QUASI-FROBENIUS RINGS

The study of quasi-Frobenius rings grew out of the theory of group representations in the 1940s and has produced an enormous body of results. This book makes no attempt to be encyclopedic but provides an elementary account of the basic facts about these rings at a level allowing researchers and graduate students to gain entry to the field. Many earlier results about self-injective rings are extended to the much wider class of mininjective rings; the methods used unify and simplify what is known in the area and so bring the reader up to current research. Sufficient background knowledge can be found in standard texts on noncommutative rings. However, appendices on Morita equivalence; on perfect, semiperfect, and semiregular rings; and on the Camps–Dicks theorem are included to make the book self-contained. After the basic results are established in Chapters 1 through 6, recent work is reviewed on three open problems in the field (the Faith conjecture, the FGF-conjecture, and the Faith–Menal conjecture). Some new results are provided and new and old methods for attacking these problems are outlined in an easily accessible format.

W. K. Nicholson is Professor of Mathematics at the University of Calgary.

M. F. Yousif is Professor of Mathematics at The Ohio State University.

CAMBRIDGE TRACTS IN MATHEMATICS

158 Quasi-Frobenius Rings

QUASI-FROBENIUS RINGS

W. K. NICHOLSON
University of Calgary

M. F. YOUSIF
The Ohio State University

CAMBRIDGE
UNIVERSITY PRESS

PUBLISHED BY THE PRESS SYNDICATE OF THE UNIVERSITY OF CAMBRIDGE
The Pitt Building, Trumpington Street, Cambridge, United Kingdom

CAMBRIDGE UNIVERSITY PRESS
The Edinburgh Building, Cambridge CB2 2RU, UK
40 West 20th Street, New York, NY 10011-4211, USA
477 Williamstown Road, Port Melbourne, VIC 3207, Australia
Ruiz de Alarcón 13, 28014 Madrid, Spain
Dock House, The Waterfront, Cape Town 8001, South Africa

http://www.cambridge.org

First published 2003

Printed in the United Kingdom at the University Press, Cambridge

Typeface Times 10/13 pt. *System* LaTeX 2_ε [TB]

A catalog record for this book is available from the British Library.

Library of Congress Cataloging in Publication Data
Nicholson, W. Keith.
Quasi-Frobenius rings / W. K. Nicholson, M. F. Yousif.
p. cm. – (Cambridge tracts in mathematics ; 158)
Includes bibliographical references and index.
ISBN 0-521-81593-2
1. Quasi-Frobenius rings. I. Yousif, M. F. (Mohamed F.) II. Title. III. Series.
QA251.5 .N53 2003
512′.4 – dc21 2002041003

ISBN 0 521 81593 2 hardback

For Kathleen and Eman

Contents

to throughout the book. Chapter 4 varies the theme of Chapters 2 and 3 and deals with the *min-CS rings* in which every simple right ideal is essential in a direct summand.

Two important subclasses of mininjective rings are introduced in Chapters 5 and 6. The right *principally injective* rings (for which every linear map from a principal right ideal to the ring is given by multiplication) are described in Chapter 5 and are shown to be closely related to the right FP-injective rings. This motivates the study of the *FP rings* [semiperfect, right FP-injective with essential right (or left) socle] as a generalization of the well-known class of pseudo-Frobenius rings. A ring is called right *simple injective* if every linear map with simple image from a right ideal to the ring is given by multiplication. These rings are investigated in Chapter 6 and are used to study *dual rings* (for which every one-sided ideal is an annihilator) and right *Ikeda–Nakayama* rings [for which $\mathtt{l}(A \cap B) = \mathtt{l}(A) + \mathtt{l}(B)$ for all right ideals A and B of R, where $\mathtt{l}(X)$ denotes the left annihilator].

A ring is called a right *C2 ring* if every right ideal that is isomorphic to a direct summand is itself a direct summand. In Chapter 7 the FGF-conjecture is shown to be closely related to these right C2 rings: A ring is quasi-Frobenius if every matrix ring over it is a C2 ring and every 2-generated right module embeds in a free module. This implies several important results in the literature and leads to a reformulation of the conjecture: The FGF-conjecture is true if and only if every right FGF ring is a right C2 ring. More recently, extensive work on the conjecture has been carried out by Gómez Pardo and Guil Asensio. They show that a right FGF ring is quasi-Frobenius if it is a right CS ring (every right ideal is essential in a direct summand). This in turn stems from their more general result: Every right Kasch, right CS ring has a finitely generated essential right socle, generalizing (and adapting the proof of) a well-known theorem of Osofsky in the right self-injective case.

The Faith–Menal conjecture is investigated in Chapter 8; in Chapter 9 a generic example is constructed to study the Faith conjecture and provide a source of examples of many of the rings studied in the book.

Of course a book like this rests on the research of many mathematicians, and it is a pleasure to acknowledge all these contributions. Special thanks go to Esperanza Sánchez Campos who gave the entire manuscript a thorough reading, made many useful suggestions, and caught a multitude of typographical errors. In addition, we thank Joanne Longworth for many consultations about the computer. We also acknowledge the support of the Ohio State University, the University of Calgary, NSERC Grant A-8075, and a Killam Resident Fellowship at the University of Calgary. Finally, we thank our families for their constant support during the time this book was being written.

1
Background

To make this monograph as self-contained as possible, this preliminary chapter contains basic characterizations of quasi-Frobenius and pseudo-Frobenius rings, together with the necessary background material. We assume familiarity with the basic facts of noncommutative ring theory, and we refer the reader to the texts by Anderson and Fuller [1] or Lam [131] for the relevant information. However, we make frequent use of facts about semiperfect, perfect, and semiregular rings and about Morita equivalence, often without comment. All these results are derived in the Appendices, again to make the book self-contained.

Throughout this book all rings considered are associative with unity and all R-modules are unital. We write $J = J(R)$ for the Jacobson radical of R and $M_n(R)$ for the ring of $n \times n$ matrices over R. Right and left modules are denoted M_R and ${}_RM$ respectively, and we write module homomorphisms opposite the scalars. If M is an R-module, we write $Z(M)$, $soc(M)$ and $M^* = hom_R(M, R)$ respectively, for the singular submodule, the socle, and the dual of M. The uniform (Goldie) dimension of a module M will be referred to simply as the dimension of M and will be denoted $dim(M)$. For a ring R, we write

$$soc(R_R) = S_r, \quad soc({}_RR) = S_l, \quad Z(R_R) = Z_r, \quad \text{and} \quad Z({}_RR) = Z_l.$$

The notations $N \subseteq^{max} M$, $N \subseteq^{ess} M$, and $N \subseteq^{sm} M$ mean that N is a maximal, (essential, and small) submodule of M, respectively, and we write $N \subseteq^{\oplus} M$ if N is a direct summand of M. Right annihilators will be denoted as

$$\mathtt{r}(Y) = \mathtt{r}_X(Y) = \{x \in X \mid yx = 0 \text{ for all } y \in Y\},$$

with a similar definition of left annihilators, $\mathtt{l}_X(Y) = \mathtt{l}(Y)$. Multiplication maps $x \mapsto ax$ and $x \mapsto xa$ will be denoted $a\cdot$ and $\cdot a$ respectively. If π is a property of modules, we say that M is a π *module* if it has the property π and that the ring R is a right π *ring* if R_R is a π module (with a similar convention on the left).

1.1. Injective Modules

Injective modules are closely related to essential extensions. If $K \subseteq M$ are modules, recall that K is called an *essential submodule* of M (and $K \subseteq M$ is called an *essential extension*) if $K \cap X \neq 0$ for every submodule $X \neq 0$ of M. This state of affairs is denoted $K \subseteq^{ess} M$. We begin with a lemma, which will be referred to throughout the book, that collects many basic properties of essential extensions.

Lemma 1.1. *Let M denote a module.*

(1) If $K \subseteq N \subseteq M$ then $K \subseteq^{ess} M$ if and only if $K \subseteq^{ess} N$ and $N \subseteq^{ess} M$.
(2) If $K \subseteq^{ess} N \subseteq M$ and $K' \subseteq^{ess} N' \subseteq M$ then $K \cap K' \subseteq^{ess} N \cap N'$.
(3) If $\alpha : M \to N$ is R-linear and $K \subseteq^{ess} N$, then $\alpha^{-1}(K) \subseteq^{ess} M$, where $\alpha^{-1}(K) = \{m \in M \mid \alpha(m) \in K\}$.
(4) Let $M = \oplus_{i \in I} M_i$ be a direct sum where $M_i \subseteq M$ for each i, and let $K_i \subseteq M_i$ for each i. Then $\oplus_{i \in I} K_i \subseteq^{ess} M$ if and only if $K_i \subseteq^{ess} M_i$ for each i.

Proof. (1) and (2). These are routine verifications.

(3). Let $0 \neq X \subseteq M$; we must show that $X \cap \alpha^{-1}(K) \neq 0$. This is clear if $\alpha(X) = 0$ since then $X \subseteq \alpha^{-1}(K)$. Otherwise, $\alpha(X) \cap K \neq 0$ by hypothesis, say $0 \neq \alpha(x) \in K$, $x \in X$. Then $0 \neq x \in X \cap \alpha^{-1}(K)$.

(4). Write $K = \oplus_{i \in I} K_i$, and assume that $K_i \subseteq^{ess} M_i$ for each i. Then $K \subseteq^{ess} M$ if and only if $mR \cap K \neq 0$ for each $0 \neq m \in M$. Since m lies in a finite direct sum of the M_i, it suffices to prove (4) when I is finite, and hence (by induction) when $|I| = 2$. Let $\pi_1 : M_1 \oplus M_2 \to M_1$ be the projection with $ker(\pi_1) = M_2$. Then $K_1 \oplus M_2 = \pi_1^{-1}(K_1) \subseteq^{ess} M_1 \oplus M_2$ by (3). Similarly, $M_1 \oplus K_2 \subseteq^{ess} M_1 \oplus M_2$, and (4) follows from (2) because $K_1 \oplus K_2 = (K_1 \oplus M_2) \cap (M_1 \oplus K_2)$. □

This book is concerned with injective modules and their generalizations, and the main properties of these modules are derived in this section. A module E_R is called *injective* if whenever $0 \to N \overset{\alpha}{\to} M$ is R-monic, every R-linear map $\beta : N \to E$ factors in the form $\beta = \gamma \circ \alpha$ for some R-linear map

$$\begin{array}{ccc} 0 \to & N & \overset{\alpha}{\to} M \\ & \beta\downarrow & \overset{\gamma}{\swarrow} \\ & E & \end{array}$$

$\gamma : M \to E$. These modules admit a characterization that we will use repeatedly in the following.

Lemma 1.2. *A module E is injective if and only if, whenever $K \subseteq M$, every R-linear map $\beta : K \to E$ extends to an R-linear map $\gamma : M \to E$.*

Proof. The condition clearly holds if E is injective. Conversely, if $N \xrightarrow{\alpha} M$ is R-monic, the map $\alpha' : \alpha(N) \to N$ is well defined by $\alpha'(\alpha(n)) = n$ for $n \in N$. Then, given $\beta : N \to E$, the map $\beta \circ \alpha' : \alpha(N) \to E$ extends to $\gamma : M \to E$ by hypothesis, and one checks that $\gamma \circ \alpha = \beta$. □

Corollary 1.3. *If $E = \Pi_i E_i$ is a direct product of modules, then E is injective if and only if each E_i is injective.*

Proof. Let $E_i \xrightarrow{\sigma_i} E \xrightarrow{\pi_j} E_j$ be the canonical maps. If E is injective, and if $K \subseteq M$ and $\beta : K \to E_i$ are given, there exists $\gamma : M \to E$ such that $\gamma = \sigma_i \circ \beta$ on K. Then $\pi_i \circ \gamma : M \to E_i$ extends β, proving that E_i is injective by Lemma 1.2. Conversely, if each E_i is injective, let $\alpha : K \to E$, where $K \subseteq M$. For each i, there exists $\gamma_i : M \to E_i$ extending $\pi_i \circ \alpha$. If $\gamma : M \to E$ is defined by $\gamma(m) = \langle \gamma_i(m) \rangle$ for each $m \in M$, then γ extends α because $x = \langle \pi_i(x) \rangle$ for each $x \in M$. It follows that E is injective by Lemma 1.2. □

Surprisingly, to prove that a module E is injective, it is enough to verify the condition in Lemma 1.2 when $M = R$.

Lemma 1.4 (Baer Criterion). *A right R-module E is injective if and only if, whenever $T \subseteq R$ is a right ideal, every map $\gamma : T \to E$ extends to $R \to E$, that is, $\gamma = c\cdot$ is multiplication by an element $c \in E$.*

Proof. The condition is clearly necessary. To prove sufficiency, let $K \subseteq M$ be modules and let $\beta : K \to E$. In this case, let $\mathcal{F}$ denote the set of pairs (K', β') such that $K \subseteq K' \subseteq M$ and $\beta' : K' \to E$ extends β. By Zorn's lemma, let (K'', β'') be a maximal member of $\mathcal{F}$. We must show that $K'' = M$. If not, let $m \in M - K''$, let $T = \{r \in R \mid mr \in K''\}$ – a right ideal, and define $\lambda : T \to E$ by $\lambda(r) = \beta''(mr)$. By hypothesis there exists $\hat{\lambda} : R \to E$ extending λ, and we use it to define $\hat{\beta} : K'' + mR \to E$ by $\hat{\beta}(y + mr) = \beta''(y) + \hat{\lambda}(r)$, where $y \in K''$ and $r \in R$. This is well defined because $y + mr = 0$ implies that $mr \in K''$ and so $\hat{\lambda}(r) = \lambda(r) = \beta''(mr) = \beta''(-y) = -\beta''(y)$. Since $\hat{\beta}$ is R-linear and extends β'' this contradicts the maximality of (K'', β'') in $\mathcal{F}$. □

It is a routine matter to show that an (additive) abelian group X is injective as a $\mathbb{Z}$-module if and only if it is *divisible*, that is, $nX = X$ for any $0 \neq n \in \mathbb{Z}$. Examples include $\mathbb{Q}$ and the Prüfer group $\mathbb{Z}_{p^\infty}$ for any prime p. Divisible groups can be used to construct injective modules over any ring. The second part of the next lemma was discovered by Baer in 1940.

Lemma 1.5. *Let R be a ring. Then the following hold:*

(1) If Q is a divisible group then $E_R = hom_{\mathbb{Z}}(R, Q)$ is an injective right R-module.

*(2) (**Baer**) Every module M_R embeds in an injective right module.*

Proof. (1). If $\lambda \in E$ and $a \in R$, E becomes a right R-module via $(\lambda \cdot a)(r) = \lambda(ar)$ for all $r \in R$. Now let $\gamma : T \to E_R$ be R-linear, where T is a right ideal of R. By Lemma 1.4 we must extend γ to $R_R \to E_R$. Define $\theta : T \to Q$ by $\theta(t) = [\gamma(t)](1)$. Then θ is a $\mathbb{Z}$-morphism; so, since ${}_{\mathbb{Z}}Q$ is injective, let $\hat{\theta} : R \to Q$ be a $\mathbb{Z}$-morphism extending θ. Since $\hat{\theta} \in E$, define $\hat{\gamma} : R \to E$ by $\hat{\gamma}(a) = \hat{\theta} \cdot a$ for all $a \in R$. One verifies that $\hat{\gamma}$ is R-linear, and we claim that it extends γ; that is, $\hat{\gamma}(t) = \gamma(t)$ for all $t \in T$. If $r \in R$, we have

$$[\hat{\gamma}(t)](r) = [\hat{\theta} \cdot t](r) = \hat{\theta}(tr) = \theta(tr) = [\gamma(tr)](1) = [\gamma(t) \cdot r](1) = [\gamma(t)](r)$$

because γ is R-linear and $\gamma(t) \in E$. Hence $\hat{\gamma}(t) = \gamma(t)$, as required.

(2). Given M_R, let $\varphi : \mathbb{Z}^{(I)} \to M$ be $\mathbb{Z}$-epic for some set I, so that ${}_{\mathbb{Z}}M \cong \mathbb{Z}^{(I)}/K \subseteq \mathbb{Q}^I/K$, where $K = ker(\varphi)$. Write $Q = \mathbb{Q}^I/K$ and note that Q is divisible. Since $M_R \cong hom(R_R, M_R)$ via $m \mapsto m\cdot$, we get

$$M_R \cong hom_R(R_R, M_R) \subseteq hom_{\mathbb{Z}}(R, M) \hookrightarrow hom_{\mathbb{Z}}(R, Q).$$

Since $E_R = hom_{\mathbb{Z}}(R, Q)$ is injective by (1), this proves (2). □

Corollary 1.6. *A module E is injective if and only if every monomorphism $\sigma : E \to M$ splits, that is, $\sigma(E) \subseteq^{\oplus} M$.*

Proof. If $\sigma : E \to M$ is monic there exists $\gamma : M \to E$ such that $\gamma \circ \sigma = 1_E$. Then $M = \sigma(E) \oplus ker(\gamma)$. The converse is clear from Lemma 1.5 because direct summands of injective modules are injective. □

Before proceeding, we need another basic property of essential extensions. If K is a submodule of a module M, it is a routine application of Zorn's lemma to see that there exist submodules C of M maximal with respect to $K \cap C = 0$. Such a submodule C is called a *complement*[1] of K in M. Thus $K \subseteq^{ess} M$ if and only if 0 is a complement of K.

[1] It is sometimes called an *intersection complement*, or *relative complement*.

Lemma 1.7 (Essential Lemma). *Let $K \subseteq M$ be modules. If C is any complement of K in M then the following hold:*

(1) $K \oplus C \subseteq^{ess} M$.
(2) $(K \oplus C)/C \subseteq^{ess} M/C$.

Proof. (1). Let X be a nonzero submodule of M; we must show that $X \cap (K \oplus C) \neq 0$. This is clear if $X \subseteq C$. Otherwise the maximality of C shows that $K \cap (X + C) \neq 0$, say $0 \neq k = x + c$ with the obvious notation. Hence $x \in X \cap (K \oplus C)$, and $x \neq 0$ because $K \cap C = 0$.

(2). Let $Y/C \cap (K \oplus C)/C = 0$. If $Y \neq C$ then $Y \cap K \neq 0$ by the choice of C, say $0 \neq a \in Y \cap K$. Then $a + C \in Y/C \cap (K \oplus C)/C = 0$ so $a \in C$. But then $0 \neq a \in C \cap K = 0$, which is a contradiction. □

Given any module M, an R-monomorphism $M \overset{\sigma}{\to} E$ is called an *injective hull* (*injective envelope*) of M if E is injective and $\sigma(M) \subseteq^{ess} E$. The following result is a famous theorem that traces back to Baer, to Eckmann and Schopf, and to Shoda.

Theorem 1.8 (Baer/Eckmann–Schopf/Shoda). *Let M_R be a module.*

(1) M has an injective hull.
(2) If $M \overset{\sigma_1}{\to} E_1$ and $M \overset{\sigma_2}{\to} E_2$ are two injective hulls there exists an isomorphism $\tau : E_1 \to E_2$ such that $\sigma_2 = \tau \circ \sigma_1$.

Proof. (1). By Lemma 1.5 let $M \subseteq Q_R$ where Q_R is injective, and, by Zorn's lemma, let E be maximal such that $M \subseteq^{ess} E \subseteq Q$. Then let $C \subseteq Q$ be maximal such that $E \cap C = 0$; it suffices to show that $E \oplus C = Q$ (so E is injective). By Lemma 1.7 we have $E \cong (E \oplus C)/C \subseteq^{ess} Q/C$. Define $\sigma : (E \oplus C)/C \to Q$ by $\sigma(x + C) = x$ if $x \in E$. Since Q is injective, σ extends to $\hat{\sigma} : Q/C \to Q$. Then $\hat{\sigma}$ is monic because $ker(\hat{\sigma}) \cap (E \oplus C)/C = 0$, and so $im(\sigma) = \hat{\sigma}((E \oplus C)/C) \subseteq^{ess} \hat{\sigma}(Q/C)$. Since $M \subseteq^{ess} E = im(\sigma)$ it follows that $E \subseteq^{ess} \hat{\sigma}(Q/C)$, and so $E = \hat{\sigma}(Q/C)$ by the maximality of E. But then $\hat{\sigma}(Q/C) = E = \hat{\sigma}((E \oplus C)/C)$ and we conclude that $Q = E \oplus C$ because $\hat{\sigma}$ is monic. This is what we wanted.

(2). The given map τ exists because E_2 injective. Moreover, τ is monic because $ker(\tau) \cap \sigma_1(M) = 0$ (since σ_2 is monic) and $\sigma_1(M) \subseteq^{ess} E_1$. Hence $\tau(E_1) \subseteq^{\oplus} E_2$ by Corollary 1.6. But $\tau(E_1) \subseteq^{ess} E_2$ because $\sigma_2(M) = \tau\sigma_1(M) \subseteq \tau(E_1)$ and $\sigma_2(M) \subseteq^{ess} E_2$ by hypothesis. It follows that τ is onto and so is an isomorphism. □

Hence we are entitled to speak of *the* injective hull of a module M and to denote it by $E(M)$. We will usually assume that $M \subseteq E(M)$; so, for example, we have $E(\mathbb{Z}) = \mathbb{Q}$ and $E(\mathbb{Z}_{p^n}) = \mathbb{Z}_{p^\infty}$ for any prime p and $n \geq 2$. The assumption that $M \subseteq E(M)$ is justified by the following result.

Lemma 1.9. *Let $\sigma : M \to E(M)$ be an injective hull of the module M. If $M \subseteq G$, where G is any injective module, there exists a copy $E \cong E(M)$ inside G such that $M \subseteq^{ess} E \subseteq^{\oplus} G$.*

Proof. As G is injective, there exists $\tau : E(M) \to G$ such that $m = \tau\sigma(m)$ for every $m \in M$. Since $ker(\tau) \cap \sigma(M) = 0$ it follows that τ is monic, and we are done by Corollary 1.6 with $E = \tau[E(M)]$. □

Lemma 1.9 will be used frequently in the following, usually without comment. In particular, let $M = \oplus_{i=1}^n M_i$ be a direct sum of modules, and let $M \subseteq E(M)$. By Lemma 1.9 we can choose a copy of $E(M_i)$ such that $M_i \subseteq^{ess} E(M_i) \subseteq E(M)$ for each i. One verifies that $E(M_1) \cap E(M_2) = 0$, so (by Lemma 1.1) $M_1 \oplus M_2 \subseteq^{ess} E(M_1) \oplus E(M_2)$. Continuing inductively, we conclude that $\Sigma_{i=1}^n E(M_i)$ is direct and that $M = \oplus_{i=1}^n M_i \subseteq^{ess} \oplus_{i=1}^n E(M_i)$. Since $\oplus_{i=1}^n E(M_i)$ is injective (Corollary 1.3) we have proved the following:

Proposition 1.10. *If $M = \oplus_{i=1}^n M_i$ is a finite direct sum of modules then $E(\oplus_{i=1}^n M_i) = \oplus_{i=1}^n E(M_i)$.*

1.2. Relative Injectivity

Let M and G denote right R-modules. We say that G is M-injective if, for any submodule $X \subseteq M$, every R-linear map $\beta : X \to G$ can be extended to an R-linear map $\hat{\beta} : M \to G$, equivalently (see the proof of Lemma 1.2) if, for every

$$\begin{array}{ccc} X & \hookrightarrow & M \\ \beta\downarrow & \swarrow_{\hat{\beta}} & \\ G & & \end{array}$$

monomorphism $\sigma : X \to M$ there exists $\lambda : M \to G$ such that $\beta = \lambda \circ \sigma$. Thus G is injective if and only if it is M-injective for every module M, equivalently (by the Baer criterion) if G is R-injective. The proof of Corollary 1.3 gives

Lemma 1.11. *Let $G = \Pi_{i \in I} G_i$ and M be modules. Then G is M-injective if and only if G_i is M-injective for each $i \in I$.*

Lemma 1.12. *If G is M-injective and $N \subseteq M$ then G is both N-injective and (M/N)-injective.*

Proof. Given $X \xrightarrow{\beta} G$, where $X \subseteq N$, extend β to $\hat{\beta} : M \to G$ by hypothesis. Then the restriction $\hat{\beta}_{|N} : N \to G$ extends β, so G is N-injective. Now let $\alpha : X/N \to G$, $N \subseteq X \subseteq M$, and let $\pi : X \to X/N$ be the coset map. Then $\alpha \circ \pi : X \to G$ extends to $\lambda : M \to G$ by hypothesis. Hence $\hat{\alpha} : M/N \to G$ is well defined by $\hat{\alpha}(m + N) = \lambda(m)$, and $\hat{\alpha}$ extends α. This shows that G is (M/N)-injective. □

Note that if G is both N- and (M/N)-injective it does not follow that G is M-injective. Indeed, there is a monomorphism $\mathbb{Z}_p \xrightarrow{\sigma} \mathbb{Z}_{p^2}$ of abelian groups, given by $\sigma(n + p\mathbb{Z}) = pn + p^2\mathbb{Z}$ for all $n \in \mathbb{Z}$. Let $G = \mathbb{Z}_p$ and $N = im(\sigma)$. Then G is both N- and $(\mathbb{Z}_{p^2}/N)$-injective (because N and $\mathbb{Z}_{p^2}/N$ are simple), but G is not $\mathbb{Z}_{p^2}$-injective because any map $\lambda : \mathbb{Z}_{p^2} \to \mathbb{Z}_p$ satisfies $\lambda(N) = 0$. However, we do have

Lemma 1.13 (Azumaya's Lemma). *If G and $M = M_1 \oplus \cdots \oplus M_n$ are modules, then G is M-injective if and only if G is M_i-injective for each $i = 1, 2, \ldots, n$.*

Proof. If G is M-injective, then G is M_i-injective for each i by Lemma 1.12. Conversely, if G is M_i-injective for each i, let $\beta : X \to G$ be R-linear, where $X \subseteq M$. As in the proof of Lemma 1.4, let (C, β^*) be maximal such that $X \subseteq C \subseteq M$ and $\beta^* : C \to G$ extends β. We show $C = M$ by proving that $M_i \subseteq C$ for each i. By hypothesis there exists $\alpha_i : M_i \to G$ such that $\alpha_i = \beta^*$ on $M_i \cap C$. Construct $\beta_i : M_i + C \to G$ by $\beta_i(m_i + c) = \alpha_i(m_i) + \beta^*(c)$ for all $m_i \in M_i$ and $c \in C$. Then β_i is well defined because $\alpha_i = \beta^*$ on $M_i \cap C$, and β_i extends β because $X \subseteq C$ and β^* extends β. Hence $M_i + C = C$ by the maximality of (C, β^*), so $M_i \subseteq C$, as required. □

It is not surprising that there is a characterization of when G is M-injective in terms of the injective hulls $E(G)$ and $E(M)$.

Lemma 1.14. *A module G is M-injective if and only if $\lambda(M) \subseteq G$ for all R-linear maps $\lambda : E(M) \to E(G)$.*

Proof. If the condition holds, let $\beta : X \to G$ be R-linear, where $X \subseteq M$. Since $E(G)$ is injective there exists $\hat{\beta} : E(M) \to E(G)$ extending β. But $\hat{\beta}(M) \subseteq G$ by hypothesis, so the restriction $\hat{\beta}_{|M} : M \to G$ extends β.

Conversely, assume that G is M-injective, and let $\lambda : E(M) \to E(G)$ be R-linear. We must show that $\lambda(M) \subseteq G$. If $X = \{x \in M \mid \lambda(x) \in G\}$ then the restriction $\lambda_{|X} : X \to G$ extends to $\mu : M \to G$. Hence it suffices to show that $(\lambda - \mu)(M) = 0$. Since $G \subseteq^{ess} E(G)$, it is enough to show that $G \cap (\lambda - \mu)(M) = 0$. But if $g = (\lambda - \mu)(m)$, where $g \in G$ and $m \in M$, then $\lambda(m) = \mu(m) + g \in G$, so $m \in X$. This means that $\lambda(m) = \mu(m)$ by the definition of μ. Hence $g = \lambda(m) - \mu(m) = 0$, as required. □

A module M is called *quasi-injective* if it is M-injective, that is, if every map $\beta : X \to M$, where X is a submodule of M, extends to an endomorphism of M. Clearly every injective or semisimple module is quasi-injective, but the converse is false (for example, $\mathbb{Z}_4$ is quasi-injective as a $\mathbb{Z}$-module, as we shall see).

Lemma 1.14 leads to an important characterization of quasi-injective modules. We say that a submodule $K \subseteq M$ is *fully invariant* in M if $\lambda(K) \subseteq K$ for every $\lambda \in end(M)$. Then taking $G = M$ in Lemma 1.14 gives immediately

Lemma 1.15 (Johnson–Wong Lemma). *A module is quasi-injective if and only if M is fully invariant in its injective hull $E(M)$.*

Thus, for example, $\mathbb{Z}_{p^n}$ is quasi-injective as a $\mathbb{Z}$-module for any prime p because it is fully invariant in its injective hull $\mathbb{Z}_{p^\infty}$.

Corollary 1.16. *Let M be a quasi-injective module. If $E(M) = \oplus_{i \in I} K_i$, then $M = \oplus_{i \in I}(M \cap K_i)$.*

Proof. Let $m = \Sigma_{i=1}^n k_i \in M$, where each $k_i \in K_i$. If $\pi_i : E(M) \to E(M)$ is the projection onto K_i, then $k_i = \pi_i(m) \in \pi_i(M) \subseteq M$ by Lemma 1.15, so $k_i \in M \cap K_i$. Hence $M \subseteq \oplus_{i \in I}(M \cap K_i)$; the other inclusion is clear. □

If p is a prime, the $\mathbb{Z}$-module $\mathbb{Q} \oplus \mathbb{Z}_p$ is not quasi-injective even though $\mathbb{Q}$ is injective and $\mathbb{Z}_p$ is simple. (The coset map $\mathbb{Z} \to \mathbb{Z}_p$ does not extend to $\mathbb{Q} \oplus \mathbb{Z}_p \to \mathbb{Z}_p$ because there is no nonzero map $\mathbb{Q} \to \mathbb{Z}_p$.) Hence the direct sum of two quasi-injective modules need not be quasi-injective. However, we do have the following lemma:

Lemma 1.17. *If M is quasi-injective so also is every direct summand N.*

Proof. If $M = N \oplus N'$ and $\beta : X \to N$ is R-linear, where $X \subseteq N$, then β extends to $\hat{\beta} : M \to M$ by hypothesis. If $\pi : M \to N$ is the projection with kernel N', then $\lambda = (\pi \circ \hat{\beta})_{|N}$ is in $end(N)$ and extends β. □

The next result uses Lemma 1.15 to identify when a finite direct sum of quasi-injective modules is again of the same type.

Proposition 1.18. *Let $M_1, \ldots, M_n$ be modules and write $E_i = E(M_i) \supseteq M_i$ for each i. The following are equivalent:*

(1) $M_1 \oplus \cdots \oplus M_n$ is quasi-injective.
(2) $\lambda(M_i) \subseteq M_j$ for all R-linear maps $\lambda : E_i \to E_j$.

Proof. Let $M_j \xrightarrow{\sigma_j} \oplus_k M_k \xrightarrow{\pi_i} M_i$ denote the canonical maps, and write $E = E(\oplus_k M_k) = \oplus_k E_k$.

(1)⇒(2). Given (1) and $\lambda : E_i \to E_j$, let $m_i \in M_i$. We have $\pi_j \circ \sigma_j = 1_{E_j}$ for each j, so $\lambda(m_i) = (\pi_j\sigma_j\lambda\pi_i\sigma_i)(m_i) = \pi_j(\sigma_j\lambda\pi_i)(\sigma_i m_i) \in M_j$ because $(\sigma_j\lambda\pi_i)(\oplus_k M_k) \subseteq \oplus_k M_k$ by (1) and Lemma 1.15.

(2)⇒(1). Given $\lambda : \oplus_k E_k \to \oplus_k E_k$, we must show (by Lemma 1.15) that $\lambda(\oplus_k M_k) \subseteq \oplus_k M_k$. Let $\bar{m} = m_1 + \cdots + m_n \in \oplus_k M_k$. Since $\Sigma_k \sigma_k \pi_k = 1_E$, we compute

$$\pi_j\lambda(\bar{m}) = \pi_j\lambda(\Sigma_k\sigma_k\pi_k\bar{m}) = \Sigma_k(\pi_j\lambda\sigma_k)(\pi_k\bar{m}) = \Sigma_k(\pi_j\lambda\sigma_k)(m_k) \in \oplus_k M_k$$

because $(\pi_j\lambda\sigma_k)(M_k) \subseteq M_j$ for all j and k by (2). □

Thus, for example, $\mathbb{Z}_n$ is quasi-injective as a $\mathbb{Z}$-module for each $n \in \mathbb{Z}$. In fact, $\mathbb{Z}_n = \mathbb{Z}_{p_1^{n_1}} \oplus \cdots \oplus \mathbb{Z}_{p_k^{n_k}}$ for distinct primes p_i, each $\mathbb{Z}_{p_i^{n_i}}$ is quasi-injective, and $hom_{\mathbb{Z}}(\mathbb{Z}_{p^\infty}, \mathbb{Z}_{q^\infty}) = 0$ if p and q are distinct primes.

Corollary 1.19. *A module M is quasi-injective if and only if M^n is quasi-injective.*

1.3. Continuous Modules

In his work on continuous rings, Utumi identified three conditions on a ring that are satisfied if the ring is self-injective. The analogs of these conditions for a module M are as follows:

(1) M satisfies the C1-*condition* if every submodule of M is essential in a direct summand of M.[2] (Note that we regard the zero submodule as essential in itself.)

(2) M satisfies the C2-*condition* if every submodule that is isomorphic to a direct summand of M is itself a direct summand.

[2] This condition is also referred to as the CS-*condition* because it is equivalent to the requirement that every complement submodule is a direct summand (complement submodules are also called *closed* submodules). We return to this topic in the following section.

(3) M satisfies the C3-*condition* if, whenever N and K are submodules of M with $N \subseteq^{\oplus} M$, $K \subseteq^{\oplus} M$, and $N \cap K = 0$, then $N \oplus K \subseteq^{\oplus} M$.

A ring R is called a right C1 *ring* (respectively C2 *ring*, C3 *ring*) if the module R_R has the corresponding property.

If M is an indecomposable module then M is a C3 module; M is a C1 module if and only if it is *uniform* (that is $X \cap Y \neq 0$ for all submodules $X \neq 0$ and $Y \neq 0$) and M is a C2 module if and only if monomorphisms in $end(M)$ are isomorphisms. The $\mathbb{Z}$-modules $\mathbb{Z}_2$ and $\mathbb{Z}_8$ each satisfy the C1-, C2- and C3-conditions, but their direct sum $N = \mathbb{Z}_2 \oplus \mathbb{Z}_8$ is not a C1 module because, writing $S = \mathbb{Z}_2 \oplus 0$ and $K = \mathbb{Z}(1 + 2\mathbb{Z}, 2 + 8\mathbb{Z})$, we see that K is contained in only two direct summands N and $S \oplus K$ and is essential in neither. Moreover, N is not a C2 module because the non-summand $0 \oplus \mathbb{Z}(4 + 8\mathbb{Z})$ is isomorphic to the summand $\mathbb{Z}_2 \oplus 0$. Hence a direct sum of C1 modules, or C2 modules, may not inherit the same property.

As an abelian group, $\mathbb{Z}$ satisfies both the C1- and C3-conditions, but it is not a C2 module. However, if F is a field let $R = \left[\begin{smallmatrix} F & V \\ 0 & F \end{smallmatrix}\right]$, where $V = F \oplus F$. If $e = \left[\begin{smallmatrix} 1 & 0 \\ 0 & 0 \end{smallmatrix}\right]$ then $eR = \left[\begin{smallmatrix} F & V \\ 0 & 0 \end{smallmatrix}\right]$ is indecomposable (in fact $eRe \cong F$) and is a C2 module because monomorphisms are epic, but it is not a C1 module because it is not uniform.

Example 1.20. Let $R = \left[\begin{smallmatrix} F & F \\ 0 & F \end{smallmatrix}\right]$, where F is a field. Then R is a right and left C1 ring, but neither a left nor right C2 ring.

Proof. We have $J = \left[\begin{smallmatrix} 0 & F \\ 0 & 0 \end{smallmatrix}\right] \cong e_{12}R$ (where e_{ij} is the matrix unit), so R is not right C2 because J_R is not a direct summand of R_R. Similarly, R is not left C2. To see that R is right C1, let $T \neq 0$ be a right ideal. If $T \nsubseteq S_r = \left[\begin{smallmatrix} 0 & F \\ 0 & F \end{smallmatrix}\right]$ then $T = e_{11}R$ or $T = R$, so T is a summand. If $T = S_r$ then $T \subseteq^{ess} R_R$ because R is right artinian. So we may assume that $dim_F(T) = 1$, say $T = xR$, $x \in S_r$. If $x^2 = x \neq 0$ we are done. Otherwise $x \in J$, so $T = J$ and one verifies that $T \subseteq^{ess} e_{11}R = \left[\begin{smallmatrix} F & F \\ 0 & 0 \end{smallmatrix}\right]$. Hence R is right C1; similarly R is right C2. □

Lemma 1.21. *The C2-condition implies the C3-condition.*

Proof. Let $N \subseteq^{\oplus} M$ and $K \subseteq^{\oplus} M$ satisfy $N \cap K = 0$; we must show that $N \oplus K \subseteq^{\oplus} M$. Write $M = N \oplus N'$, and let $\pi : M \to N'$ be the projection with $ker(\pi) = N$. If $k \in K$ and $k = n + n'$, $n \in N$, $n' \in N'$, then $\pi(k) = n'$ and it follows that $N \oplus K = N \oplus \pi(K)$. Hence we show that $N \oplus \pi(K) \subseteq^{\oplus} M$. Since $\pi_{|K} : K \to M$ is monic we have $\pi(K) \subseteq^{\oplus} M$ by the C2-condition. Since $\pi(K) \subseteq N'$, it follows that $N' = \pi(K) \oplus W$ for some submodule W and hence that $M = N \oplus \pi(K) \oplus W$. Thus M satisfies the C3-condition. □

A module is called *continuous* if it satisfies both the C1- and C2-conditions, and a module is called *quasi-continuous* if it satisfies the C1- and C3-conditions, and R is called a right *continuous ring* (right *quasi-continuous ring*) if R_R has the corresponding property. As the terminology suggests, every continuous module is quasi-continuous (by Lemma 1.21). Clearly every injective or semisimple module is continuous; in fact:

Proposition 1.22. *Every quasi-injective module is continuous.*

Proof. Let M be quasi-injective. If $N \subseteq M$ then $E(M)$ contains a copy of $E(N) = E$, and $E(M) = E \oplus G$ for some submodule G because E is injective. But then Corollary 1.16 shows that $M = (M \cap E) \oplus (M \cap G)$. Moreover, $N \subseteq^{ess} E$, so $N \subseteq^{ess} (M \cap E)$. This shows that M has the $C1$-property.

Now suppose that $N \cong P \subseteq^{\oplus} M$. Since M is M-injective, it follows from Lemma 1.11 that P is also M-injective and hence that N is M-injective. But then the identity map $1_N : N \to N$ extends to $\lambda : M \to N$, and it follows that $M = N \oplus ker(\lambda)$. This proves C2. □

The following lemma is a useful connection between essential submodules and singular modules.

Lemma 1.23. *If $K \subseteq^{ess} M$ are modules then M/K is singular, that is, $Z(M/K) = M/K$.*

Proof. If $K \subseteq^{ess} M$ and $m \in M$, we must show that $\mathbf{r}_R(m + K) \subseteq^{ess} R_R$, that is, $bR \cap \mathbf{r}_R(m + K) \neq 0$ for every $0 \neq b \in R$. This is clear if $mb = 0$. Otherwise, we have $mbR \cap K \neq 0$ by hypothesis, say $0 \neq mba \in K$, $a \in R$. But then $0 \neq ba \in bR \cap \mathbf{r}_R(m + K)$, as required. □

We can now prove two important results about endomorphism rings. A ring R is called *semiregular*[3] if R/J is (von Neumann) regular and idempotents lift modulo J, equivalently (by Lemma B.40 in Appendix B) if, for each $a \in R$ there exists $e^2 = e \in aR$ such that $(1 - e)a \in J$. We are going to prove that the endomorphism ring S of a continuous module M_R is semiregular and $J(S) = \{\alpha \in S \mid ker(\alpha) \subseteq^{ess} M\}$. We will need the following lemma.

Lemma 1.24. *Given M_R, write $S = end(M)$ and $\bar{S} = S/J(S)$, and assume S is semiregular and $J(S) = \{\alpha \in S \mid ker(\alpha) \subseteq^{ess} M\}$.*

(1) If $\pi^2 = \pi$ and $\tau^2 = \tau$ in S satisfy $\bar{\pi}\bar{S} \cap \bar{\tau}\bar{S} = 0$ then $\pi M \cap \tau M = 0$.
(2) If M satisfies the C3-condition and $\Sigma_{i \in I}\bar{\pi}_i\bar{S}$ is direct in $\bar{S}$, where $\pi_i^2 = \pi_i \in S$ for each i, then $\Sigma_{i \in I}\pi_i M$ is direct in M.

[3] These rings are also called *F-semiperfect* in the literature.

(3) *If M is quasi-continuous and $\Sigma_{i\in I}\pi_i M$ is direct in M, where $\pi_i^2 = \pi_i \in S$ for each i, then $\Sigma_{i\in I}\bar{\pi}_i\bar{S}$ is direct in $\bar{S}$.*

Proof. (1). We begin with a simplifying adjustment.

Claim 1. We may assume that $\bar{\tau}\bar{\pi} = 0$.

Proof. As $\bar{S}$ is regular, let $\bar{\pi}\bar{S} \oplus \bar{\tau}\bar{S} \oplus T = \bar{S}$, with T a right ideal of $\bar{S}$. Let $\bar{\eta}^2 = \bar{\eta}$ be such that $\bar{\tau}\bar{S} = \bar{\eta}\bar{S}$ and $\bar{\pi}\bar{S} \oplus T = (1-\bar{\eta})\bar{S}$. By hypothesis, we may assume that $\eta^2 = \eta$ in S. Note that $\bar{\eta}\bar{\pi} = 0$. Then $\gamma = \tau + \tau\eta(1-\tau)$ satisfies $\gamma^2 = \gamma$, $\gamma\tau = \tau$, and $\tau\gamma = \gamma$, so $\tau M = \gamma M$ and $\bar{\tau}\bar{S} = \bar{\gamma}\bar{S}$. Moreover, $\bar{\gamma} = \bar{\eta}$ because $\bar{\tau}\bar{\eta} = \bar{\eta}$ and $\bar{\eta}\bar{\tau} = \bar{\tau}$. Hence $\bar{\gamma}\bar{\pi} = \bar{\eta}\bar{\pi} = 0$, so replacing τ by γ proves Claim 1.

By Claim 1 we have $\tau\pi \in J(S)$, so writing $K = ker(\tau\pi)$, we have $K \subseteq^{ess} M$ by hypothesis. It follows that $\pi K \subseteq^{ess} \pi M$ (if $0 \neq X \subseteq \pi M$ then $\pi x = x$ for each $x \in X$, so $0 \neq X \cap K \subseteq X \cap \pi K$). However, $\tau\pi K = 0$, so $\pi K \subseteq ker(\tau)$, whence $\pi K \cap \tau M = 0$. This in turn implies that $\pi M \cap \tau M = 0$ because $\pi K \subseteq^{ess} \pi M$.

(2). It is enough to show that $\Sigma_{i\in F}\pi_i M$ is direct for any finite subset $F \subseteq I$. Write $F = \{1, \ldots, n\}$ and proceed by induction on n. If $n = 1$ there is nothing to prove, and if $n = 2$ then (2) follows from (1). Assume inductively that $\pi_1 M + \cdots + \pi_n M$ is a direct sum, $n \geq 1$. Then the C3-condition implies that $\pi_1 M \oplus \cdots \oplus \pi_n M = \pi M$ for some $\pi^2 = \pi \in S$.

Claim 2. $\pi S = \pi_1 S + \cdots + \pi_n S$.

Proof. We have $\pi\pi_i = \pi_i$ for each i (because $\pi_i M \subseteq \pi M$), so $\Sigma_{i=1}^n \pi_i S \subseteq \pi S$. For each $i = 1, \ldots, n$, let $\rho_i : \pi M \to \pi_i M$ be the projection, so that $\pi\rho_i = \rho_i$ for each i and $\pi = \Sigma_{i=1}^n \rho_i\pi = \Sigma_{i=1}^n \tau_i$, where we define $\tau_i = \rho_i\pi$ for each i. Then $\pi\tau_i = \tau_i$ for each i, and so $\pi S = \Sigma_{i=1}^n \tau_i S = \Sigma_{i=1}^n \pi_i\tau_i S \subseteq \Sigma_{i=1}^n \pi_i S$. This proves Claim 2.

By Claim 2 we have $\bar{\pi}\bar{S} = \bar{\pi}_1\bar{S} \oplus \cdots \oplus \bar{\pi}_n\bar{S}$, so $\bar{\pi}\bar{S} \cap \bar{\pi}_{n+1}\bar{S} = 0$. But then (1) implies that $\pi M \cap \pi_{n+1}M = 0$. Since $\pi_1 M \oplus \cdots \oplus \pi_n M = \pi M$, this shows that $\pi_1 M \oplus \cdots \oplus \pi_n M \oplus \pi_{n+1} M$ is a direct sum, as required.

(3). For each $i \in I$ let C_i be a closure of $\oplus_{j\neq i}\pi_j M$, so that $\oplus_{j\neq i}\pi_j M \subseteq^{ess} C_i$. It follows that $\pi_i M \cap C_i = 0$. But C_i is a direct summand of M by the C1-condition, so the C3-condition implies that $M = \pi_i M \oplus C_i \oplus N_i$ for some submodule $N_i \subseteq M$. So, for each $i \in I$, let $\tau_i^2 = \tau_i \in S$ satisfy $\tau_i M = \pi_i M$ and $ker(\tau_i) = C_i \oplus N_i$. Then $\pi_i\tau_i = \tau_i$ and $\tau_i\pi_i = \pi_i$, and so $\tau_i S = \pi_i S$. Furthermore, $\tau_i\pi_j = 0$ for all $j \neq i$ because $\pi_j M \subseteq C_i \subseteq ker(\tau_i)$. But then

$\tau_i\tau_j = \tau_i(\pi_j\tau_j) = (\tau_i\pi_j)\tau_j = 0$ whenever $i \neq j$, so the τ_i are orthogonal. Thus $\Sigma_i\bar{\pi}_i\bar{S} = \Sigma_i\bar{\tau}_i\bar{S}$ is direct in $\bar{S}$ because the $\bar{\tau}_i$ are also orthogonal. □

We can now prove an important result about the endomorphism ring of a continuous (or quasi-injective) module.

Theorem 1.25. *Let M_R be a continuous module with $S = end(M_R)$. Then:*

(1) S is semiregular and $J(S) = \{\alpha \in M \mid ker(\alpha) \subseteq^{ess} M\}$.
(2) $S/J(S)$ is right continuous.
(3) If M is actually quasi-injective, $S/J(S)$ is right self-injective.

Proof. (1). Write $\Delta = \{\alpha \in S \mid ker(\alpha) \subseteq^{ess} M\}$. It is a routine exercise to show that Δ is a left ideal of S; it is also a right ideal using (3) of Lemma 1.1. If $\alpha \in \Delta$ the fact that $ker(\alpha) \cap ker(1-\alpha) = 0$ means that $ker(1-\alpha) = 0$. Hence $(1-\alpha)M \cong M$, so, by C2, $(1-\alpha)M \subseteq^{\oplus} M$. But $ker(\alpha) \subseteq (1-\alpha)M$, so it follows that $(1-\alpha)M = M$. Hence $1-\alpha$ is a unit in S, and it follows that $\Delta \subseteq J(S)$.

Let $\alpha \in S$ and (by C1) let $ker(\alpha) \subseteq^{ess} P$ where $P \oplus Q = M$. Then $\alpha Q \cong Q$, so (by C2) let $\alpha Q \oplus W = M$. Then $\beta \in S$ is well defined by $\beta(\alpha q + w) = q$, $q \in Q$, $w \in W$. If $\pi^2 = \pi \in S$ satisfies $\pi M = Q$, then $\beta\alpha\pi = \pi$. Define $\tau = \alpha\pi\beta$. Then $\tau^{\,2} = \tau \in \alpha S$ and $(1-\tau)\alpha = \alpha - \alpha\pi\beta\alpha$ is in Δ because $ker(\alpha - \alpha\pi\beta\alpha) \supseteq ker(\alpha) \oplus Q$ and $ker(\alpha) \oplus Q \subseteq^{ess} P \oplus Q = M$ (by Lemma 1.1). It follows that S/Δ is regular and hence that $J(S) \subseteq \Delta$. This proves that $J(S) = \Delta$ and so S is semiregular by Lemma B.40. This proves (1).

In preparation for the proof of (2) and (3), let T be a right ideal of $\bar{S}$ and, by Zorn's lemma, choose a family $\{\bar{\pi}_i\bar{S} \mid i \in I\}$ of nonzero, principal right ideals of $\bar{S}$ maximal such that $\bar{\pi}_i\bar{S} \subseteq T$ for each i and $\Sigma_i\bar{\pi}_i\bar{S}$ is direct. Since $\bar{S}$ is regular, we may assume that each $\bar{\pi}_i$ is an idempotent; since idempotents lift modulo $J(S)$ we may further assume that $\pi_i^2 = \pi_i$ in S. Then $\Sigma_i\pi_i M$ is direct by Lemma 1.24.

(2). Let $\oplus_i\pi_i M \subseteq^{ess} \pi M$, where $\pi^2 = \pi \in S$ (by C1). Since $\pi_i M \subseteq \pi M$ for each i, we have $\bar{\pi}_i\bar{S} \subseteq \bar{\pi}\bar{S}$, so $\oplus_i\bar{\pi}_i\bar{S} \subseteq \bar{\pi}\bar{S}$.

Claim. $\oplus_i\bar{\pi}_i\bar{S} \subseteq^{ess} \bar{\pi}\bar{S}$.

Proof. Suppose that $\bar{\eta}\bar{S} \cap (\oplus_i\bar{\pi}_i\bar{S}) = 0$, where $\bar{\eta} \in \bar{\pi}\bar{S}$. As before, we may assume that $\eta^2 = \eta$ in S. Thus $\eta M \cap (\oplus_i\pi_i M) = 0$ by Lemma 1.24. Since $\bar{\pi}\bar{\eta} = \bar{\eta}$, we have $(\pi\eta - \eta) \in J(S) = \Delta$ and so $(\pi\eta - \eta)K = 0$ for some $K \subseteq^{ess} M$. But this implies that $\eta K \subseteq \pi M$, and it follows that $\eta K = 0$ because $(\oplus_i\pi_i M) \subseteq^{ess} \pi M$. Hence $\eta \in J(S)$, so $\bar{\eta} = 0$. This proves the Claim.

We can now show that $T \subseteq^{ess} \bar{\pi}\bar{S}$. Since $\oplus_i \bar{\pi}_i \bar{S} \subseteq T$ it suffices (by the Claim) to show that $T \subseteq \bar{\pi}\bar{S}$. Let $0 \neq \bar{\tau} \in T$. Since $\oplus_i \bar{\pi}_i \bar{S} \subseteq T$ it follows that $\bar{\tau}\bar{S} \cap (\oplus_i \bar{\pi}_i \bar{S}) \subseteq^{ess} \bar{\tau}\bar{S}$. Also, $\bar{\tau}\bar{S} \cap (\oplus_i \bar{\pi}_i \bar{S}) \subseteq \bar{\tau}\bar{S} \cap \bar{\pi}\bar{S} \subseteq \bar{\tau}\bar{S}$, so we have $\bar{\tau}\bar{S} \cap \bar{\pi}\bar{S} \subseteq^{ess} \bar{\tau}\bar{S}$. But $\bar{\tau}\bar{S} \cap \bar{\pi}\bar{S}$ is generated by an idempotent ($\bar{S}$ is a regular ring) so it follows that $\bar{\tau}\bar{S} \cap \bar{\pi}\bar{S} = \bar{\tau}\bar{S}$. In particular, $\bar{\tau} \in \bar{\pi}\bar{S}$, as required. Thus $\bar{S}$ satisfies C1 and (2) follows ($\bar{S}$ satisfies C2 because it is regular by (1)).

(3). Let $\mathfrak{f} : T \to \bar{S}$ be $\bar{S}$-linear, where T is a right ideal of $\bar{S}$; we must show that $\mathfrak{f} = \bar{\delta}\cdot$ is left multiplication by an element $\bar{\delta}$ of $\bar{S}$. By Lemma 1.7 we may assume that $T \subseteq^{ess} \bar{S}_{\bar{S}}$. Write $\mathfrak{f}(\bar{\pi}_i) = \bar{\sigma}_i$, where $\sigma_i \in S$. Since M is quasi-injective the map $\oplus_{i\in I} \pi_i M \to M$ given by $(\Sigma_i \pi_i m_i) \mapsto \Sigma_i \sigma_i(\pi_i m_i)$ extends to $\delta \in S$. Hence $\delta(\pi_i m) = \sigma_i(\pi_i m)$ for every $m \in M$, so $\delta\pi_i = \sigma_i\pi_i$ in S for each i. But then $\bar{\delta}\bar{\pi}_i = \bar{\sigma}_i\bar{\pi}_i = \mathfrak{f}(\bar{\pi}_i)\bar{\pi}_i = \mathfrak{f}(\bar{\pi}_i)$ for each i, and so $\bar{\delta}\cdot$ agrees with $\mathfrak{f}$ on $\oplus_i \bar{\pi}_i \bar{S}$. This implies that $\bar{\delta}\cdot = \mathfrak{f}$ because $\oplus_i \bar{\pi}_i \bar{S} \subseteq^{ess} \bar{S}_{\bar{S}}$ and $Z(\bar{S}_{\bar{S}}) = 0$ (because $\bar{S}$ is regular). This proves (3). □

If M_R is continuous and $S = end(M_R)$ then $Z(S_S) \subseteq J(S)$ because S is semiregular (Lemma B.40), and we have equality if M is free (see Lemma 7.43). If we specialize Theorem 1.25 to the case of a right continuous ring, we obtain one of the original results about continuity.

Theorem 1.26 (Utumi's Theorem). *If R is right continuous, then R is semiregular, $Z_r = J$, and R/J is right continuous.*

Proof. This is Theorem 1.25 where $M = R_R$, except that $J = \{\alpha \in R \mid ker(\alpha\cdot) \subseteq^{ess} R_R\}$. But $ker(\alpha\cdot) = \mathrm{r}(\alpha)$, so this shows that $J = Z_r$. □

1.4. Quasi-Continuous Modules

If $K \subseteq M$ are modules, recall that a complement C of K in M is a submodule of M maximal with respect to $K \cap C = 0$ and that in this case $K \oplus C \subseteq^{ess} M$ by the Essential lemma (Lemma 1.7). A submodule C of M is said to be *closed* in M if C is the complement of some submodule of M. These submodules will play an important role in what follows, and the following characterizations will be referred to frequently.

Proposition 1.27. *The following conditions are equivalent for a submodule C of a module M:*

(1) C is closed in M.
(2) If $C \subseteq^{ess} N \subseteq M$ then $C = N$.
(3) If $C \subseteq N \subseteq^{ess} M$ then $N/C \subseteq^{ess} M/C$.
(4) If D is any complement of C in M then C is a complement of D in M.

Proof. (1)⇒(2). Let C be a complement of $K \subseteq M$. Given N as in (2), it suffices to show that $K \cap N = 0$. But if $K \cap N \neq 0$ then, as $C \subseteq^{ess} N$, we have $0 \neq C \cap (K \cap N) = C \cap K$, which is a contradiction.

(2)⇒(3). If (3) fails let $C \subseteq N \subseteq^{ess} M$ and assume that $(X/C) \cap (N/C) = 0$, where $X/C \neq 0$. By (2), C is not essential in X, say $Y \cap C = 0$ with $0 \neq Y \subseteq X$. Then $Y \cap N = (Y \cap X) \cap N = Y \cap (X \cap N) = Y \cap C = 0$, which is a contradiction because $N \subseteq^{ess} M$.

(3)⇒(4). If D is a complement of C, let $D \cap N = 0$, where $C \subseteq N$; we must show that $C = N$. We have $D \oplus C \subseteq^{ess} M$ by Lemma 1.7, so $(D \oplus C)/C \subseteq^{ess} M/C$ by (3). Hence it suffices to show that $(N/C) \cap [(D \oplus C)/C] = 0$. But if $n + C = d + C$, $n \in N$, $d \in D$, then $n - d \in C \subseteq N$, so $d \in N \cap D = 0$.

(4)⇒(1). If D is any complement of C in M, then C is a complement of D by (4). □

It is a routine matter to verify that every direct summand of a module M is closed in M and that M is semisimple if and only if every submodule is closed in M (using Lemma 1.7 and (2) of Proposition 1.27). We regard the zero submodule as closed and as essential in itself.

If K is a proper submodule of M, Zorn's lemma provides modules C maximal with respect to the condition $K \subseteq^{ess} C \subseteq M$; these maximal essential extensions of K are closed in M by Proposition 1.27 and will be referred as *closures* of K in M. Direct summands of M have a unique closure in M; if M is nonsingular we can say more, and the result will be needed later.

Lemma 1.28. *Let M_R be a nonsingular module; that is, $Z(M) = 0$. Then every submodule $K \subseteq M$ has a unique closure $\hat{K}$ in M given by*

$$\hat{K} = \{m \in M \mid mI \subseteq K \text{ for some } I \subseteq^{ess} R_R\}.$$

Proof. Let m and m_1 be in $\hat{K}$, say $mI \subseteq K$ and $m_1 I_1 \subseteq K$, where I and I_1 are essential in R_R. Then $m \pm m_1 \in \hat{K}$ because $I \cap I_1 \subseteq^{ess} R_R$, and $ma \in \hat{K}$ for all $a \in R$ because $a^{-1}I = \{r \in R \mid ar \in I\} \subseteq^{ess} R$ by Lemma 1.1. It is clear that $K \subseteq \hat{K}$; we claim that $K \subseteq^{ess} \hat{K}$. For if $0 \neq m \in \hat{K}$, say $mI \subseteq K$, $I \subseteq^{ess} R_R$, then $mI \neq 0$ because $m \notin Z(M)$ and so $0 \neq mI \subseteq K \cap mR$.

Hence it remains to prove that if $K \subseteq^{ess} C \subseteq M$ then $C \subseteq \hat{K}$. But if $c \in C$ then $K \subseteq^{ess} (cR + K)$, so $(cR + K)/K$ is singular by Lemma 1.23. Hence $(c + K)I = 0$ for some $I \subseteq^{ess} R_R$, whence $cI \subseteq K$ and $c \in \hat{K}$. □

If M is a C1 module and $C \subseteq M$ is a closed submodule, then C is essential in a direct summand of M and so *is* a summand of M by (2) of Proposition 1.27. Conversely, if every closed submodule of M is a summand, then every submodule is essential in a summand (any closure), so M is C1. Thus C1 modules are often referred to as *CS modules*, and a ring R is called a right CS *ring* if R_R is a CS module. We have seen that the direct sum of two C1 modules need not be C1 (see the discussion preceding Example 1.20). However, we are going to prove that the C1-condition is inherited by direct summands, and we need the fact that closure is transitive.

Lemma 1.29. *If $C \subseteq D \subseteq M$, where C is closed in D and D is closed in M, then C is closed in M.*

Proof. Let C' be a complement of C in D, and let D' be a complement of D in M. Then $C \oplus C' \subseteq^{ess} D$ and $D \oplus D' \subseteq^{ess} M$, so Proposition 1.27 shows that

$$\frac{C \oplus C'}{C} \subseteq^{ess} \frac{D}{C} \quad \text{and} \quad \frac{D \oplus D'}{D} \subseteq^{ess} \frac{M}{D}. \qquad (*)$$

Claim. $\frac{C+C'+D'}{C} \subseteq^{ess} \frac{M}{C}$.

Proof. Observe first that $(C + C') \cap (C + D') = C$ (in fact, if $c + c' = c_1 + d'$ then $c + c' - c_1 = d' \in D \cap D' = 0$). Hence (*) and Lemma 1.1 give

$$\frac{C + C' + D'}{C} = \frac{C + C'}{C} \oplus \frac{C + D'}{C} \subseteq^{ess} \frac{D}{C} \oplus \frac{C + D'}{C} = \frac{D + D'}{C},$$

so it suffices to show that $\frac{D+D'}{C} \subseteq^{ess} \frac{M}{C}$. To this end, consider $\alpha : \frac{M}{C} \to \frac{M}{D}$ given by $\alpha(m + C) = m + D$. Then $\frac{D+D'}{C} = \alpha^{-1}(\frac{D \oplus D'}{D}) \subseteq^{ess} \frac{M}{C}$ by (*) and Lemma 1.1, proving the Claim.

Now suppose that $C \subseteq^{ess} N \subseteq M$; we must show that $C = N$. We have $C \cap (C' + D') = 0$ (because $c = c' + d'$ means $c - c' = d' \in D \cap D' = 0$), so $N \cap (C' + D') = 0$. Hence $\frac{N}{C} \cap \frac{C+C'+D'}{C} = 0$, and the Claim applies. □

Since a module satisfies the C1-condition if and only if every closed submodule is a direct summand, and since direct summands are closed, it is clear by Lemma 1.29 that every direct summand of a C1 module is again a C1 module. This proves the first part of the following result; the rest is a routine verification.

Proposition 1.30. *Each of the conditions C1, C2, and C3 is inherited by direct summands. In particular, any summand of a (quasi-) continuous module is again (quasi-) continuous.*

The following theorem presents some important characterizations of quasi-continuous modules and clarifies their relationship to quasi-injective modules (see Lemma 1.15 and Corollary 1.16).

Theorem 1.31. *The following conditions on a module M are equivalent:*

(1) M is quasi-continuous.
(2) If C and D are complements of each other then $M = C \oplus D$.
(3) $\tau(M) \subseteq M$ for every $\tau^2 = \tau \in end[E(M)]$.
(4) If $E(M) = \oplus_{i \in I} E_i$ then $M = \oplus_{i \in I}(M \cap E_i)$.

Proof. Write $E(M) = E$ for convenience.

(1)⇒(2). Given C and D as in (2), each is closed by Proposition 1.27, so each is a direct summand by C1. But $C \cap D = 0$, so $C \oplus D = M$ by C3.

(2)⇒(3). Given $\tau^2 = \tau : E \to E$, write $K = M \cap \tau(E)$ and $N = M \cap (1 - \tau)(E)$. Thus $K \cap N = 0$, so let $C \supseteq K$ be a complement of N in M. Thus $C \cap N = 0$, so let $D \supseteq N$ be a complement of C in M. Since C is closed it is also a complement of D (by Proposition 1.27), so $M = C \oplus D$ by (2). Let $\pi : M \to C$ be the projection with kernel D. Because $M \subseteq^{ess} E$, it suffices to show that $M \cap (\tau - \pi)(M) = 0$. But if $m = (\tau - \pi)(x)$, where m and x are in M, then $\tau(x) = m + \pi(x) \in M \cap \tau(E) = K \subseteq C$. Hence $(1 - \tau)(x) \in M \cap (1 - \tau)(E) = N \subseteq D$ so, since $x = \tau(x) + (1 - \tau)(x)$, we have $\pi(x) = \tau(x)$ by the definition of π. Hence $m = 0$, proving (3).

(3)⇒(4). Let $m \in M$, say $m \in E_1 \oplus \cdots \oplus E_n$, and let $\tau_1, \ldots, \tau_n$ be orthogonal idempotents in $end(E)$ with $\tau_i(E) = E_i$ for each i. Then $\tau_i(M) \subseteq M$ for each i by (3), so $m = \Sigma_{i=1}^n \tau_i(m) \in \oplus_{i=1}^n (M \cap E_i)$. It follows that $M \subseteq \oplus_{i \in I}(M \cap E_i)$; the other inclusion is clear.

(4)⇒(1). If $K \subseteq M$ write $E = E(K) \oplus G$. Then $M = (M \cap E(K)) \oplus (M \cap G)$ by (4), and $K \subseteq^{ess} M \cap E(K)$. Thus M satisfies C1. To prove C3, let $K_1 \subseteq^{\oplus} M$ and $K_2 \subseteq^{\oplus} M$, where $K_1 \cap K_2 = 0$. We must show that $(K_1 \oplus K_2) \subseteq^{\oplus} M$. For each i, choose an injective hull $E_i = E(K_i)$ such that $K_i \subseteq E_i \subseteq E$. Then $E_1 \cap E_2 = 0$ because $K_i \subseteq^{ess} E_i$ for each i. Since $E_1 \oplus E_2$ is injective, we get $E = E_1 \oplus E_2 \oplus H$ for some H. Then (4) gives $M = (M \cap E_1) \oplus (M \cap E_2) \oplus (M \cap H)$, so it suffices to show that $K_i = M \cap E_i$ for each i. But this follows because $K_i \subseteq^{ess} M \cap E_i$ (since $K_i \subseteq^{ess} E_i$), and $K_i \subseteq^{\oplus} M \cap E_i$ (since $K_i \subseteq^{\oplus} M$). □

Corollary 1.32. *If $M = K \oplus N$ is quasi-continuous, then K is N-injective.*

Proof. If $X \subseteq N$ and $\alpha : X \to K$ is R-linear, we must extend α to $N \to K$. Put $Y = \{x - \alpha(x) \mid x \in X\}$. Then $Y \cap K = 0$, so let $C \supseteq Y$ be a complement of K in M. Since K is closed in M it is a complement of C by Proposition 1.27, so $M = K \oplus C$ by Theorem 1.31. Let $\pi : M \to K$ be the projection with $ker(\pi) = C$. Then $Y \subseteq ker(\pi)$, so $\pi(x) = \pi[\alpha(x)] = \alpha(x)$ for any $x \in X$. Thus the restriction of π to N extends α. □

Theorem 1.33. *The following conditions are equivalent for a module $M = M_1 \oplus \cdots \oplus M_n$:*

(1) M is quasi-continuous.
(2) Each M_i is quasi-continuous and M_i is M_j-injective for all $i \neq j$.

Proof. (1)⇒(2). Given (1), each M_i is quasi-continuous by Proposition 1.30. If $i \neq j$ then $M_i \oplus M_j$ is quasi-continuous by the same Proposition, so M_i is M_j-injective by Corollary 1.32.

(2)⇒(1). If $N = M_2 \oplus \cdots \oplus M_n$ then N is M_1-injective by Lemma 1.11 and M_1 is N-injective by Lemma 1.13 (Azumaya's lemma). Hence we may assume that $n = 2$. In that case write $E = E(M)$, and choose $E_i = E(M_i) \subseteq E$ for each i. Then

$$E = E(M) = E(M_1 \oplus M_2) = E(M_1) \oplus E(M_2) = E_1 \oplus E_2.$$

So let $\tau^2 = \tau \in end(E)$; by Theorem 1.31 we must show that $\tau(M) \subseteq M$. We can represent τ as a matrix $\tau = \begin{bmatrix} \tau_{11} & \tau_{12} \\ \tau_{21} & \tau_{22} \end{bmatrix}$, where $\tau_{ij} : E_j \to E_i$. Hence $\tau M_1 = \tau_{11} M_1 + \tau_{21} M_1$ and $\tau M_2 = \tau_{12} M_2 + \tau_{22} M_2$. Moreover, $\tau_{21} M_1 \subseteq M_2$ by Lemma 1.14 because M_2 is M_1-injective, and similarly $\tau_{12} M_2 \subseteq M_1$. Hence it suffices to show that $\tau_{11} M_1 \subseteq M_1$ and $\tau_{22} M_2 \subseteq M_2$; we verify the former.

Since $\tau^2 = \tau$ we have $\tau_{11} = \tau_{11}^2 + \tau_{12}\tau_{21}$. For convenience, write $\alpha = \tau_{11}$ and $\beta = 1 - \tau_{11}$, so $\alpha\beta = \beta\alpha = \alpha - \alpha^2 = \beta - \beta^2 = \tau_{12}\tau_{21}$ in $end(E_1)$. Write $K = ker(\alpha\beta)$.

Claim 1. $K = \alpha K \oplus \beta K$.

Proof. If $x \in \alpha K \cap \beta K$ then $\alpha x \in \alpha\beta K = 0$, so $x = x - \alpha x = \beta x \in \beta\alpha K = 0$. Thus $\alpha K \cap \beta K = 0$. We have $\alpha K \subseteq ker(\beta) \subseteq ker(\alpha\beta) = K$ and (similarly) $\beta K \subseteq K$, so $\alpha K \oplus \beta K \subseteq K$. Since $\alpha + \beta = 1$, Claim 1 is proved.

Now, since E_1 is injective choose injective hulls $E(\alpha K) \subseteq E_1$ and $E(\beta K) \subseteq E_1$. By Claim 1, $K = \alpha K \oplus \beta K \subseteq E(\alpha K) \oplus E(\beta K)$, and $E(\alpha K) \oplus E(\beta K)$ is a direct summand of E_1. Hence there exist orthogonal idempotents μ and ν in $end(E_1)$ such that $\alpha K \subseteq \mu E_1$ and $\beta K \subseteq \nu E_1$. It follows that $\mu(\alpha K) = \alpha K$, $\nu(\beta K) = \beta K$, and $\mu(\beta K) = 0 = \nu(\alpha K)$.

Claim 2. The restriction $\alpha_{|\beta\mu E_1}$ is monic.

Proof. Observe first that $\mu K = \mu(\alpha K \oplus \beta K) = \alpha K$, so $K \cap \mu E_1 \subseteq \mu K = \alpha K \subseteq K \cap \mu E_1$. Hence $K \cap \mu E_1 = \alpha K \subseteq ker(\beta)$. Now suppose that $x \in ker(\alpha) \cap \beta\mu E_1$, say $x = \beta\mu e_1$ where $e_1 \in E_1$. Then $\alpha\beta(\mu e_1) = 0$, so $\mu e_1 \in ker(\alpha\beta) = K$. Then $\mu e_1 \in K \cap \mu E_1 \subseteq ker(\beta)$ and so $0 = \beta\mu e_1 = x$. This proves Claim 2.

Write $\iota : \beta\mu E_1 \to E_1$ for the inclusion map in the diagram. Since E_1 is injective there exists $\lambda \in end(E_1)$ such that $\beta\mu = \lambda\alpha\beta\mu$ on E_1. Observe that

$$\begin{array}{ccc} 0 \to \beta\mu E_1 & \overset{\alpha}{\to} & E_1 \\ \iota\downarrow & \swarrow \lambda & \\ E_1 & & \end{array}$$

$\mu(M_1) \subseteq M_1$ because M_1 is quasi-continuous (by Theorem 1.31) and $\mu^2 = \mu \in end(E_1)$, and observe that $\tau_{ij}(M_j) \subseteq M_i$ for $i \neq j$ because M_i is M_j-injective. So we have

$$\beta\mu M_1 = \lambda\alpha\beta\mu M_1 \subseteq \lambda\alpha\beta M_1 = \lambda\tau_{12}\tau_{21} M_1 \subseteq (\lambda\tau_{12}) M_1 \subseteq M_1.$$

Similarly, $\alpha\nu M_1 \subseteq M_1$, so

$$\alpha M_1 = \alpha(\mu + \nu) M_1 \subseteq \alpha\mu M_1 + \alpha\nu M_1 = (1 - \beta)\mu M_1 + \alpha\nu M_2 \subseteq M_1.$$

Since $\alpha = \tau_{11}$, this is what we wanted. □

We have verified that the following implications hold for modules:

semisimple ⇒ quasi-injective ⇒ continuous ⇒ quasi-continuous ⇒ CS.

The next example shows that none of the reverse implications are true, even for the module R_R where R is a ring.

Example 1.34.

(1) An infinite product of fields is a regular self-injective ring that is not semisimple.
(2) If F is a field, $R = \begin{bmatrix} F & F \\ 0 & F \end{bmatrix}$ is a left and right artinian, right CS ring that is not right quasi-continuous.
(3) The integers are a commutative noetherian quasi-continuous ring that is not continuous.

(4) If F_i is a field and $K_i \subseteq F_i$ is a proper subfield for $i \geq 1$, let R denote the set of all sequences in $\prod F_i$ with almost all entries in K_i. Then R is a regular continuous ring that is not self-injective.

We now give a characterization of right self-injective rings in terms of quasi-continuity that will be referred to later. Here a ring R is called right *self-injective* if R_R is an injective R-module.

Theorem 1.35. *The following conditions on a ring R are equivalent:*

(1) R is right self-injective.

(2) $R \oplus R$ is continuous (quasi-continuous) as a right R-module.

(3) $M_2(R)$ is right continuous (quasi-continuous).

(4) $M_n(R)$ is right continuous (quasi-continuous) for all $n \geq 1$.

(5) $M_n(R)$ is right self-injective for all $n \geq 1$.

Proof. (1)⇔(2) follows from Corollary 1.32, (1)⇒(5) because right self-injectivity is a Morita-invariant property, and (5)⇒(4)⇒(3) is clear. Given (3), write $S = M_2(R)$, so that $S = e_{11}S \oplus e_{22}S$, where e_{ii} is the matrix unit. Then S is right quasi-continuous by (3), so $e_{ii}S$ is $e_{jj}S$-injective for $i \neq j$ by Corollary 1.32. As $e_{11}S \cong e_{22}S$, $e_{11}S$ is $e_{jj}S$-injective for all j. Hence $e_{11}S$ is S-injective by Lemma 1.13 and so is injective as a right S-module by Lemma 1.4. Similarly, $e_{22}S$ is injective and it follows that S is right self-injective. Now (1) follows, again because right self-injectivity is a Morita invariant. □

1.5. Quasi-Frobenius Rings

A ring is called *quasi-Frobenius*[4] (or a *QF ring*) if it is left and right artinian and left and right self-injective. Examples include the following:

(1) semisimple, artinian rings;
(2) group algebras FG, where F is a field and G is a finite group; and
(3) rings R/aR, where a is a nonzero, nonunit in a (commutative) principal ideal domain R.

These rings have a natural duality between the right and left ideals (see Theorem 1.50) and are perhaps the most interesting class of nonsemisimple rings. We

[4] There *are* Frobenius rings. They are the quasi-Frobenius rings R in which $S_r \cong (R/J)_R$ and $S_l \cong {}_R(R/J)$, and they arose earlier in the work of Nakayama on duality. The unfortunate terminology is standard in the literature.

will present several basic characterizations of quasi-Frobenius rings in this book; in this section we derive some of the classical characterizations of these rings (Theorem 1.50). This requires some preliminary work that is not without interest in itself.

For convenience, we call a right ideal T of a ring R *extensive* if every R-linear map $\alpha : T \to R$ extends to $R \to R$, that is, if $\alpha = a\cdot$ is left multiplication by an element $a \in R$. Thus R is right self-injective if every right ideal is extensive. We now give a useful result that encompasses many special situations arising in injectivity proofs. It will be used several times throughout this book.

Lemma 1.36. *Let T and T' denote right ideals of a ring R.*

(1) If $T + T'$ is extensive then $\mathtt{l}(T \cap T') = \mathtt{l}(T) + \mathtt{l}(T')$.
(2) Conversely, if $\mathtt{l}(T \cap T') = \mathtt{l}(T) + \mathtt{l}(T')$ and $\alpha : T + T' \to R$ is an R-linear map such that the restrictions $\alpha_{|T}$ and $\alpha_{|T'}$ are given by left multiplication, then α is also given by left multiplication.

Proof. (1). If $b \in \mathtt{l}(T \cap T')$ then $\alpha : T + T' \to R_R$ is well defined by $\alpha(t + t') = bt$, so $\alpha = a\cdot$ for some $a \in R$ by hypothesis. Then $b - a \in \mathtt{l}(T)$ and $a \in \mathtt{l}(T')$, so $b = (b - a) + a \in \mathtt{l}(T) + \mathtt{l}(T')$. Hence $\mathtt{l}(T \cap T') \subseteq \mathtt{l}(T) + \mathtt{l}(T')$; the other inclusion always holds.

(2). Let $\alpha = b\cdot$ on T and let $\alpha = c\cdot$ on T'. Then $b - c \in \mathtt{l}(T \cap T') = \mathtt{l}(T) + \mathtt{l}(T')$, say $b - c = d - d'$, where $dT = 0 = d'T'$. Put $a = b - d = c - d'$. Then $at = (b - d)t = bt = \alpha(t)$ for all $t \in T$ and $at' = (c - d')t' = ct' = \alpha(t')$ for all $t' \in T'$. It follows that $\alpha = a\cdot$, as required. □

Lemma 1.36 is often applied when $T + T' = R$, in which case $\mathtt{l}(T \cap T') = \mathtt{l}(T) \oplus \mathtt{l}(T')$.

A natural generalization of the right self-injective rings is to consider the right *F-injective* rings, where we require only that every *finitely generated* right ideal is extensive. We can now characterize these rings.

Lemmaa 1.37 (Ikeda–Nakayama Lemma). *A ring R is right F-injective if and only if it satisfies the following two conditions:*

(a) $\mathtt{l}(T \cap T') = \mathtt{l}(T) + \mathtt{l}(T')$ for all finitely generated right ideals T and T'.
(b) $\mathtt{lr}(a) = Ra$ for all $a \in R$.

Proof. Assume that R is right F-injective. Then $T + T'$ is extensive for all finitely generated right ideals T and T', so (a) holds by Lemma 1.36. Next, if

$b \in \mathtt{lr}(a)$, the map $\gamma : aR \to R$ is well defined by $\gamma(ar) = br$, so $\gamma = c\cdot$ for some $c \in R$. Hence $b = \gamma(a) = ca \in Ra$, so $\mathtt{lr}(a) \subseteq Ra$; the other inclusion always holds.

Conversely, if the conditions hold and $\alpha : T \to R_R$ is given, induct on n, where $T = \Sigma_{i=1}^n t_i R$. If $n = 1$ then $\alpha(t_1) \in \mathtt{lr}(t_1) = Rt_1$, say $\alpha(t_1) = at_1$. It follows that $\alpha = a\cdot$, so α extends. In general, the restrictions of α to t_1R and to $T' = \Sigma_{i=2}^n t_i R$ both extend to R by induction, so α extends by Lemma 1.36. □

Corollary 1.38. *If R is right self-injective, then*

(1) $\mathtt{l}(T \cap T') = \mathtt{l}(T) + \mathtt{l}(T')$ for all right ideals T and T' of R and
(2) $\mathtt{lr}(L) = L$ for all finitely generated left ideals L of R.

Proof. (1) follows by Lemma 1.36. To prove (2), write $L = \Sigma_{i=1}^n Ra_i$. Then (1) gives $\mathtt{lr}(L) = \mathtt{l}[\cap_{i=1}^n \mathtt{r}(a_i)] = \Sigma_{i=1}^n \mathtt{lr}(a_i) = \Sigma_{i=1}^n Ra_i = L$, using (1) and Lemma 1.37. □

The following companion to the Ikeda–Nakayama lemma also follows from Lemma 1.36 and will be needed later.

Lemma 1.39. *Assume that $\mathtt{lr}(a) = Ra$ for all $a \in R$ and that*

$$\mathtt{l}(T_0 \cap T) = \mathtt{l}(T_0) + \mathtt{l}(T)$$

for all right ideals T_0 and T of R with T_0 finitely generated. Then every R-linear map $\alpha : T \to R$ with finitely generated image extends to $R \to R$.

Proof. Since $\alpha(T)$ is finitely generated, we have $T = T_0 + ker(\alpha)$, where $T_0 = t_1R + \cdots + t_nR$. Since $\mathtt{lr}(t_1) = Rt_1$ by hypothesis, the restriction of α to t_1R extends to R (see the proof of Lemma 1.37). Hence, by induction, the restriction of α to T_0 extends, as does the restriction to $ker(\alpha)$. So α extends by Lemma 1.36. □

Rings R in which $\mathtt{lr}(a) = Ra$ for all a (as in Lemma 1.37) will be extensively studied in Chapter 5. The condition that $\mathtt{lr}(L) = L$ for a left ideal L arises frequently (see Corollary 1.38), and we characterize it (for right ideals) in the next result. A right R-module is called *torsionless* if it can be embedded in a product of copies of R_R.

Lemma 1.40. *If T is a right ideal of R then $\mathtt{rl}(T) = T$ if and only if R/T is torsionless.*

Proof. If $\sigma : R/T \to R^I$ is monic, let $\sigma(1+T) = \langle a_i \rangle$ and write $A = \{a_i \mid i \in I\}$. Then $T = \mathtt{r}(A)$, so $\mathtt{rl}(T) = \mathtt{rlr}(A) = \mathtt{r}(A) = T$. Conversely, if $T = \mathtt{r}\{a_i \mid i \in I\}$, the map $R \to R^I$ given by $r \mapsto \langle a_i r \rangle$ has kernel T. □

Before proceeding with our discussion, we introduce a class of modules that will play a basic role in much of what we do. A module C is said to *cogenerate* a module M if M can be embedded in a direct product C^I of copies of C, and C_R is called a *cogenerator* if it cogenerates every right module. The following lemma is useful.

Lemma 1.41. *A module C_R is a cogenerator if and only if*

$$\cap\{ker(\lambda) \mid \lambda : M \to C\} = 0$$

for all modules M_R.

Proof. If $\sigma : M \to C^I$ is monic and $\pi_i : C^I \to C$ are the projections, then $\cap\{ker(\pi_i \circ \sigma) \mid i \in I\} = 0$. Conversely, if $\cap\{ker(\lambda) \mid \lambda : M \to C\} = 0$ let $I = hom_R(M, C)$ and define $\sigma : M \to C^I$ by $\sigma(m) = \langle \lambda(m) \rangle_{\lambda \in I}$. Then σ is R-monic. □

The cogenerators are dual to the *generators* (that is, the modules G such that every module is an image of a direct sum $G^{(I)}$ for some set I). The ring R is clearly a projective generator, so it is not surprising that the injective cogenerators play an important role in the category of modules. The following lemma gives an important characterization of them.

Lemma 1.42. *If E_R is an injective module, then E is a cogenerator if and only if every simple right module can be embedded in E.*

Proof. If K_R is simple and $\sigma : K \to E^I$ is monic then $\pi_i \circ \sigma \neq 0$ for some projection $\pi_i : E^I \to E$. Hence $K \hookrightarrow E$. Conversely, given M_R write $N = \cap\{ker(\lambda) \mid \lambda : M \to E\}$; it suffices (by Lemma 1.41) to show that $N = 0$. But if $N \neq 0$, let $0 \neq n \in N$ and choose $X \subseteq^{max} nR$. By hypothesis let $\sigma : nR/X \to E$ be monic so, as E is injective, extend σ to $\hat{\sigma} : M/X \to E$. Then define $\lambda : M \to E$ by $\lambda(m) = \hat{\sigma}(m+X)$. As $n \in N$ we have $0 = \lambda(n) = \hat{\sigma}(n+X) = \sigma(n+X)$, whence $n \in X$, which is a contradiction. □

Proposition 1.43. *Let $\{K_i \mid i \in I\}$ be a system of distinct representatives of the simple right R-modules, and write $C = \oplus_{i \in I} E(K_i)$. Then:*

(1) C_R is a cogenerator.
(2) C_R embeds in every cogenerator.

Proof. (1). The injective module $E = \Pi_{i \in I} E(K_i)$ is a cogenerator by Lemma 1.42, and (1) follows because $E \hookrightarrow C^I$.

(2). If D is any cogenerator, let $\sigma : E(K_i) \to D^J$ be monic. Then $\pi_j[\sigma(K_i)] \neq 0$ for some projection $\pi_j : D^J \to D$, $j \in J$, so there exists $\gamma : E(K_i) \to D$ such that $\gamma(K_i) \neq 0$. Thus $K_i \nsubseteq ker(\gamma)$, so $K_i \cap ker(\gamma) = 0$ because K_i is simple. But then γ is monic because $K_i \subseteq^{ess} E(K_i)$. Thus $E(K_i)$ embeds in D for all i. If we regard $E(K_i) \subseteq D$, then $\Sigma_{i \in I} K_i$ is direct because the K_i are pairwise nonisomorphic, and so $\Sigma_{i \in I} E(K_i)$ is direct. It follows that $C \hookrightarrow D$. □

Because of Proposition 1.43, the module $C = \oplus_{i \in I} E(K_i)$ in that result is called a *minimal cogenerator* for the category of right R-modules.

A ring R is called a right *Kasch* ring (or simply *right Kasch*) if every simple right module K embeds in R_R, equivalently if R_R cogenerates K. Every semisimple artinian ring is right and left Kasch, and a local ring R is right Kasch if and only if $S_r \neq 0$ because R has only one simple right module. However, the right Kasch condition is a strong requirement on a ring R [for example, every right Kasch ring with $Z_r = 0$ is semisimple because every maximal right ideal is an annihilator (see the next proposition) and so is a direct summand].

Proposition 1.44. *The following are equivalent for a ring R:*

(1) R is right Kasch.
(2) $hom(M, R_R) \neq 0$ for every finitely generated right R-module M.
(3) $\mathtt{l}(T) \neq 0$ for every proper (respectively maximal) right ideal T of R.
(4) $\mathtt{rl}(T) = T$ for every maximal right ideal T of R.
(5) $E(R_R)$ is a cogenerator.

Proof. (1)⇒(2). This is because every finitely generated module has a simple image.

(2)⇒(3). We may assume that T is maximal. If $0 \neq \sigma : R/T \to R_R$, and if $\sigma(1 + T) = a$, then $0 \neq a \in \mathtt{l}(T)$.

(3)⇒(4). Given (3), $T \subseteq \mathtt{rl}(T) \neq R$, and (4) follows.

(4)⇒(5). If T is a maximal right ideal of R, let $0 \neq a \in \mathtt{l}(T)$ by (4). Then $\gamma : R/T \to R$ is well defined by $\gamma(r + T) = ar$. Since $T \subseteq \mathtt{r}(a) \neq R$, we have $T = \mathtt{r}(a)$, which shows that γ is monic. Thus $R/T \hookrightarrow R \subseteq E(R)$, and (5) follows by Lemma 1.42.

(5)⇒(1). If K_R is simple let $\sigma : K \to E(R)$ be monic by (5). Then $\sigma(K) \cap R \neq 0$ because $R \subseteq^{ess} E(R)$, so $\sigma(K) \subseteq R$ because $\sigma(K)$ is simple. □

It follows that every quasi-Frobenius ring is left and right Kasch because, as we show later in Theorem 1.50, $\mathtt{rl}(T) = T$ and $\mathtt{lr}(L) = L$ hold for all right ideals T and all left ideals L in a quasi-Frobenius ring.

Corollary 1.45. *A right self-injective ring R is right Kasch if and only if $\mathtt{rl}(T) = T$ for every (maximal) right ideal T of R.*

Proof. If R is right Kasch then $R = E(R_R)$ is a cogenerator by Proposition 1.44, so Lemma 1.40 applies. The converse is by Proposition 1.44. □

We now relate the Kasch condition to other ring-theoretic properties that will recur later. Call a ring R a right *C2 ring* if R_R satisfies the C2-condition.

Proposition 1.46. *Let R be any ring. Then*

$$R \text{ is left Kasch} \quad \Rightarrow \quad R \text{ is a right } C2 \text{ ring} \quad \Rightarrow \quad Z_r \subseteq J.$$

Proof. Suppose R is left Kasch. If aR is isomorphic to a summand of R, $a \in R$, it suffices to show that $Ra \subseteq^{\oplus} {}_RR$ (then a is a regular element, so $aR \subseteq^{\oplus} R_R$ too). Since aR is projective, let $\mathtt{r}(a) = (1-e)R$, $e^2 = e$. Then $a = ae$, so $Ra \subseteq Re$, and we claim that $Ra = Re$. If not let $Ra \subseteq M \subseteq^{max} Re$. By the Kasch hypothesis let $\sigma : Re/M \to {}_RR$ be monic and write $c = (e + M)\sigma$. Then $ec = c$ and (since $ae = a \in M$) $c \in \mathtt{r}(a) = (1-e)R$. It follows that $c = ec = 0$ and hence that $e \in M$ since σ is monic. This contradiction shows that $Ra = Re$, as required.

Now assume that R is a right C2 ring and let $a \in Z_r$. Since we have $\mathtt{r}(a) \cap \mathtt{r}(1-a) = 0$ it follows that $\mathtt{r}(1-a) = 0$, whence $(1-a)R \cong R$. By hypothesis $(1-a)R \subseteq^{\oplus} R$, whence $R(1-a) \subseteq^{\oplus} R$, say $R(1-a) = Rg$, $g^2 = g$. It follows that $1 - g \in \mathtt{r}(1-a) = 0$, so $R(1-a) = R$. Since $a \in Z_r$ was arbitrary, this means that $Z_r \subseteq J$. □

The converses of both implications in Proposition 1.46 are false even for commutative rings. If F is a field then $F \times F \times F \times \cdots$ is a C2 ring (it is regular), but it is not Kasch (as already mentioned, right Kasch rings R with

$Z_r = 0$ are artinian). The commutative local ring $\mathbb{Z}_{(2)}$ has $Z_r = 0 \subseteq J$ but it is not a C2 ring (it is a domain that is not a division ring).

We will require the following useful result. Recall that a submodule K of M is called a *small* submodule of M (written $V \subseteq^{sm} M$) if $K + X = M$, with X a submodule, implies that $X = M$.

Lemma 1.47 (Nakayama's Lemma). *If M_R is finitely generated, then*

(1) if $MJ = M$ then $M = 0$ and
(2) $MJ \subseteq^{sm} M$.

Proof. (1). If M is principal, say $M = mR$, then $M = MJ = mJ$. It follows that $m(1 - a) = 0$ for some $a \in J$, so $m = 0$. In general, if $M = m_1R + m_2R + \cdots + m_nR$ and $M = MJ$, then $M = m_1J + m_2J + \cdots + m_nJ$, so let $m_1 = m_1a + y$, where $a \in J$ and $y \in m_2J + \cdots + m_nJ$. It follows that $M = m_2J + \cdots + m_nJ$, so $M = 0$ by induction.

(2). If $MJ + X = M$, where X is a submodule of M, then $(M/X)J = M/X$. Hence (2) follows from (1). □

Recall that a ring R is called *semiperfect* if R/J is semisimple and idempotents lift modulo J, and that R is called *semiprimary* if it it semiperfect and J is nilpotent. These rings are discussed at length in Appendix B, where the properties used in the next three results are derived.

Lemma 1.48. *Suppose that R is a semiperfect ring in which $S_l \subseteq^{ess} R_R$. Then R is left Kasch.*

Proof. Let L be a maximal left ideal of R. Because R is semiperfect, choose $e^2 = e \in R$ such that $(1 - e) \in L$ and $Re \cap L \subseteq J$. Then $\mathrm{r}(Re \cap L) \supseteq \mathrm{r}(J) \supseteq S_l$, so $\mathrm{r}(Re \cap L)$ is essential in R_R by hypothesis. In particular $0 \neq eR \cap \mathrm{r}(Re \cap L) = \mathrm{r}[R(1 - e) \oplus (Re \cap L)] = \mathrm{r}(L)$. Hence R is a left Kasch ring. □

The hypotheses of the next result identify a class of rings called pseudo-Frobenius, and we return to them later.

Lemma 1.49. *Assume that R is a right self-injective, semiperfect ring with the property that $S_r \subseteq^{ess} R_R$. Then R is right and left Kasch.*

Proof. Let $e_1, \ldots, e_n$ be basic idempotents in R, so that R has exactly n isomorphism classes of simple right modules. Hence each e_iR is an indecomposable injective module and so is uniform. Write $S_i = soc(e_iR) = S_r \cap e_iR$. Then $S_i \neq 0$ because $S_r \subseteq^{ess} R_R$, so $S_i \subseteq^{ess} e_iR$ and hence S_i is simple. Thus $e_iR = E(S_i)$ is the injective hull of S_i for each i. But then $S_i \cong S_j$ implies $e_iR \cong e_jR$, whence $i = j$ because the e_i are basic. Thus R contains n nonisomorphic simple right ideals, and it follows that R is right Kasch.

To show that R is left Kasch, it is enough (by Lemma 1.48) to show that $S_r \subseteq S_l$. So let kR be a minimal right ideal of R; we show that $k \in S_l$ by proving that Rk is simple. If $0 \neq b \in Rk$, we must show that $k \in Rb$. If $b = ak$, $a \in R$, define $\gamma : kR \to bR$ by $\gamma(kr) = a(kr) = br$. Then γ is an isomorphism because kR is simple and $ak \neq 0$, so, since R is right self-injective, let $\gamma^{-1} = c\cdot$, $c \in R$. Thus $k = \gamma^{-1}(b) = cb \in Rb$, as required. □

The next theorem contains characterizations of the quasi-Frobenius rings that are part of the folklore of the subject. If b is a class of modules, we say that a module M has the *ascending chain condition* (ACC) on modules in b if $C_1 \subseteq C_2 \subseteq \cdots \subseteq M$, $C_i \in b$, implies $C_n = C_{n+1} = \cdots$ for some n similarly, we have the *descending chain condition* (DCC). If M is a module, let $lat(M)$ denote the lattice of submodules of M.

Theorem 1.50. *The following are equivalent for a ring R:*

(1) R is quasi-Frobenius.
(2) R is left or right artinian, and R is left or right self-injective.
(3) R is left or right noetherian, and R is left or right self-injective.
(4) R has ACC on left or right annihilators, and R is left or right self-injective.
(5) R is right and left noetherian, $\mathtt{rl}(T) = T$ for all right ideals T, and $\mathtt{lr}(L) = L$ for all left ideals L.

In this case the maps

$$f : lat({}_RR) \to lat(R_R) \quad \text{and} \quad g : lat(R_R) \to lat({}_RR)$$

given by $f(L) = \mathtt{r}(L)$ and $g(T) = \mathtt{l}(T)$ are mutually inverse lattice anti-isomorphisms.

Proof. (1)⇒(2)⇒(3)⇒(4) are obvious, and the last sentence follows from the annihilator conditions in (5).

(4)⇒(5). We may assume that R is right self-injective.

Case 1: R Has ACC on Left Annihilators

Then R has ACC on finitely generated left ideals L (by Corollary 1.38), so R is left noetherian. Hence $L = \mathtt{lr}(L)$ for all left ideals, again by Corollary 1.38. Furthermore, $\mathtt{r}(J) \subseteq \mathtt{r}(J^2) \subseteq \cdots$ terminates, say $\mathtt{r}(J^n) = \mathtt{r}(J^{n+1}) = \cdots$. Hence $J^n = \mathtt{lr}(J^n) = \mathtt{lr}(J^{n+1}) = J^{n+1} = JJ^n$, so $J^n = 0$ by Nakayama's lemma (Lemma 1.47). Since R is right self-injective, it is semiregular by Theorem 1.26 and hence semiperfect (being left noetherian), and so it is semiprimary. But then the Hopkins–Levitzki theorem[5] shows that R is left artinian. It remains to show that $\mathtt{rl}(T) = T$ for all right ideals T of R [then R is right noetherian because if $T_1 \subseteq T_2 \subseteq \cdots$ are right ideals then $\mathtt{l}(T_1) \supseteq \mathtt{l}(T_2) \supseteq \cdots$ terminates]. By Corollary 1.45 it suffices to show that R is right Kasch, and this follows by Lemma 1.49 because $S_r \subseteq^{ess} R_R$ (R is semiprimary).

Case 2: R Has ACC on Right Annihilators

Suppose $L_1 \supseteq L_2 \supseteq \cdots$ are finitely generated left ideals in R. Then $\mathtt{r}(L_1) \subseteq \mathtt{r}(L_2) \subseteq \cdots$, so $\mathtt{r}(L_m) = \mathtt{r}(L_{m+1}) = \cdots$ for some m by hypothesis. Hence $L_m = L_{m+1} = \cdots$ by Corollary 1.38. In particular R has DCC on principal left ideals and so is right perfect (see Theorem B.39). Furthermore, J is nilpotent. Indeed, $\mathtt{r}(J) \subseteq \mathtt{r}(J^2) \subseteq \cdots$ implies $\mathtt{r}(J^n) = \mathtt{r}(J^{n+1}) = \cdots$ for some n by hypothesis, and we claim that $J^n = 0$. For if not then $\mathtt{r}(J^n) \neq R$, so (as R is right perfect) $R/\mathtt{r}(J^n)$ contains a minimal left module $X/\mathtt{r}(J^n)$. But then $JX \subseteq \mathtt{r}(J^n)$, so $X \subseteq \mathtt{r}(J^{n+1}) = \mathtt{r}(J^n)$, which is a contradiction. So $J^n = 0$.

Hence $S_r \subseteq^{ess} R_R$, so R is right Kasch by Lemma 1.49. It follows that $\mathtt{rl}(T) = T$ for every right ideal T by Corollary 1.45. In particular every right ideal T is a right annihilator, so R is right noetherian by hypothesis; hence R is right artinian by the Hopkins–Levitzki theorem. This in turn shows that R is left noetherian. [If $L_1 \subseteq L_2 \subseteq \cdots$ are finitely generated left ideals then $\mathtt{r}(L_1) \supseteq \mathtt{r}(L_2) \supseteq \cdots$, so $\mathtt{r}(L_k) = \mathtt{r}(L_{k+1}) = \cdots$ for some k because R is right artinian, whence $L_m = L_{m+1} = \cdots$ by Corollary 1.38.] Finally, every left ideal L is finitely generated (R is left noetherian), so $\mathtt{lr}(L) = L$ by Corollary 1.38.

(5)⇒(1). If $T_1 \supseteq T_2 \supseteq \cdots$ are right ideals, then the chain $\mathtt{l}(T_1) \subseteq \mathtt{l}(T_2) \subseteq \cdots$ terminates because R is left noetherian, and hence $T_1 \supseteq T_2 \supseteq \cdots$ terminates because $\mathtt{rl}(T_i) = T_i$ for each i. Thus R is right artinian. Now observe

[5] The Hopkins–Levitzki theorem states that a semiprimary ring is right artinian if and only if it is right noetherian. In fact, if R is semiprimary, any module is artinian if and only if it is noetherian.

that condition (5) shows that f and g (in the last statement of the theorem) are mutually inverse lattice anti-isomorphisms. In particular, $\mathtt{l}(T \cap T') = \mathtt{l}(T) + \mathtt{l}(T')$ for all right ideals T and T'. Hence R is right F-injective by Lemma 1.37 and (5), and so is right self-injective (being right noetherian). An analogous argument shows that R is left artinian and left self-injective. $\square$

1.6. Pseudo-Frobenius Rings

Before proving the next theorem, we develop another useful condition that a ring is right artinian. It is clear that a module M is finitely generated if and only if $M = \Sigma_{i \in I} N_i$ implies that $M = \Sigma_{i \in F} N_i$ for some finite subset $F \subseteq I$. Dually, M is called *finitely cogenerated*[6] if $\cap_{i \in I} K_i = 0$, $K_i \subseteq M$, implies that $\cap_{i \in F} K_i = 0$ for some finite subset $F \subseteq I$. These modules admit several other useful characterizations.

Lemma 1.51. *The following are equivalent for a module M:*

(1) M is finitely cogenerated.
(2) If M embeds in a direct product $\Pi_{i \in I} N_i$ then M embeds in $\Pi_{i \in F} N_i$ for some finite subset $F \subseteq I$.
(3) M is an essential extension of an artinian submodule.
(4) $soc(M)$ is finitely generated and essential in M.

Proof. (1)⇒(2). If $\sigma : M \to \Pi_{i \in I} N_i$ is monic, write $K_j = ker(\pi_j \circ \sigma)$, where $\pi_j : \Pi_{i \in I} N_i \to N_j$ is the projection. Then $\cap_{j \in I} K_j = 0$, so $\cap_{j \in F} K_j = 0$ for some finite set $F = \{i_1, \ldots, i_n\} \subseteq I$. Then $m \mapsto (\pi_{i_1}\sigma(m), \ldots, \pi_{i_n}\sigma(m))$ embeds M in $\Pi_{i \in F} N_i$.

(2)⇒(3). Let $\{S_i \mid i \in I\}$ be a set of representatives of the simple modules, and write $E = \Pi_{i \in I} E(S_i)$. Then M embeds in a direct product of copies of E by Lemma 1.42 and Proposition 1.43, so it embeds in $\Pi_{i \in F} E(S_i)$ for a finite set F by (2). Since this latter module enjoys the property in (3), so does M.

(3)⇒(4). Let $A \subseteq^{ess} M$, where A is artinian, and write $S = soc(M)$. Then $S \subseteq A$ (every simple submodule of M is in A), so S is artinian and hence is finitely generated. We have $S \subseteq^{ess} M$ because $A \subseteq^{ess} M$ and A is artinian.

(4)⇒(1). Let $\cap_{i \in I} K_i = 0$, $K_i \subseteq M$, and let $\mathcal{F}$ denote the set of all finite intersections of these K_i; we must show that $\mathcal{F}$ contains the zero submodule. If not, write $S = soc(M)$. Since S is artinian, let $S \cap X_0$ be a minimal member

[6] These modules are also called *finitely embedded*.

of $\{S \cap X \mid X \in \mathcal{F}\}$. If $X \in \mathcal{F}$ is arbitrary then $X \cap X_0 \in \mathcal{F}$, and it follows that $S \cap X_0 \subseteq S \cap X$ by minimality. Hence $S \cap X_0 \subseteq \cap_{X \in \mathcal{F}}(S \cap X) = S \cap (\cap_{X \in \mathcal{F}} X) = S \cap (\cap_{i \in I} K_i) = 0$, which is a contradiction because $S \subseteq^{ess} M$. □

Since a module is noetherian if and only if every submodule is finitely generated, the following result shows again that "finitely cogenerated" plays a role dual to the notion of "finitely generated."

Lemma 1.52 (Vámos Lemma). *A module is artinian if and only if every factor module is finitely cogenerated.*

Proof. The condition is clearly necessary. So assume that M/N is finitely cogenerated for all submodules N.

Claim. If M is finitely cogenerated and $N_1 \supseteq N_2 \supseteq \cdots$ are nonzero submodules of M, then $\cap_i N_i \neq 0$.

Proof. Write $S_i = N_i \cap soc(M)$ so that $S_1 \supseteq S_2 \supseteq \cdots$ in $soc(M)$. But $soc(M)$ is artinian (by hypothesis), so $S_k = S_{k+1} = \cdots$ for some k. Hence $(\cap_i N_i) \cap soc(M) = \cap_i S_i = S_k$, and $S_k \neq 0$ because $soc(M) \subseteq^{ess} M$. Thus $\cap_i N_i \neq 0$, proving the Claim.

Now assume that $N_1 \supseteq N_2 \supseteq \cdots$ are submodules of M; we must show that $N_k = N_{k+1} = \cdots$ for some k. If $N = \cap_i N_i$ we have $N_1/N \supseteq N_2/N \supseteq \cdots$ in M/N and $\cap_i (N_i/N) = 0$. Since M/N is finitely cogenerated by hypothesis, the Claim shows that $N_k/N = 0$ for some k. Hence $N = N_k \supseteq N_{k+1} \supseteq \cdots \supseteq N$, and we are done. □

Note that $\mathbb{Z}/N$ is finitely cogenerated for every subgroup $N \neq 0$, so it is not enough in Lemma 1.52 to require that M/N is finitely cogenerated for every *proper* submodule N.

The following two lemmas about projective modules, while interesting in themselves, will be needed in the proof of Theorem 1.56. Recall that the *radical* of a module M is defined to be the intersection $rad(M) = \cap\{N \mid N \subseteq^{max} M\}$ of all maximal submodules of M. We define $rad(M) = M$ if M has no maximal submodules. Clearly, $rad(R_R) = J = rad({}_R R)$, and $MJ \subseteq rad(M)$ for every module M. If M_R is a finitely generated module and $MJ = M$ then Nakayama's lemma (Lemma 1.47) asserts that $M = 0$. The following important result is a version of this for projective modules.

Lemma 1.53. *Let P_R denote a projective module.*

(1) If $PJ = P$ then $P = 0$.
(2) $rad(P) = PJ$ so, if $P \neq 0$, then P has maximal submodules.

Proof. (1). Assume that $P \oplus Q = F$ is free with basis $\{f_i \mid i \in I\}$. For each i write $f_i = p_i + q_i$, where $p_i \in P$ and $q_i \in Q$. If $p \in P$, write $p = \Sigma_i f_i r_i = \Sigma_i p_i r_i + \Sigma_i q_i r_i$, $r_i \in R$; we show that $r_i = 0$ for each i. First, observe that $\Sigma_i q_i r_i \in P \cap Q = 0$. Since $P = PJ \subseteq FJ$, write $p_i = \Sigma_j f_j a_{ij}$, where each $a_{ij} \in J$. Hence $q_i = \Sigma_j f_j(\delta_{ij} - a_{ij})$, where $\delta_{ij} = 1$ for $i = j$ or 0 for $i \neq j$. Then

$$0 = \Sigma_i q_i r_i = \Sigma_i[\Sigma_j f_j(\delta_{ij} - a_{ij})]r_i = \Sigma_j f_j[\Sigma_i(\delta_{ij} - a_{ij})r_i],$$

so $\Sigma_i(\delta_{ij} - a_{ij})r_i = 0$ for each j. Inserting zeros where necessary, we may assume that i and j range over $\{1, 2, \ldots, n\}$, so the $n \times n$ matrix $[\delta_{ij} - a_{ij}]$ is invertible in $M_n(R)$. It follows that each $r_i = 0$, proving (1).

(2). Since $rad(\oplus_i M_i) = \oplus_i rad(M_i)$ for any modules M_i, we obtain $rad(R^{(I)}) = J^{(I)} = (R^{(I)})J$. Hence $rad(F) = FJ$ for every free module F. Now suppose that $P \oplus Q = F$ is free. Then $PJ \oplus QJ = FJ = rad(P \oplus Q) = rad(P) \oplus rad(Q)$. Since $PJ \subseteq rad(P)$ for any module P, it follows that $rad(P) = PJ$. With this, (2) follows from (1). □

Lemma 1.54. *Let $P_R \neq 0$ be projective. Then the following are equivalent:*

(1) P is not the sum of two proper submodules.
(2) $rad(P)$ is a maximal submodule of P that is small in P.
(3) $rad(P)$ is a proper submodule of P that contains all proper submodules.
(4) $end(P)$ is local.

Proof. (1)⇒(2). Lemma 1.53 shows that P *has* a maximal submodule; it is unique and small by (1). Now (2) follows.

(2)⇒(3). $rad(P) \neq P$ by Lemma 1.53. Let $X \neq P$ be a submodule of P. If $X \nsubseteq rad(P)$ then $X + rad(P) = P$ because $rad(P)$ is maximal, so $X = P$ because $rad(P)$ is small in P. This proves (3).

(3)⇒(4). Write $E = end(P)$ and $W = rad(P)$, and define $A = \{\alpha \in E \mid \alpha(P) \subseteq W\}$. Then A is an ideal of E and $(1 - \alpha)(P) \nsubseteq W$ for every $\alpha \in A$. Hence $1 - \alpha$ is epic by (3) and so has a right inverse in E because P is projective. Hence $A \subseteq J(E)$. But if $\lambda \in E - J(E)$ then $\lambda \notin A$, so (as before) λ has a right inverse in E. It follows that $E - J(E)$ consists of units, proving (4).

(4)$\Rightarrow$(1). Let $P = K + N$, where K and N are submodules. If $\phi : P \to P/K$ is the coset map, then the restriction $\phi_{|N} : N \to P/K$ is epic, so, since P is projective, there exists $\lambda : P \to N$ such that $\phi \circ \lambda = \phi$. Hence $\lambda(P) \subseteq N$ and

$$\begin{array}{ccc} & & P \\ & \lambda \swarrow & \downarrow \phi \\ N & \overset{\phi_{|N}}{\to} & P/K \to 0 \end{array}$$

$(1 - \lambda)(P) \subseteq K$, so (1) follows because one of λ and $1 - \lambda$ is a unit in $end(P)$ by (4). □

If M is a right R-module define the *trace* of M by $trace_R(M) = \Sigma_{\lambda \in M^*} \lambda(M)$, where $M^* = hom_R(M_R, R)$ denotes the *dual* of M. The trace of M is a two-sided ideal of R. Recall that a right R-module G is called a *generator* if every right module M is an image of a direct sum $G^{(I)}$ of copies of G.

Lemma 1.55. *The following conditions are equivalent for a module G_R:*

(1) G_R is a generator.
(2) $trace_R(G) = R$, equivalently $1 = \Sigma_{i=1}^n \lambda_i(g_i)$, $\lambda_i \in G^*$, $g_i \in G$.
(3) $G^n \cong R \oplus N$ for some integer n and submodule $N \subseteq G^n$.

Proof. (1)$\Rightarrow$(2). Because R_R is finitely generated, (1) gives an epimorphism $\theta : G^n \to R_R$ for some n. If $\sigma_i : G \to G^n$ are the injections, then $\lambda_i = \theta \circ \sigma_i$ is in G^* for each i and $R = \lambda_1(G) + \cdots + \lambda_n(G)$.

(2)$\Rightarrow$(3). Given (2), define $\theta : G^n \to R$ by $\theta(x_1, \ldots, x_n) = \Sigma \lambda_i(x_i)$. Then θ is epic by (2) and splits because R_R is projective. This proves (3).

(3)$\Rightarrow$(1). Given any module M_R there is an epimorphism $R^{(J)} \to M$ for some set J. Since there is an epimorphism $G^n \to R$ by (3), we obtain an epimorphism $(G^n)^{(J)} \to M$, proving (1). □

Every generator is faithful by (3) of Lemma 1.55, and the rings for which every faithful right module is a generator are called *right pseudo-Frobenius* rings (right *PF rings*). Like the quasi-Frobenius rings, these rings admit several characterizations.

Theorem 1.56 (Azumaya–Kato–Osofsky–Utumi Theorem). *The following conditions are equivalent for a ring R:*

(1) R is right pseudo-Frobenius.
(2) Every right cogenerator is a generator.

(3) R is right self-injective and S_r is finitely generated and essential in R_R.
(4) R is right self-injective and semiperfect, and S_r is essential in R_R.
(5) R_R is a cogenerator and R is left Kasch.
(6) R_R is a cogenerator and there are only a finite number of nonisomorphic simple right R-modules.

Proof. (1)⇒(2). If M is a cogenerator, then R embeds in a direct product of copies of M and so M is faithful. Hence M is a generator by (1).

(2)⇒(3). If $\{K_i \mid i \in I\}$ are distinct representatives of the simple right R-modules, Proposition 1.43 shows that $E = \oplus_{i \in I} E(K_i)$ is a cogenerator. Since E is a generator by (2), R is (isomorphic to) a direct summand of E^n for some $n \geq 1$, and so it is a direct summand of $\oplus_{i=1}^m E(K_i)$ for some $m \geq 1$. But $\oplus_{i=1}^m E(K_i)$ is injective with finitely generated essential socle, so R_R has the same property. This proves (3).

(3)⇒(4). R is semiregular by Utumi's theorem (Theorem 1.26) and contains no infinite set of orthogonal idempotents because it is right finite dimensional (by hypothesis). Hence R is semiperfect.

(4)⇒(5) and (4)⇒(6). Given (4), R is left and right Kasch by Lemma 1.49. Since R_R is injective it is a cogenerator by Proposition 1.44. This proves (5); it also gives (6) because R is semiperfect.

(5)⇒(1). Let M_R be a faithful right R-module and write $T = trace_R(M)$. By Lemma 1.55 we must show that $T = R$; since R is left Kasch it suffices to prove that $\mathtt{r}(T) = 0$. If $x \in \mathtt{r}(T)$, then $\lambda(mx) = \lambda(m)x = 0$ for every $\lambda \in M^*$ and $m \in M$, so $Mx \subseteq \cap_{\lambda \in M^*} ker(\lambda)$. But $\cap_{\lambda \in M^*} ker(\lambda) = 0$ because R_R is a cogenerator, and it follows that $x = 0$ because M is faithful.

(6)⇒(4). We begin with the following claim:

Claim. If K is a simple right R-module then $E(K)$ embeds in R_R.

Proof. Since R_R is a cogenerator, let $\sigma : E(K) \to R^I$ be a monomorphism for some index set I. If $\pi_i : R^I \to R$ is the ith projection then $\pi_i[\sigma(K)] \neq 0$ for some i, and it suffices to show that $ker(\pi_i \circ \sigma) = 0$. But if $ker(\pi_i \circ \sigma) \neq 0$ then $ker(\pi_i \circ \sigma) \cap K \neq 0$ because $K \subseteq^{ess} E(K)$, whence $K \subseteq ker(\pi_i \circ \sigma)$ because K is simple. This contradiction proves the Claim.

By hypothesis, let $K_1, \ldots, K_n$ denote a set of representatives of the simple right R-modules, and write $E_i = E(K_i)$ for each i. Then each E_i embeds in R_R by the Claim, so, being injective, E_i is a summand of R and so is projective. Moreover, E_i is indecomposable (actually uniform), so Lemma 1.54 shows that $rad(E_i)$ is a maximal, small submodule of E_i. Hence $S_i = E_i/rad(E_i)$ is simple and E_i is a projective cover of S_i. Furthermore, if $S_i \cong S_j$ then $E_i \cong E_j$

by the uniqueness of projective covers (see Corollary B.17). This means that $K_i \cong K_j$ and hence that $i = j$. It follows that $\{S_1, S_2, \ldots, S_n\}$ is a set of distinct representatives of the simple right R-modules, from which we infer that every simple right R-module has a projective cover. Thus R is a semiperfect ring by Theorem B.21.

Hence write $R = \oplus_{i=1}^{m} e_i R$, where $e_1, \ldots, e_m$ are orthogonal, primitive idempotents with $\Sigma_{i=1}^{m} e_i = 1$. If $\{e_1, \ldots, e_n\}$ are a basic set of idempotents then, since each E_i is indecomposable and projective, $E_i \cong e_{\sigma(i)}R$ for some $\sigma(i) \in \{1, 2, \ldots, n\}$. But the E_i are pairwise nonisomorphic, so it follows that $e_i R$ is an injective module with simple essential socle for each $i = 1, 2, \ldots, n$. Thus R is right self-injective and $S_r \subseteq^{ess} R_R$, proving (4). □

There is another characterization of these pseudo-Frobenius rings that will be proved in Chapter 7 as a consequence of a more general result (Corollary 7.33).

Theorem 1.57 (Osofsky's Theorem). *A ring is right pseudo-Frobenius if and only if it is a right self-injective ring that is a right cogenerator.*

Notes on Chapter 1

Divisible groups were introduced by Baer [14], who essentially proved that every module can be embedded in an injective module. The general concept of an injective module is due to Eckmann and Schopf [50], and the existence of an injective hull seems to have been discovered independently by Eckmann and Schopf and by Shoda [204]. A good survey of the properties of injective modules can be found in Sharpe and Vámos [203].

Quasi-injective modules were introduced and characterized by Johnson and Wong [113], and relative injectivity in general was studied by various authors including Harada [90, 94] and Azumaya [10].

The C1-, C2-, and C3-conditions were identified by Utumi [219] for rings, and were extended to the module case by Jeremy [108], Takeuchi [214], and Mohammed and Bouhy [146]. An account of continuous and quasi-continuous modules can be found in Mohamed and Müller [147].

The fact that closures are unique in a nonsingular module (Lemma 1.28) is due to Johnson [112].

The term "Kasch ring" honors F. Kasch.

Quasi-Frobenius rings were first investigated by Nakayama [150], who was interested in the natural duality between right and left ideals. Theorem 1.50, with its characterizations of quasi-Frobenius rings, is associated with the names Faith, Ikeda, Nakayama, and Eilenberg (see Faith [54]). It was first stated in this generality by Faith [54], where condition (4) was introduced.

The extension of Nakayama's lemma to projective modules (Lemma 1.53) is due to Bass [16].

Left and right pseudo-Frobenius rings (in Theorem 1.56) are called rings with perfect duality in the literature. They were introduced and investigated independently by Azumaya [11], Kato [121, 122], Utumi [221], and Osofsky [182] as generalizations of quasi-Frobenius rings where the chain conditions are dropped.

2

Mininjective Rings

A surprising amount of information about right self-injective rings can be obtained by studying a much larger class of rings, the right mininjective rings. A ring is called right mininjective if every isomorphism between two simple right ideals is given by left multiplication. The basic general facts about these rings are derived in this chapter, and this work serves as a basis for the study of two important subclasses: the P-injective rings in Chapter 5 and the simple injective rings in Chapter 6.

After giving several examples (including a right mininjective ring that is not left mininjective), we show that mininjectivity is a Morita invariant and that "min" versions of the C2- and C3-conditions hold. (The "min" version of the C1-condition is studied in Chapter 4.) Surprisingly, under a mild commutativity condition, a ring R is right minininjective if and only if its right socle is square-free, and every factor ring of R is right mininjective if and only if R has a distributive lattice of right ideals. In general, it is shown that the right socle of any right mininjective ring R is contained in the left socle (in fact, if kR is a simple right ideal then Rk is also simple). This remarkable fact is used repeatedly throughout the book. If a right mininjective ring is semiregular and has essential right socle, we show that the right singular ideal equals the Jacobson radical, extending the situation for right self-injective rings.

Right mininjective rings admit another important characterization: A ring is right mininjective if and only if the dual of every simple right module is simple or zero. This result is used to give a simple proof of Ikeda's theorem: R is quasi-Frobenius if and only if it is right and left artinian and right and left mininjective. The chapter concludes with an investigation of the relationship between mininjectivity and the condition that every minimal left ideal is an annihilator.

2.1. Definition and Examples

If R is a ring, a module M_R is called *mininjective* if, for every simple right ideal K of R, each R-linear map $\gamma : K \to M_R$ extends to $\bar{\gamma} : R \to M$; that is, $\gamma = m\cdot$ is multiplication by some $m \in M$ [in fact $m = \bar{\gamma}(1)$].

Recall that a right ideal T in a ring R is called *extensive* if every R-morphism $\gamma : T \to R_R$ can be extended to $R_R \to R_R$, equivalently if $\gamma = c\cdot$ is left multiplication by an element $c \in R$. Thus, R is right mininjective if and only if every simple right ideal K is extensive.

Lemma 2.1. *The following are equivalent for a ring R:*

(1) R is right mininjective.
(2) If kR is simple, $k \in R$, then $\mathtt{lr}(k) = Rk$.
(3) If kR is simple and $\mathtt{r}(k) \subseteq \mathtt{r}(a)$, $k, a \in R$, then $Ra \subseteq Rk$.
(4) If kR is simple and $\gamma : kR \to R$ is R-linear, $k \in R$, then $\gamma(k) \in Rk$.

Proof. (1)⇒(2). Always $Rk \subseteq \mathtt{lr}(k)$. If $a \in \mathtt{lr}(k)$ then $\mathtt{r}(k) \subseteq \mathtt{r}(a)$, so $\gamma : kR \to R$ is well defined by $\gamma(kr) = ar$. Thus $\gamma = c\cdot$ for some $c \in R$ by (1), whence $a = \gamma(k) = ck \in Rk$.

(2)⇒(3). If $\mathtt{r}(k) \subseteq \mathtt{r}(a)$ then $a \in \mathtt{lr}(k)$, so $a \in Rk$ by (2).

(3)⇒(4). If $\gamma(k) = a$ then $\mathtt{r}(k) \subseteq \mathtt{r}(a)$, so $a \in Rk$ by (3).

(4)⇒(1). Let $\gamma : kR \to R_R$ and use (4) to write $\gamma(k) = ck$, $c \in R$. Then $\gamma = c\cdot$, proving (1). □

In practice, it is enough to verify conditions (2), (3), or (4) in Lemma 2.1 for simple right ideals $kR \subseteq J$. Indeed, the proof of (1)⇒(2)⇒(3)⇒(4) goes through as before in this case. If $\gamma : kR \to R$ is R-linear, where $kR \nsubseteq J$ is simple, then $(kR)^2 \neq 0$, so[1] $kR = eR$ for some idempotent $e = e^2 \in R$, and γ extends.

We note in passing that the proof of Lemma 2.1 actually gives equivalent conditions that *any* principal right ideal kR is extensive.

Note further that a simple right ideal K need not be extensive even if it isomorphic to an extensive right ideal. For example, let $R = \begin{bmatrix} D & D \\ 0 & D \end{bmatrix}$, where D is a division ring and write $k = \begin{bmatrix} 0 & 1 \\ 0 & 0 \end{bmatrix}$ and $e = \begin{bmatrix} 0 & 0 \\ 0 & 1 \end{bmatrix}$. Then $kR \cong eR$ and $e^2 = e$, but kR is not extensive by Lemma 2.1 because $\mathtt{lr}(k) \neq Rk$. Thus R is not right mininjective even though it has a projective, homogeneous right socle.

[1] If $kK \neq 0$, $k \in K$, then $kK = K$ because K is simple, say $ke = k$, where $0 \neq e \in K$. Hence $e^2 - e \in K_0 = \{a \in K \mid ka = 0\}$, so $e^2 = e$ because $K_0 \neq K$ and K is simple. It follows that $K = eR$, again by simplicity. This result is sometimes called *Brauer's lemma*.

Clearly every right self-injective ring is right mininjective, as is every ring in which every simple right ideal is a direct summand. Thus every ring with zero right socle is right mininjective, and every semiprime ring (in particular R/J) is right and left mininjective because every simple right or left ideal is a direct summand. In particular, the ring $\mathbb{Z}$ of integers is a commutative, noetherian mininjective ring that is not self-injective.

Example 2.2 follows easily from (2) of Lemma 2.1.

Example 2.2. A direct product $R = \Pi R_i$ of rings is right mininjective if and only if R_i is right mininjective for each i.

Example 2.3. Every polynomial ring $R[x]$ is right and left mininjective.

Proof. This is because both socles of $R[x]$ are zero. For example, if $K = kR[x]$ is simple where $deg(k) = n$, then $K = x^{n+1}K$ because x^{n+1} is central, so $k \in x^{n+1}kR[x]$, which is a contradiction. □

Example 2.4. If S_r is simple as a left ideal, then R is right mininjective.

Proof. If kR is simple and $\mathtt{r}(k) \subseteq \mathtt{r}(a)$, where $k, a \in R$, we must show that $a \in Rk$. This is clear if $a = 0$; if $a \neq 0$, then $\mathtt{r}(k) = \mathtt{r}(a)$ because $\mathtt{r}(k)$ is maximal, so aR is simple too. Hence $Ra = S_r = Rk$ by hypothesis. □

The next example shows, among other things, that a right mininjective ring need not be left mininjective; it will be referred to several times in the following. Recall that a ring R is called *local* if R/J is a division ring, equivalently if R has a unique maximal right (left) ideal.

Example 2.5 (Björk Example). Let F be a field and assume that $a \mapsto \bar{a}$ is an isomorphism $F \to \bar{F} \subseteq F$, where the subfield $\bar{F} \neq F$. Let R denote the left vector space on basis $\{1, t\}$, and make R into an F-algebra by defining $t^2 = 0$ and $ta = \bar{a}t$ for all $a \in F$. Then the following are true:

(1) R is local, $R/J \cong F$, and $J^2 = 0$.
(2) $J = Rt = Ft$ is the only proper left ideal of R.
(3) R is right mininjective but not left mininjective.
(4) $X \mapsto Xt$ is a lattice isomorphism from the right $\bar{F}$-subspaces X of F to the lattice of right ideals Xt of R contained in J.
(5) R is left artinian, and the following are equivalent:
    (a) R is right artinian;
    (b) R is right noetherian;

(c) R is right finite dimensional;
(d) ${}_{\bar{F}}F$ is finite dimensional.
Furthermore, if p is a prime and $F = \mathbb{Z}_p(x)$ is the field of rational forms over $\mathbb{Z}_p$, then the map $w \longmapsto w^p$ is an isomorphism $F \to \bar{F}$, where $\bar{F} = \{w^p \mid w \in F\}$ and $dim({}_{\bar{F}}F) = p$.

Proof. (1) and (2) are routine verifications.

(3). R is right mininjective by Example 2.4. If $d \in F - \bar{F}$, define $\gamma : Ft \to R$ by $\gamma(at) = adt$. This is well defined, and it is left R-linear because $\gamma[(b + ct)at] = (ba)dt = (b + ct) \cdot \gamma(at)$. If R is left mininjective then $\gamma = \cdot c$ is right multiplication by some $c = e + ft$ in R. But then $dt = \gamma(t) = tc = te = \bar{e}t$, which implies that $d = \bar{e} \in \bar{F}$, a contradiction.

(4). Let $P \subseteq J$ be a right ideal of R, and define $X = \{x \in F \mid xt \in P\}$. Then $Xt \subseteq P$ is clear, and the reverse inclusion follows because $P \subseteq J = Ft$. This shows that the map in (4) is onto; it is routine to show that it is one-to-one and preserves sums and intersections.

(5). R is left artinian by (2); the equivalence of (a), (b), (c), and (d) follows from (4); and the last observation is a routine calculation (in fact $\{1, x, \ldots, x^{p-1}\}$) is a basis of F over $\bar{F}$. □

The next example presents a commutative, local mininjective ring with $J^3 = 0$, which has a simple, essential socle that is not artinian. This example will also be cited frequently in what follows.

Example 2.6 (Camillo Example). Let $R = F[x_1, x_2, \ldots]$, where F is a field and the x_i are commuting indeterminants satisfying the relations

$$x_i^3 = 0 \text{ for all } i, \quad x_i x_j = 0 \text{ for all } i \neq j, \quad \text{and} \quad x_i^2 = x_j^2 \text{ for all } i \text{ and } j.$$

Write $m = x_1^2 = x_2^2 = \cdots$, so that $m^2 = 0 = x_i m$ for all i. Then the following are true:

(1) R is a commutative ring with F-basis $\{1, m, x_1, x_2, \ldots\}$.
(2) R is local, $J = span_F\{m, x_1, x_2, \ldots\}$, $R/J \cong F$, and $J^3 = 0$.
(3) $Fm \subseteq A$ for every ideal $A \neq 0$ of R.
(4) $soc(R) = J^2 = Fm$ is simple and essential in R.
(5) R is mininjective with simple essential socle, but R is not noetherian.
(6) If we denote $X = span_F\{x_1, x_2, \ldots\}$, the maps

$$A \mapsto A \cap X \quad \text{and} \quad U \mapsto Fm \oplus U$$

are mutually inverse lattice isomorphisms between the lattice of all ideals $A \neq 0, R$ of R, and the lattice of all F-subspaces U of X.

Proof. (1) and (2) and (3) are routine verifications, and then (4) follows because the product of any two elements of J is in Fm. Next, R is mininjective by (4) and Example 2.4, and R is not noetherian by (6) because $Fm \subset Fm \oplus Fx_1 \subset Fm \oplus Fx_1 \oplus Fx_2 \subset \cdots$.

As to (6), observe first that $Fm \oplus U$ is an ideal for each U because $(Fm \oplus U)R \subseteq FmR + UR \subseteq Fm + (UF + mF) \subseteq Fm + U$. The composites of the maps in (6) are $U \mapsto Fm \oplus U \mapsto (Fm \oplus U) \cap X = U$ because $U \subseteq X$, and $A \mapsto A \cap X \mapsto Fm \oplus (A \cap X) = A$ by the modular law because $Fm \subseteq A \subseteq J = Fm \oplus X$ using (3). Hence the two maps are mutual inverses; they clearly preserve inclusions. Note that $J = Fm \oplus X$. □

If R is a ring and ${}_RV_R$ is a bimodule, the *trivial extension* $T(R, V)$ of R over V is the direct sum $R \oplus V$ with multiplication $(a, v)(b, w) = (ab, aw + vb)$. It is sometimes useful to view $T(R, V)$ as the set of all matrices $\left[\begin{smallmatrix} a & v \\ 0 & a \end{smallmatrix}\right]$, where $a \in R$ and $v \in V$, using matrix multiplication.

In particular, if $A = Fm \oplus U$ as in Example 2.6 then $R/A \cong T(F, V)$, where V is any F-subspace of X such that $U \oplus V = X$. We will return to this later; we observe for now that $T(F, V)$ is mininjective if and only if $dim_F(V) = 1$.

A semisimple module is said to be *square-free* if it contains at most one copy of any simple module. The condition that the right socle of a ring is square-free is in a sense dual to the requirement that the ring is right mininjective.

Lemma 2.7. *The following are equivalent for a ring R:*

(1) The right socle S_r is square-free.
(2) If kR is simple and $\gamma : kR \to R$ is R-linear then $\gamma(k) \in kR$.
(3) If kR is simple then $\mathtt{lr}(k) \subseteq kR$.

Proof. (1)⇒(2). If γ is as in (2), then $\gamma(k)R = \gamma(kR) \subseteq kR$ by (1).

(2)⇒(3). If $a \in \mathtt{lr}(k)$ then $\mathtt{r}(k) \subseteq \mathtt{r}(a)$, so $\gamma : kR \to R$ is well defined by $\gamma(kr) = ar$. By (2) $a = \gamma(k) \in kR$.

(3)⇒(1). Let $\gamma : K \to M$ be an isomorphism of simple right ideals. If $K = kR$, write $\gamma(k) = m$. Then $\mathtt{r}(k) \subseteq \mathtt{r}(m)$, so $m \in \mathtt{lr}(k) \subseteq kR = K$ by (3). Thus $M \subseteq K$, whence $M = K$. □

Note that if we ask instead that $\gamma(k) \in Rk$ in (2) of Lemma 2.7, we characterize right mininjectivity by Lemma 2.1. Observe further that if S_r is square-free then (3) of Lemma 2.7 shows that every simple right ideal kR is two-sided, that is $Rk \subseteq kR$. Moreover, the converse holds if R is right mininjective by

Lemmas 2.1 and 2.7. If we insist that $Rk = kR$, we obtain the following result, which includes a characterization of the commutative mininjective rings.

Proposition 2.8. *Assume that $Rk = kR$ whenever kR is a simple right ideal of the ring R. Then the following hold:*

(1) R is right mininjective if and only if S_r is square-free.
(2) A commutative ring is mininjective if and only if S_r is square-free.

A module M is called *distributive* if $A \cap (B + C) = (A \cap B) + (A \cap C)$ for all submodules A, B, and C of M. We need the first of the following characterizations of when this happens. If x and y are elements of M_R, write $(xR : y) = \{r \in R \mid yr \in xR\}$.

Theorem 2.9. *The following are equivalent for a module M_R:*

(1) M_R is distributive.
(2) M/N has square-free socle for every submodule $N \subseteq M$.
(3) $(xR : y) + (yR : x) = R$ for all x and y in M.

Proof. (1)⇒(2). Since images of distributive modules are distributive, we may assume $N = 0$. Suppose $\sigma : X \to Y$ is an isomorphism, where X and Y are simple submodules of M with $X \cap Y = 0$, and let $K = \{x + \sigma(x) \mid x \in X\}$. Then $K \cap Y = 0$ and $K \subseteq X + Y$, so $K = K \cap X$ by (1). It follows that $Y \subseteq X$, which is a contradiction.

(2)⇒(3). If (3) fails, let $(xR : y) + (yR : x) \subseteq T \subseteq^{max} R_R$, $x, y \in M$. Write $N = xT + yT$ and $Z = (xR + yR)/N \subseteq M/N$, so that Z has square-free socle by (2). Denote $\bar{x} = x + N$ and $\bar{y} = y + N$, and consider the map $\theta : R \to \bar{x}R$ given by $\theta(r) = \bar{x}r$. Then $T \subseteq ker\theta$, so either $\bar{x}R = 0$ or $\bar{x}R \cong R/T$. Similarly, either $\bar{y}R = 0$ or $\bar{y}R \cong R/T$. But if $\bar{x}R \cong R/T \cong \bar{y}R$ then $\bar{x}R = \bar{y}R$ by (2) and so, in any case, either $\bar{x} \in \bar{y}R$ or $\bar{y} \in \bar{x}R$. If $\bar{x} \in \bar{y}R$ then $x \in xT + yR$, say $x = xt + yr$, $t \in T$, $r \in R$. This implies that $1 - t \in (yR : x)$, so $(yR : x) \nsubseteq T$, which is a contradiction. Similarly, $\bar{y} \in \bar{x}R$ leads to a contradiction.

(3)⇒(1). Observe first that, for all $z, w \in M$, then $(zR : z + w) = (zR : w)$ and $w(zR : w) = zR \cap wR = z(wR : z)$. Hence, for all $x, y \in M$,

$$
\begin{aligned}
&(xR \cap (x + y)R) + (yR \cap (x + y)R) \\
&\qquad = (x + y)(xR : x + y) + (x + y)(yR : x + y) \\
&\qquad = (x + y)(xR : y) + (x + y)(yR : x) = (x + y)R
\end{aligned}
$$

using (3). It is a routine matter to show that this implies (1). □

Theorem 2.9 (with Proposition 2.8) gives a characterization of when a commutative ring is distributive. In fact, we get a version for duo rings, where a ring R is called a *duo ring* if every one-sided ideal is two-sided (equivalently $aR = Ra$ for all $a \in R$).

Theorem 2.10. *The following are equivalent for a duo ring R:*

(1) Every factor ring R/A is right mininjective.
(2) R_R is distributive.

Proof. Observe that the duo hypothesis is inherited by factors R/A. If (1) holds then $(R/A)_{R/A}$ has square-free socle by Proposition 2.8. Hence $(R/A)_R$ has square-free socle and (2) follows from Theorem 2.9. Conversely, (2) implies that $(R/A)_{R/A} = (R/A)_R$ has square-free socle, again by Theorem 2.9, so (1) follows from Proposition 2.8. □

Note that every regular ring R has the property that R/A is right (and left) mininjective for all ideals A, but R_R need not be distributive. For example, if F is a field then $R = M_2(F)$ does not have square-free socle, so R_R is not distributive by Theorem 2.9. Hence the duo hypothesis cannot be dispensed with in Theorem 2.10.

2.2. Morita Invariance

As for right self-injectivity, right mininjectivity turns out to be a Morita invariant. The next result gives half the proof.

Proposition 2.11. *If R is right mininjective, so is eRe for all $e^2 = e \in R$ satisfying $ReR = R$.*

Proof. Write $S = eRe$ and let $\mathrm{r}_S(k) \subseteq \mathrm{r}_S(a)$, where $k, a \in S$ and kS is a simple right ideal of S. We claim first that kR is simple in R. For if $kr \neq 0$, $r \in R$, then $krReR \neq 0$, so there exists $t \in R$ such that $0 \neq krte = (ke)rte \in kS$. Hence $k \in krteS \subseteq krR$, whence kR is simple. Thus it suffices to show that $\mathrm{r}_R(k) \subseteq \mathrm{r}_R(a)$; then $a \in Rk$ by Lemma 2.1 and so $a = ea \in eRk = Sk$, as required. So let $kx = 0$, $x \in R$, and write $1 = \sum_{i=1}^n a_i e b_i$, where $a_i, b_i \in R$. Then $k(exa_ie) = 0$ for each i, so $a(exa_ie) = 0$ by hypothesis. Hence $ax = \sum_{i=1}^n axa_ieb_i = 0$ because $a = ae$, as required. □

The proof that right mininjectivity is inherited by matrix rings requires the following result.

Lemma 2.12. *Let K be a simple right ideal of a ring R. If $dK \neq 0$ is extensive for some $d \in R$, then K is extensive.*

Proof. The map $\sigma(x)=dx$ defines an isomorphism $\sigma : K \to dK$. Given $\gamma : K \to R_R$ we have $\gamma \circ \sigma^{-1} : dK \to R$, so $\gamma \circ \sigma^{-1} = c\cdot$, where $c \in R$ by hypothesis. It follows that $\gamma = (cd)\cdot$. □

Proposition 2.13. *A ring R is right mininjective if and only if the ring $M_n(R)$ is right mininjective for all (some) $n \geq 1$.*

Proof. If $S = M_n(R)$ is right mininjective, so is $R \cong e_{11}Se_{11}$ by Proposition 2.11 because $Se_{11}S = S$ (here e_{ij} denotes the matrix unit). Conversely, assume that R is right mininjective. By Proposition 2.11 it suffices[2] to do the case $n = 2$. Let $\bar{k}S$ be a simple right ideal of $S = M_2(R)$; we must show that $\mathtt{lr}(\bar{k}) = S\bar{k}$. If row i of $\bar{k}$ is nonzero then $e_{1i}\bar{k} \neq 0$, so, by Lemma 2.12, we may assume that $\bar{k} \in e_{11}S$. In this case, if column j of $\bar{k}$ is nonzero then $\bar{k}e_{j1} \neq 0$, so $\bar{k}S = \bar{k}e_{j1}S$. Thus we may assume that $\bar{k} \in e_{11}Se_{11}$, so write $\bar{k} = \left[\begin{smallmatrix} k & 0 \\ 0 & 0 \end{smallmatrix}\right]$, $k \in R$. Then kR is simple, so $Rk = \mathtt{lr}(k)$ by Lemma 2.1. But then $\mathtt{r}_S(\bar{k}) = \left[\begin{smallmatrix} \mathtt{r}(k) & \mathtt{r}(k) \\ R & R \end{smallmatrix}\right]$, whence $\mathtt{lr}_S(\bar{k}) = \left[\begin{smallmatrix} \mathtt{lr}(k) & 0 \\ \mathtt{lr}(k) & 0 \end{smallmatrix}\right] = \left[\begin{smallmatrix} Rk & 0 \\ Rk & 0 \end{smallmatrix}\right] = S\bar{k}$, as required. □

If M_R is the free right R-module of rank χ, and if we view $end(M_R)$ as the set of $\chi \times \chi$ column finite matrices over R, the proof of Proposition 2.13 goes through to give

Corollary 2.14. *If R is right mininjective, so also is the endomorphism ring of any free right R-module.*

Combining Propositions 2.11 and 2.13 with Theorem A.20 gives

Theorem 2.15. *Right mininjectivity is a Morita invariant.*

The next theorem gives another characterization of right mininjectivity based on the following "relative" version of the concept. If e is an idempotent in R, a right R-module M_R is called *eR-mininjective* if R-morphisms $\gamma : K \to M$ extend to $eR \to M$ whenever $K \subseteq eR$ is a simple right ideal, equivalently if $\gamma = m\cdot$ for some $m \in M$. Clearly, M is mininjective if and only if it is R-mininjective.

[2] Choose $2^k \geq n$. Since $T = M_{2^k}(R)$ is right mininjective by hypothesis, and since $M_n(R) \cong eTe$, where $e = \left[\begin{smallmatrix} I_n & 0 \\ 0 & 0 \end{smallmatrix}\right]$, it suffices by Proposition 2.11 to show that $TeT = T$. But $e_{11} \in TeT$ by block multiplication, and so $e_{ii} = e_{i1}e_{11}e_{1i} \in TeT$ for each i.

Lemma 2.16. *Let e, f, $e_1, \ldots, e_n$ denote idempotents in a ring R, and let M_R be a module.*

(a) If $1 = e_1 + \cdots + e_n$ then M is mininjective if and only if M is e_iR-mininjective for each i.
(b) If $eR \cong fR$ and M is eR-mininjective then M is fR-mininjective.
(c) If $M_R = \oplus_{i \in I} M_i$ then M is eR-mininjective if and only if each M_i is eR-mininjective.

Proof. (a). Let $\gamma : K \to M_R$ be R-linear, where K is a simple right ideal. Then $e_iK \neq 0$ for some i, so $\sigma(k) = e_ik$ defines an isomorphism $\sigma : K \to e_iK$. If M is e_iR-mininjective we have $\gamma \circ \sigma^{-1} = m\cdot$ for some $m \in M$, whence $\gamma = (me_i)\cdot$ and M is mininjective. The converse is clear.

(b). Let $\sigma : eR \to fR$ be an isomorphism. Given $\gamma : X \to M$, where $X \subseteq fR$ is simple, then $K = \sigma^{-1}(X) \subseteq eR$ is simple and $\gamma \circ \sigma_{|K}$ is an R-morphism $K \to M$. Hence $\gamma \circ \sigma_{|K} = m\cdot$ for some $m \in M$ by hypothesis. Write $a = \sigma^{-1}(f)$. Then $\gamma(x) = (\gamma \circ \sigma)(\sigma^{-1}x) = m\,\sigma^{-1}(x) = m(ax)$ for all $x \in X$, so $\gamma = (ma)\cdot$.

(c). Assume that each M_i is eR-mininjective. If $\gamma : K \to M$ is R-linear, where $K = kR \subseteq eR$ is simple, let $\gamma(K) \subseteq \oplus_{t=1}^{n} M_t$. If $\pi_t : M \to M_t$ is the projection for each t, then $\pi_t \circ \gamma = m_t\cdot$ for some $m_t \in M_t$ by hypothesis. If $k \in K$ and $\gamma(k) = \sum_{t=1}^{n} m_t'$, then $m_t' = \pi_t\gamma(k) = m_tk$, and it follows that $\gamma = m\cdot$, where $m = \sum_{t=1}^{n} m_t$. Hence M is eR-injective. The converse is clear. □

Proposition 2.17. *Let $1 = f_1 + \cdots + f_n$ in R, where the f_i are orthogonal idempotents, and let $e_1, \ldots, e_m$ be idempotents such that $\{e_1R, \ldots, e_mR\}$ is a complete set of representatives of $\{f_1R, \ldots, f_nR\}$. The following are equivalent:*

(1) R is right mininjective.
(2) e_iR is e_jR-mininjective whenever $1 \le i, j \le m$.

Proof. (1)⇒(2). Assume (1). Since $R = \oplus f_jR$, it follows that each f_jR is mininjective (direct summands of mininjective modules are again mininjective) and hence that each e_iR is mininjective (mininjectivity is preserved by isomorphisms). But then Lemma 2.16(a) (with $M = e_iR$) shows that each e_iR is f_jR-mininjective for each j. Now (2) follows by Lemma 2.16(b).

(2)⇒(1). Fix i where $1 \le i \le m$. Then (2) shows that e_iR is f_jR-mininjective for all j by Lemma 2.16(b). Hence e_iR is mininjective as a right R-module by Lemma 2.16(a). This holds for each i, so each f_jR is right mininjective (mininjectivity is preserved by isomorphisms). Finally, this shows that $R_R = \oplus_{j=1}^{n} f_jR$ is mininjective by Lemma 2.16(c). This proves (1). □

We now show that any right mininjective ring satisfies "min-" versions of conditions C2 and C3.

Proposition 2.18. *Let R be a right mininjective ring.*

(1) (min-C2) If K is a simple right ideal and $K \cong eR, e^2 = e$, then $K = gR$ for some $g^2 = g$.
(2) (min-C3) If $eR \neq fR$ are simple, $e^2 = e$, $f^2 = f$, then $eR \oplus fR = gR$ for some $g^2 = g$.

Proof. (1). If $\gamma : K \to eR$ is an isomorphism, let $\gamma = c\cdot$, $c \in R$. Then $cK = eR \nsubseteq J(R)$, so $K^2 \neq 0$ and (1) follows because K is simple.

(2). Observe that $eR \oplus fR = eR \oplus (1-e)fR$. If $(1-e)fR = 0$ we are done. Otherwise, $(1-e)fR \cong fR$, so $(1-e)fR = hR$, $h^2 = h$, by (1). Hence $eh = 0$ and so $g = e + h - he$ is an idempotent such that $eg = e = ge$ and $hg = h = gh$. It follows that $eR \oplus fR = eR \oplus hR = gR$. □

A module M satisfies the C1-condition if every submodule is essential in a direct summand of M. The min-C1 analogue of this condition reads as follows: *If K is a minimal submodule of M then K is essential in a direct summand of M*. However, the Björk example (Example 2.5) shows that this may not hold in a right mininjective ring. The rings that satisfy min-C1 are called *min-CS rings*, and we return to them in Chapter 4.

A ring R is called *I-finite* if it contains no infinite orthogonal family of idempotents (see Lemma B.6). This condition implies that $1 = e_1 + \cdots + e_n$, where the e_i are orthogonal, primitive idempotents. Proposition 2.18 has a nice application to I-finite mininjective rings.

Theorem 2.19. *Let R be I-finite and right mininjective. Then $R \cong R_1 \times R_2$, where R_1 is semisimple and every simple right ideal of R_2 is nilpotent.*

Proof. Let $1 = e_1 + \cdots + e_n$ where the e_i are orthogonal primitive idempotents, and where e_iR is simple if $1 \leq i \leq m$ and e_jR is not simple if $j > m$.

Claim. $e_iRe_j = 0 = e_jRe_i$ for all $1 \leq i \leq m < j \leq n$.

Proof. If $0 \neq a \in e_iRe_j$ then $a\cdot : e_jR \to e_iR$ is epic (since e_iR is simple) and so is an isomorphism (since e_jR is indecomposable), which is a contradiction as e_jR is not simple. Now suppose that $0 \neq b \in e_jRe_i$. Then $b\cdot : e_iR \to e_jR$ is monic (since e_iR is simple) and so $bR = fR$, where $f^2 = f$ by Proposition 2.18. Hence $b\cdot$ is epic (since e_jR is indecomposable) – a contradiction as before. This proves the Claim.

If $e = e_1 + \cdots + e_m$ it follows that $eR(1-e) = 0 = (1-e)Re$. Hence e is a central idempotent and $R_1 = eR = eRe$ is semisimple. It remains to show that each simple right ideal $K \subseteq (1-e)R$ is nilpotent. If not, $K = fR$, $f^2 = f$. As $K(1-e) \neq 0$, let $Ke_j \neq 0$, where $j > m$. Thus $fRe_j \neq 0$, say

$0 \neq c \in fRe_j$. Then $c\cdot : e_jR \to fR$ is an isomorphism as before; this is a contradiction because e_jR is not simple. □

Corollary 2.20. *Let R be right mininjective and I-finite with* $1 = e_1 + \cdots + e_n$, *where the* e_i *are orthogonal, primitive idempotents. Either some* e_iR *is simple or* $S_r^2 = 0$.

Proof. If no e_iR is simple then $R = R_2$ in Theorem 2.19, so $K^2 = 0$ for every simple right ideal K of R. It remains to show that $SK = 0$ for every simple right ideal S. But otherwise $SK = S$, whence $S = (SK)K = 0$, which is a contradiction. □

2.3. Minsymmetric Rings

If R is right mininjective the right socle is necessarily contained in the left socle. This surprising fact is fundamental and is contained in the following theorem, which will be used frequently.

Theorem 2.21. *Let R be a right mininjective ring, and let* $k, m \in R$.

(a) If kR *is a simple right ideal, then* Rk *is a simple left ideal.*
(b) If $kR \cong mR$ *are simple, then* $Rk \cong Rm$; *in fact* $Rk = (Rm)u$ *for some element* $u \in R$.
(c) $S_r \subseteq S_l$.

Proof. (a) and (c). If kR is simple and $0 \neq ak \in Rk$, define $\gamma = a\cdot : kR \to akR$. Then γ is an isomorphism and so, as R is right mininjective, let $\gamma^{-1} = c\cdot$, $c \in R$. Thus $k = \gamma^{-1}(ak) = cak \in Rak$, and (a) follows. Let $x \in S_r$, say $x \in k_1R \oplus \cdots \oplus k_nR$, where each k_iR is simple. Hence $k_i \in S_l$ for each i by (1), and (c) follows.

(b). If $\sigma : kR \to mR$ is an isomorphism, write $\sigma(k) = mu$, $u \in R$. Clearly $muR = mR$ is simple and $\mathbf{r}(mu) = \mathbf{r}[\sigma(k)] = \mathbf{r}(k)$. Since R is right mininjective, this gives $Rmu = Rk$ by Lemma 2.1. □

If K and M are modules with K simple, write $soc_K(M) = \Sigma\{X \subseteq M \mid X \cong K\}$ for the *homogeneous component* of M generated by K. It is fully invariant in M in the sense that $\alpha[soc_K(M)] \subseteq soc_K(M)$ for every $\alpha \in end(M)$.

Corollary 2.22. *If* kR *is a simple right ideal of a right mininjective ring* R, *then* $soc_{kR}(R_R) = RkR$ *is a simple ideal of* R *contained in* $soc_{Rk}({}_RR)$.

Proof. Write $S = soc_{kR}(R_R)$. Always $R(kR) \subseteq S$. Suppose $\sigma : kR \to M \subseteq R$ is an R-isomorphism. Then $\sigma k \in Rk$ by Lemma 2.1, so $M = (\sigma k)R \subseteq$

$(Rk)R$. Thus $S \subseteq RkR$, proving that $S = RkR$. Now let $0 \neq A \subseteq S$, with A an ideal of R. If $M \subseteq A$ is a simple right ideal then $M \cong kR$. Hence if X is any right ideal isomorphic to kR, let $\gamma : M \to X$ be an R-isomorphism. Then $\gamma = c\cdot$, $c \in R$, so $X = \gamma(M) = cM \subseteq cA \subseteq A$. It follows that $S \subseteq A$, whence $S = A$ and S is a simple ideal. Finally, $(Rk)R \subseteq soc_{Rk}({}_RR)$ always holds. □

If R is right mininjective, it follows from Corollary 2.22 that every two-sided ideal of S_r is a direct sum of simple ideals (the converse is false by Example 2.23). Moreover, if R is right and left mininjective and we write $S = S_r = S_l$, the set of left homogeneous components of S is the same as the set of right homogeneous components of S.

Example 2.23. If F is a field, the ring $R = \left[\begin{smallmatrix} F & F \\ 0 & F \end{smallmatrix}\right]$ is a two-sided artinian ring, $J = \left[\begin{smallmatrix} 0 & F \\ 0 & 0 \end{smallmatrix}\right]$, $S_r = \left[\begin{smallmatrix} 0 & F \\ 0 & F \end{smallmatrix}\right]$ and $S_l = \left[\begin{smallmatrix} F & F \\ 0 & 0 \end{smallmatrix}\right]$. Hence R is neither left nor right mininjective because $S_r \nsubseteq S_l$ and $S_l \nsubseteq S_r$.

We can now give an example of a two-sided mininjective ring with an image that is neither right nor left mininjective.

Example 2.24. If S is a subring of a ring R, define a subring C of $R^{\mathbb{N}}$ by

$$C = C[R, S] = \{(r_0, r_1, r_2, \ldots, r_n, s, s, \ldots) \mid n \geq 0, r_i \in R \text{ and } s \in S\}.$$

Then C is right mininjective if and only if R is right mininjective. Moreover, the map carrying $(r_0, r_1, r_2, \ldots, r_n, s, s, \ldots) \mapsto s$ is a ring homomorphism of C onto S. In particular, if $R = \left[\begin{smallmatrix} F & F \\ F & F \end{smallmatrix}\right]$ and $S = \left[\begin{smallmatrix} F & F \\ 0 & F \end{smallmatrix}\right]$, where F is a field, then C is a right and left mininjective ring with zero Jacobson radical that has an image S that is neither left nor right mininjective by Example 2.23.

Proof. Assume that R is right mininjective and that $\gamma : K \to C$ is C-linear, where K is a simple right ideal of C. Since $K \neq 0$, let $\bar{k} = (0, 0, \ldots, k, 0, 0, \ldots) \in K$, where $0 \neq k \in R$. Then kR is simple and $K = \bar{k}C$. If k is in the mth component of $\bar{k}$, let $\gamma_0 = \pi_m \gamma \sigma_m : kR \to R$, where $R \overset{\sigma_m}{\to} C \overset{\tau_m}{\to} R$ are the canonical maps. Then $\gamma_0 = c\cdot$ for some $c \in R$ by hypothesis, whence $\gamma = \sigma_m(c)\cdot$, as is easily verified. Hence C is right mininjective; the converse is a routine verification. Finally, if $R = \left[\begin{smallmatrix} F & F \\ F & F \end{smallmatrix}\right]$ then $J(C) = 0$ because every nonzero one-sided ideal of C contains a nonzero idempotent. □

Note that Example 2.24 shows that any subring of a right mininjective ring is also an image of a right mininjective ring.

Motivated by Theorem 2.21, we call a ring R *right minsymmetric* if kR simple, $k \in R$, implies that Rk is simple. These rings all have the property that $S_r \subseteq S_l$. This is a large class of rings containing all commutative rings, all right mininjective rings (by Theorem 2.21), and hence every semiprime ring. This proves the first half of

Example 2.25. Every semiprime ring R is right and left minsymmetric, but not conversely.

Proof. The converse fails for any commutative mininjective ring that is not semiprime (see Example 2.6). Note that if $e = \begin{bmatrix} 0 & 0 \\ 0 & 1 \end{bmatrix}$ in the ring $R = \begin{bmatrix} F & F \\ 0 & F \end{bmatrix}$, where F is a field, then eR is simple but Re contains a nilpotent ideal. □

The following characterization of right minsymmetric rings has some independent interest and will be referred to later.

Proposition 2.26. *The following are equivalent for a ring R:*

(1) R is right minsymmetric.
(2) If kR is a simple right ideal, then $\mathtt{l}[kR \cap \mathtt{r}(a)] = \mathtt{l}(k) + Ra$ for all elements $a \in R$.

Proof. (1)⇒(2). Assume kR is simple and let $a \in R$. If $ak = 0$ then $kR \cap \mathtt{r}(a) = kR$ and $\mathtt{l}(k) + Ra = \mathtt{l}(k)$, and (2) follows. If $ak \neq 0$ then $\mathtt{l}(k) + Ra = R$ [because $\mathtt{l}(k)$ is maximal by (1)] and $kR \cap \mathtt{r}(a) = 0$ (because kR is simple), and again (2) follows.

(2)⇒(1). If kR is simple, let $a \notin \mathtt{l}(k)$. Then $kR \cap \mathtt{r}(a) = 0$, so $\mathtt{l}(k) + Ra = R$ by (2). This shows that $\mathtt{l}(k)$ is maximal, proving (1). □

Utumi's theorem shows that every right self-injective ring satisfies $J = Z_r$, but this fails if R is merely right mininjective: The localization $\mathbb{Z}_{(p)}$ of $\mathbb{Z}$ at the prime p is a commutative local ring that is mininjective (it has no simple ideals) and in which $Z_r = 0$ (it is a domain) and $J = p\mathbb{Z}_{(p)} \neq 0$. However, as we shall see in Chapter 5, $J = Z_r$ does hold for right principally injective rings (that is, every principal right ideal is extensive).

In general, necessary and sufficient conditions that $J = Z_r$ are elusive. We have $Z_r \subseteq J$ (and $Z_l \subseteq J$) if R is semiregular because then every one-sided ideal not contained in J contains a nonzero idempotent (see Lemma B.40). On the other hand, if $S_l \subseteq^{ess} R_R$ then $J \subseteq Z_r$ because $S_l \subseteq \mathtt{r}(J) \subseteq \mathtt{r}(a)$ for all $a \in J$. Combining these observations we obtain conditions that $Z_r = J$.

Proposition 2.27. *If R is a right mininjective, semiregular ring in which $S_r \subseteq^{ess} R_R$, then $J = Z_r$.*

Proof. We have $S_r \subseteq S_l$ because R is right mininjective, so $S_l \subseteq^{ess} R_R$ by hypothesis. Hence $J \subseteq Z_r$ by the preceding discussion. We have $Z_r \subseteq J$ because R is semiregular (again by the preceding discussion). □

2.4. Duality

Right mininjective rings have another characterization that is related to duality. If M_R is a right R-module, recall that the *dual* $M^* = hom_R(M_R, R)$ of M is a left module via $(r\lambda)(m) = r \cdot \lambda(m)$ for all $r \in R$, $\lambda \in M^*$, and $m \in M$.

Lemma 2.28. *If $M = mR$ is a principal right R-module and $T = \mathtt{r}(m)$ then $M^* \cong \mathtt{l}(T) = \mathtt{lr}(m)$ as left R-modules.*

Proof. If $b \in \mathtt{l}(T)$, the map $\lambda_b : M \to R$ is well defined by $\lambda_b(mr) = br$. Then $b \mapsto \lambda_b$ is a monomorphism $\mathtt{l}(T) \to M^*$ of left R-modules, and it is onto because, if $\lambda \in M^*$, then $\lambda = \lambda_b$, where $b = \lambda(m) \in \mathtt{l}(T)$. □

With this we can give some useful duality characterizations of mininjectivity.

Theorem 2.29. *The following are equivalent for a ring R:*

(1) R is right mininjective.
(2) M^* is simple or zero for every simple right R-module M_R.
(3) $\mathtt{l}(T)$ is simple or zero for every maximal right ideal T of R.
(4) K^* is simple for every simple right ideal K of R.

Proof. (1)⇒(2). Let M_R be simple. If $M^* = 0$ there is nothing to prove. Otherwise, let $0 \neq \delta \in M^*$; we must show that $M^* = R\delta$. Observe first that $\delta : M \to \delta(M)$ is an isomorphism. Given $\gamma \in M^*$ we have $\delta(M) \xrightarrow{\delta^{-1}} M \xrightarrow{\gamma} R$, so $\gamma \circ \delta^{-1} = a\cdot$ for some $a \in R$ by (1). It follows that $\gamma = a\delta$, proving (2).

(2)⇒(3). If T is a maximal right ideal then $R/T = (1+T)R$ is simple and $\mathtt{r}(1+T) = T$. Hence $\mathtt{l}(T) \cong (R/T)^*$ by Lemma 2.28, so (2) applies.

(3)⇒(4). If $K = kR$ is a simple right ideal, write $T = \mathtt{r}(k)$. Then T is maximal, so $K^* \cong \mathtt{l}(T)$ is simple or zero by (3) and Lemma 2.28. But $K^* \neq 0$ because it contains the inclusion map.

(4)⇒(1). Given $\gamma : K \to R$, where $K = kR$ is a simple right ideal, let $\iota : K \to R$ be the inclusion. Then $K^* = R\iota$ by (4), so $\gamma = c\iota$ for some $c \in R$. It follows that $\gamma = c\cdot$ as required. □

Note that this gives a different proof of the fact that if kR is simple in a right mininjective ring R, then Rk is simple (see Theorem 2.21). Indeed, $\mathtt{lr}(k) \cong K^*$ by Lemma 2.28, so $\mathtt{lr}(k)$ is simple by Theorem 2.29. Since $Rk \subseteq \mathtt{lr}(k)$, it follows that $Rk = \mathtt{lr}(k)$ is simple.

We conclude this section by applying Theorem 2.29 to give an important characterization of quasi-Frobenius rings, due originally to Ikeda in 1952.

Theorem 2.30 (Ikeda's Theorem). *The following conditions are equivalent for a ring R:*

(1) R is quasi-Frobenius.
(2) R is two-sided mininjective and two-sided artinian.[3]

Proof. Given (2), it suffices by Theorem 1.50 to show that $\mathtt{rl}(T) = T$ for every right ideal T of R [by symmetriy $\mathtt{lr}(L) = L$ for each left ideal L].

Claim. *If $K \subseteq T$ are right ideals and T/K is simple, then $\mathtt{l}(K)/\mathtt{l}(T)$ is zero or simple.*

Proof. If $a \in \mathtt{l}(K)$ then $\lambda_a : T/K \to R_R$ is well defined by $\lambda_a(t + K) = at$. Hence $a \mapsto \lambda_a$ is a homomorphism $\mathtt{l}(K) \to (T/K)^*$ with kernel $\mathtt{l}(T)$, and the claim follows from Theorem 2.29 because R is right mininjective.

Now let T be a right ideal of R, and (as R is right artinian) let $0 = T_0 \subset T_1 \subset \cdots \subset T_n = R$ be a composition series for R that contains T. We show that $\mathtt{rl}(T_i) = T_i$ for each i. Taking left annihilators gives the series

$$R = \mathtt{l}(T_0) \supseteq \mathtt{l}(T_1) \supseteq \cdots \supseteq \mathtt{l}(T_n) = 0, \qquad (*)$$

so the Claim implies that $length({}_RR) \leq n = length(R_R)$. The other inequality is proved similarly, and so $length(R_R) = length({}_RR)$. But then the Jordan–Hölder theorem (and the Claim) show that (*) is actually a composition series for ${}_RR$ [and that the inclusions in (*) are strict]. Now repeat the process on (*), taking right annihilators to get a composition series $0 = \mathtt{rl}(T_0) \subset \mathtt{rl}(T_1) \subset \cdots \subset \mathtt{rl}(T_n) = R$. Since $T_1 \subseteq \mathtt{rl}(T_1)$, this gives $T_1 = \mathtt{rl}(T_1)$, whence $\mathtt{rl}(T_1) = T_1 \subset T_2 \subseteq \mathtt{rl}(T_2)$ gives $T_2 = \mathtt{rl}(T_2)$. Continue in this way to get $T_i = \mathtt{rl}(T_i)$ for each i, as required. □

Note that it is essential that R be two-sided mininjective in (2) of Theorem 2.30: The Björk example (Example 2.5) is a local ring R with $J^2 = 0$ that is

[3] It is enough that R is only right artinian; in fact it is sufficient that R has the ACC on right annihilators and $S_r \subseteq^{ess} R_R$ (see Theorem 3.31 in Chapter 3).

two-sided artinian and right mininjective, but not left mininjective (and hence not quasi-Frobenius).

2.5. The Kasch Condition

In Theorem 2.29 we showed that a ring is right mininjective if and only if M^* is simple or zero for every simple right module M_R. Recall from Proposition 1.44, that a ring is right Kasch if and only if the left annihilator of every maximal right ideal is nonzero. Combining this with Theorem 2.29 gives the following characterization of the right mininjective, right Kasch rings.

Theorem 2.31. *The following are equivalent for a ring R:*

(1) R is right mininjective and right Kasch.
(2) M^* is simple for every simple right module M_R.
(3) $\mathtt{l}(T)$ is simple for every maximal right ideal T of R.

In this case every nonzero left annihilator in R contains a simple left ideal.

Proof. The equivalence of (1), (2), and (3) follows at once from Proposition 1.44 and from the equivalence of (1), (2), and (3) in Theorem 2.29.

Now suppose that $0 \neq L = \mathtt{l}(X)$ is a left annihilator, where $X \subseteq R$. We may assume that X is a right ideal and so, as $X \neq R$, let $X \subseteq T$, where T is a maximal right ideal. Thus $\mathtt{l}(T) \subseteq \mathtt{l}(X) = L$, and $\mathtt{l}(T)$ is simple by (3). □

In a quasi-Frobenius ring the maps $T \mapsto \mathtt{l}(T)$ and $L \mapsto \mathtt{r}(L)$ are mutually inverse, inclusion reversing bijections between the right ideals T and the left ideals L. This result is the beginning of the theory of Morita duality. The next theorem identifies a condition under which a weakened form of this duality holds in a right mininjective, right Kasch ring.

Theorem 2.32. *Let R be a right mininjective, right Kasch ring, and consider the map*

$$\theta : T \mapsto \mathtt{l}(T)$$

from the set of maximal right ideals T of R to the set of minimal left ideals of R. Then the following conditions hold:

(a) θ is one-to-one.
(b) θ is a bijection if and only if $\mathtt{lr}(K) = K$ for all minimal left ideals K of R. In this case the inverse map is given by $K \mapsto \mathtt{r}(K)$.

Proof. (a). If T is a maximal right ideal, then $\mathtt{l}(T)$ is simple by Theorem 2.31, so θ is defined. Since $T \subseteq \mathtt{rl}(T) \neq R$, we have $T = \mathtt{rl}(T)$ because T is maximal. Now (a) follows.

(b). If θ is onto, every minimal left ideal K is an annihilator, so $\mathtt{lr}(K) = K$. Conversely, assume that $\mathtt{lr}(K) = K$ for all minimal left ideals K.

Claim. If K is a minimal left ideal then $\mathtt{r}(K)$ is maximal.

Proof. Let $\mathtt{r}(K) \subseteq T$, where T is maximal. Then $K = \mathtt{lr}(K) \supseteq \mathtt{l}(T) \neq 0$ by the right Kasch hypothesis, so $K = \mathtt{l}(T)$ because K is simple. Thus $\mathtt{r}(K) = \mathtt{rl}(T) \supseteq T$, whence $\mathtt{r}(K) = T$, proving the Claim.

By the Claim we have a map φ given by $K \mapsto \mathtt{r}(K)$ that we assert is the inverse of θ. Indeed, $\varphi \circ \theta$ carries $T \mapsto \mathtt{l}(T) \mapsto \mathtt{rl}(T) = T$ by the calculation in (a), whereas $\theta \circ \varphi$ carries $K \mapsto \mathtt{r}(K) \mapsto \mathtt{lr}(K) = K$ by hypothesis. This completes the proof of (b). □

2.6. Minannihilator Rings

Motivated by Theorem 2.32(b), we call a ring R a *left minannihilator* ring if every minimal left ideal K of R is an annihilator, equivalently if $\mathtt{lr}(K) = K$. This condition is clearly satisfied if $K = Re$, $e^2 = e$, so it is enough to have $\mathtt{lr}(K) = K$ for all simple left ideals contained in J.

Examples.

(1) Every semiprime ring is two-sided minannihilator.
(2) A commutative ring is mininjective if and only if it is a minannihilator ring by Lemma 2.1.
(3) Every two-sided mininjective ring is two-sided minannihilator. For if $K = Rk$ is simple then kR is simple because R is left mininjective (Theorem 2.21), so $\mathtt{lr}(K) = K$ because R is right mininjective (Lemma 2.1). So R is left minannihilator; it is right minannihilator by symmetry.
(4) The Björk example (Example 2.5) is left minannihilator (it has only one proper left ideal) but it is not right minannihilator. Indeed, in the notation of Example 2.5, $tF = \bar{F}t$ is a simple right ideal but $\mathtt{rl}(tF) = \mathtt{r}(J) = J = Ft \neq \bar{F}t$.

The following result and its corollaries reveal the close connection between right mininjective rings and left minannihilator rings. A right mininjective ring R that is *left* minsymmetric (Rk simple, $k \in R$, implies kR simple) is a left minannihilator ring by Lemma 2.1. The converse requires the right minsymmetric condition.

Proposition 2.33. *The following are equivalent for a left minannihilator ring R:*

(1) R is right mininjective.
(2) R is right minsymmetric.
(3) $S_r \subseteq S_l$.

Proof. (1)⇒(2) and (2)⇒(3) always hold (see Theorem 2.21). Given (3), let kR be simple. Then $k \in S_l$ by (3), so let $Rk \supseteq Rm$, where Rm is simple. Thus $\mathtt{r}(k) \subseteq \mathtt{r}(m)$, so $\mathtt{r}(k) = \mathtt{r}(m)$ because $\mathtt{r}(k)$ is maximal. Since R is left minannihilator, $Rk \subseteq \mathtt{lr}(Rk) = \mathtt{lr}(Rm) = Rm \subseteq Rk$ because Rm is simple, so $Rk = \mathtt{lr}(Rk) = \mathtt{lr}(k)$. This proves (1) by Lemma 2.1. □

Together with Theorem 2.21, Proposition 2.33 gives

Corollary 2.34. *A ring R is left and right mininjective if and only if $S_r = S_l$ and R is a left and right minannihilator ring.*

The proof that (3)⇒(1) in Proposition 2.33 yields

Corollary 2.35. *Suppose that R is a left minannihilator ring in which S_l is essential in ${}_RR$ (for example if R is right perfect). Then R is right mininjective.*

2.7. Universally Mininjective Rings

Recall that a module M_R is called mininjective if, for every simple right ideal K of R, each R-linear map $\gamma : K \to M_R$ extends to $\bar{\gamma} : R \to M$; that is, $\gamma = m\cdot$ is multiplication by some $m \in M$. The rings for which every right module is mininjective admit several characterizations.

Theorem 2.36. *The following conditions are equivalent for a ring R:*

(1) Every right R-module is mininjective.
(2) Every principal right R-module is mininjective.
(3) $K^2 \neq 0$ for every simple right ideal K of R.
(4) $S_r \cap J = 0$.
(5) R is right mininjective and S_r is projective as a right R-module.

Proof. (1)⇒(2). This is clear.

(2)⇒(3). If $K = kR$ is simple, $k \in R$, we have an R-isomorphism $\gamma : kR \to R/\mathtt{r}(k)$ given by $\gamma(ka) = a + \mathtt{r}(k)$. By (2), γ is left multiplication by $c + \mathtt{r}(k)$ for some $c \in R$. Thus $ck + \mathtt{r}(k) = \gamma(k) = 1 + \mathtt{r}(k)$, whence $kck = k$. If $e = kc$ then $0 \neq e^2 = e \in K$.

(3)⇒(4). This is clear.

(4)⇒(5). If K is a simple right ideal then $K = eR$, $e^2 = e$, by (4).

(5)⇒(1). If $\gamma : K \to M_R$ is R-linear, where K is a simple right ideal, then $K \cong eR$, where $e^2 = e$, because S_r is projective. Since R is right mininjective, it follows that $eR = cK$ for some $c \in R$. Hence $K \nsubseteq J$, so $K = fR$ for some $f^2 = f \in R$. It follows that $\gamma = m\cdot$, where $m = \gamma(f)$. □

Call a ring R right *universally mininjective* if it satisfies the conditions in Theorem 2.36. Clearly each ring with zero right socle (hence every polynomial ring) is right universally mininjective, and every semiprime ring is both right and left universally mininjective. In contrast, a right universally mininjective ring R with essential right socle is semiprime by (4) of Theorem 2.36, and if R is right Kasch it is semisimple artinian (every simple right module is projective). Although every right universally mininjective ring has projective right socle, the converse is false: If F is a field, the ring $\left[\begin{smallmatrix} F & F \\ 0 & F \end{smallmatrix}\right]$ has both socles projective but is neither right nor left mininjective.

A direct product of rings is right universally mininjective if and only if each factor is right universally mininjective. With only minor variations, the proof of Theorem 2.15 goes through to prove that being right universally mininjective is a Morita invariant property of rings.

Finally, if R is I-finite, this discussion and Theorem 2.19 give

Theorem 2.37. *If R is I-finite then R is right universally mininjective if and only if $R \cong R_1 \times R_2$, where R_1 is semisimple artinian and R_2 has zero right socle.*

Notes on Chapter 2

In the artinian case, the right mininjective rings were introduced by Ikeda [102] in 1952 and were studied by Dieudonné [45] and later by Harada [91, 92]. The related minannihilator condition that every left ideal is an annihilator was studied in artinian rings by Dieudonné [45], Storrer [212], and Björk [21]. The "relative" version of mininjectivity stems from Harada [94].

Example 2.5 traces back to Björk [21] in 1970 and was mentioned by Rutter [199]. Example 2.6 is an adaptation of an example of Camillo [26].

Distributive modules are investigated in Cohn [39]. In Theorem 2.9, the equivalence of (1) and (2) is due to Camillo [25], and the equivalence of (1) and (3) is due to Stephenson [209].

I-finite rings have the property that 1 is a finite sum of orthogonal primitive idempotents. It is an open question whether the converse holds.

In Theorem 2.29 characterizing mininjectivity in terms of duality, the equivalence of (1) and (2) was first noted for artinian rings by Björk [21] in 1970. It leads to Ikeda's characterization [102] of quasi-Frobenius rings (Theorem 2.30), originally proved in 1952.

Theorem 2.37 was proved by Gordon [88] with a different proof.

3

Semiperfect Mininjective Rings

Only in the semiperfect case does the full power of the mininjective hypothesis comes into focus. The reason is that we can get combinatorial information about the simple right and left ideals because a semiperfect ring has only a finite number of simple right modules that are represented in the ring by basic idempotents. We remind the reader that all the facts needed about semiperfect rings are developed in Appendix B.

We begin by characterizing the semiperfect, right mininjective rings, and we use the result to give easy proofs of two well-known characterizations of quasi-Frobenius rings: R is quasi-Frobenius if and only if it is left and right artinian, both $soc(Re)$ and $soc(eR)$ are simple for every local idempotent e in R, and either $S_r = S_l$ or R is right and left Kasch. The following class of rings is important in extending these results. A ring R is called a right minfull ring if it is semiperfect, right mininjective, and $soc(eR) \neq 0$ for every local idempotent $e \in R$. We show that these right minfull rings are a Morita invariant class that exhibits many of the basic properties of quasi-Frobenius rings. In particular, they are right and left Kasch, and $S_r = S_l$ if and only if $\mathtt{lr}(K) = K$ for every minimal left ideal K that is contained in Re for some local idempotent e. With this we obtain the following result: If R is a semilocal, left and right mininjective ring with ACC on right annihilators in which $S_r \subseteq^{ess} R_R$, then R is quasi-Frobenius. This is an improvement of Ikeda's theorem (where R is left and right artinian and left and right mininjective).

If R is a semiperfect ring with basic idempotents $\{e_1, \ldots, e_n\}$, a permutation σ of $\{1, 2, \ldots n\}$ is called a Nakayama permutation for R if $soc(Re_{\sigma i}) \cong Re_i/Je_i$ and $soc(e_iR) \cong e_{\sigma i}R/e_{\sigma i}J$ for each $i = 1, 2, \ldots, n$. We show that every right and left minfull ring admits a Nakayama permutation, and that a semiperfect ring R with a Nakayama permutation is right and left minfull if $S_r \subseteq^{ess} R_R$ (or $S_l \subseteq^{ess} {}_RR$). This leads to an easy proof of a characterization of quasi-Frobenius rings due to Nakayama: A ring R is quasi-Frobenius if and only if it is left and right artinian and has a Nakayama permutation.

These results suggest the following definition: A ring R is called a right min-PF ring if it is a semiperfect, right mininjective ring in which $S_r \subseteq^{ess} R_R$ and $\mathtt{lr}(K) = K$ for every simple left ideal $K \subseteq Re$ for some local idempotent e. As the name suggests every right PF ring is right (and left) min-PF, and the right min-PF rings form a Morita invariant class. We show that a right min-PF ring with ACC on right annihilators is left artinian, and so we conclude that a right and left min-PF ring with ACC on right annihilators is quasi-Frobenius. In particular, this settles the $J^2 = 0$ case of the Faith conjecture by showing that, if R is a semiprimary, right and left mininjective ring with $J^2 = 0$, then R is quasi-Frobenius.

3.1. Basic Properties

The following property of idempotents will be used frequently in this section. Recall that a ring R is called *semilocal* if R/J is semisimple artinian.

Lemma 3.1. *Let e be an idempotent in any ring R. Then:*

(1) $(eR/eJ)^* \cong \mathtt{l}(J)e$.
(2) If R is semilocal then $(eR/eJ)^* \cong S_r e$.

Proof. We have $eR/eJ = mR$, where $m = e + eJ$, so (by Lemma 2.28) we have $(eR/eJ)^* \cong \mathtt{l}(T)$, where $T = \mathtt{r}(m)$. Here $T = J + (1 - e)R$, so $\mathtt{l}(T) = \mathtt{l}(J) \cap Re = \mathtt{l}(J)e$. This proves (1), and (2) follows because $S_r = \mathtt{l}(J)$ whenever R is semilocal. □

Our first application of Lemma 3.1 is to derive the following characterizations of when a semiperfect ring is right mininjective or right Kasch. They will be referred to repeatedly later. We use the fact that if $\sigma : N \to M$ is an isomorphism of right R-modules then ${}_R M^* \cong {}_R N^*$ via $\lambda \mapsto \lambda \circ \sigma$. Recall that an idempotent e in a ring R is called a *local idempotent* if eRe is a local ring, and that R is semiperfect if and only if $1 = e_1 + e_2 + \cdots + e_n$ where the e_i are orthogonal, local idempotents.

Theorem 3.2. *Let R be a semiperfect ring.*

(1) R is right mininjective if and only if $S_r e$ is simple or zero for each local idempotent $e \in R$.
(2) R is right Kasch if and only if $S_r e \neq 0$ for each local idempotent $e \in R$.

Proof. Let M_R denote a simple module. Since R is semiperfect we have $Me \neq 0$ for some local idempotent e, say $me \neq 0$, where $m \in M$. Then the map $x \mapsto mex$ from $eR \to M$ induces an R-isomorphism $eR/eJ \to M$ (see Proposition B.2). Hence a module M_R is simple if and only if $M_R \cong eR/eJ$ for some local idempotent e. Moreover, $(eR/eJ)^* \cong S_r e$ by Lemma 3.1 because R is semilocal. Now (2) follows, and (1) restates Theorem 2.29. □

Note that, since $S_r \subseteq \mathtt{l}(J)$ in any ring, semiperfect or not, the fact that $S_r e \subseteq \mathtt{l}(J)e \cong (eR/eJ)^*$ shows that $S_r e$ is simple or zero for each local idempotent e in *any* right mininjective ring. Furthermore, the converse is true in any ring R that has "enough" local idempotents in the sense that, for every simple right module M, $Me \neq 0$ for some local idempotent e (equivalently, if $M \cong eR/eJ$ for some local idempotent e).

The local case of Theorem 3.2 gives a converse to Example 2.4.

Corollary 3.3. *A local ring R is right mininjective if and only if S_r is zero or simple as a left ideal.*

In the artinian case, Theorem 3.2 gives an easy proof of the following characterizations of quasi-Frobenius rings, which are part of the folklore of the subject.

Theorem 3.4. *The following conditions are equivalent for a right and left artinian ring R:*

(1) R is quasi-Frobenius.
(2) R is right and left Kasch, and both $soc(Re)$ and $soc(eR)$ are simple for every local idempotent $e \in R$.
(3) $S_r = S_l$ and both $soc(Re)$ and $soc(eR)$ are simple or zero for every local idempotent $e \in R$.

Proof. (1)⇒(2). Given (1), R is right Kasch because $\mathtt{rl}(T) = T$ for every maximal right ideal T (Theorem 1.50). Similarly, R is left Kasch. If $e = e^2 \in R$ is local then eR is uniform [being injective by (1) and indecomposable]. Hence, $soc(eR)$ is essential in eR (it is nonzero because R is right artinian). It follows that $soc(eR)$ is simple; similarly $soc(Re)$ is simple.

(2)⇒(3). Since R is right Kasch, we have $S_r e \neq 0$ for every local idempotent e by Theorem 3.2. Hence $S_r e$ contains a simple left ideal (R is left artinian), so $soc(Re) \subseteq S_r e$ by (2). Thus if $1 = e_1 + \cdots + e_n$, where the e_i are local, orthogonal idempotents, then $S_l = \oplus_i \, soc(Re_i) \subseteq \oplus_i \, S_r e_i \subseteq S_r$. Similarly, $S_r \subseteq S_l$.

(3)⇒(1). If $e^2 = e$ is local then $S_r e = S_r \cap Re = S_l \cap Re = soc(Re)$ using (3). Hence R is right mininjective by Theorem 3.2. Similarly, R is left mininjective. Since R is two-sided artinian, it is quasi-Frobenius by Ikeda's theorem (Theorem 2.30). □

Note that the proof of (2)⇒(3) in Theorem 3.4 requires only that $S_l \subseteq^{ess} {}_RR$ and $S_r \subseteq^{ess} R_R$ and so holds for any left and right perfect ring. Moreover, the proof that (3)⇒(1) yields

Proposition 3.5. *If R is a semiperfect ring in which $S_r \subseteq S_l$ and $soc(Re)$ is simple or zero for all local idempotents e, then R is right mininjective.*

Before proceeding, we need a lemma about local idempotents, which will be used several times. It is based on the following module-theoretic observation. Suppose $M = P \oplus Q$ is a direct sum of modules. Then the maps

$$X \mapsto X \oplus Q \qquad \text{and} \qquad Y \mapsto Y \cap P \tag{*}$$

are mutually inverse lattice isomorphisms between the lattice $\{X \mid X \subseteq P\}$ of all submodules of P and the lattice $\{Y \mid Q \subseteq Y \subseteq M\}$ of all submodules of M that contain Q. The verification is a routine computation using the modular law, and we apply the result as follows:

Lemma 3.6. *An idempotent e in a ring R is local if and only if $J + R(1 - e)$ is the unique maximal left ideal of R that contains $R(1 - e)$. Furthermore,*

$$\frac{R}{J + R(1-e)} \cong \frac{Re}{Je}$$

in this case.

Proof. Apply (*) to the decomposition $R = Re \oplus R(1 - e)$ to conclude that Je is the unique maximal submodule of Re (equivalently, that e is local) if and only if $J + R(1 - e)$ is the unique maximal submodule of ${}_RR$ that contains $R(1 - e)$. Since $J + R(1 - e) = Je \oplus R(1 - e)$, we have

$$\frac{R}{J + R(1-e)} = \frac{Re \oplus R(1-e)}{Je \oplus R(1-e)} \cong \frac{Re}{Je}. \qquad \square$$

We can now begin our investigation of the structure of an arbitrary semiperfect, right mininjective ring. The next theorem collects some basic properties of these rings, including some important criteria that the two socles are equal.

Note that, if R is *any* right mininjective, right Kasch ring, it follows from Theorem 2.31 that Re contains a simple left ideal for every $e^2 = e \neq 0$. If R is semiperfect, we can say more.

Theorem 3.7. *Let R be a semiperfect, right mininjective ring. Then:*

(1) S_r is semisimple and artinian as a left R-module.
(2) If $0 \neq k \in soc(eR)$, where $e^2 = e$ is local, then Rk is simple.
(3) If R is right Kasch, the following conditions are equivalent:
  (a) $S_r = S_l$.
  (b) $\mathbf{lr}(K) = K$ whenever $e^2 = e \in R$ is local and $K \subseteq Re$ is a simple left ideal.
  (c) $soc(Re) = S_r e$ for every local $e^2 = e \in R$.
  (d) $soc(Re)$ is simple for every local $e^2 = e \in R$.

Proof. If $1 = e_1 + \cdots + e_n$ where the e_i are local idempotents in R then $S_r = \Sigma_i S_r e_i$, so statement (1) is clear from Theorem 3.2. Turning to (2), we let $0 \neq k \in soc(eR)$. Then $R(1-e) \subseteq \mathbf{l}(k)$, and $J \subseteq \mathbf{l}(R)$ because $soc(eR) \subseteq S_r \subseteq S_l$ (using Theorem 2.21). Hence $J + R(1-e) \subseteq \mathbf{l}(k) \neq R$, and so $\mathbf{l}(k) = J + R(1-e)$ is maximal by Lemma 3.6, and (2) follows. To prove (3), assume that R is right Kasch.

(a)⇒(b). Suppose that $K \subseteq Re$ is a simple left ideal, where $e^2 = e \in R$ is local. We have $KJ = 0$ by (a), so $\mathbf{r}(K) \supseteq J + (1-e)R$. As $J + (1-e)R$ is maximal by Lemma 3.6, it follows that $\mathbf{r}(K)$ is maximal. Hence $\mathbf{lr}(K)$ is simple by Theorem 2.31, and (b) follows because $K \subseteq \mathbf{lr}(K)$.

(b)⇒(c). Let $K \subseteq Re$ be a simple left ideal, so that $\mathbf{r}(K) \supseteq (1-e)R$. It follows that $\mathbf{r}(K) \subseteq J + (1-e)R$ because $J + (1-e)R$ is the unique maximal right ideal containing $(1-e)R$ by Lemma 3.6. But then (b) gives $K = \mathbf{lr}(K) \supseteq \mathbf{l}[J + (1-e)R] = \mathbf{l}(J) \cap Re = S_r \cap Re = S_r e$, again because R is semilocal. But $S_r e \neq 0$ by Theorem 3.2, so $K = S_r e$ because K is simple. This proves (c).

(c)⇒(d). This follows by Theorem 3.2 because R is right mininjective and right Kasch.

(d)⇒(a). Given (d) we have $soc(Re) = S_r e$ for each local $e^2 = e$ by Theorem 3.2. Let $1 = e_1 + \cdots + e_n$, where the e_i are orthogonal local idempotents. Then $S_l = \oplus_i soc(Re_i) = \oplus_i S_r e_i \subseteq S_r$. Since $S_r \subseteq S_l$ in any right mininjective ring, (a) follows. □

The localization $\mathbb{Z}_{(p)} = \{\frac{m}{n} \mid p \text{ does not divide } n\}$ of $\mathbb{Z}$ at the prime p is a commutative, local, mininjective ring in which conditions (a), (b), and (c) in Theorem 3.7 are all true, but $\mathbb{Z}_{(p)}$ is not Kasch (it is a domain).

If we insist that R is (left and right) mininjective in Theorem 3.7, we get the following result.

Corollary 3.8. *Let R be a semiperfect, two-sided mininjective, two-sided Kasch ring. Then R is a two-sided minannihilator ring, $S_r = S_l$ (denoted S), and $soc(eR) = eS$ and $soc(Re) = Se$ are both simple whenever $e^2 = e$ is local.*

We can now characterize the commutative, semiperfect, mininjective rings.

Proposition 3.9. *The following are equivalent for a commutative, semiperfect ring R:*

(1) R is mininjective.
(2) $soc(eR)$ is simple or zero for all local idempotents e.
(3) R is a finite product of local rings whose socles are simple or zero.

Proof. (1)⇒(2). Assume that $soc(eR) \neq 0$. If $K, M \subseteq eR$ are simple R-submodules, they are simple eRe-submodules and so are eRe-isomorphic (as e is local). But then they are R-isomorphic, and so $M = cK$, $c \in R$, by (1). Hence $M = K$, proving (2).

(2)⇒(3). If $1 = e_1 + \cdots + e_m$ in R, where the e_i are orthogonal, local idempotents, then $R \cong \Pi_i e_i R e_i$. Hence (3) follows because $soc[(eR)_{eRe}] = soc[(eRe)_R]$.

(3)⇒(1). If $R \cong R_1 \times \cdots \times R_n$ as in (3) then each R_i has square-free socle and so is mininjective by Proposition 2.8. Hence R is mininjective by Example 2.2. □

Note that the localization $\mathbb{Z}_{(p)}$ of $\mathbb{Z}$ at the prime p is a commutative, local, mininjective ring that has zero socle.

The simplicity of $soc(eR)$ for each local idempotent e plays an important role in the theory of semiperfect mininjective rings as we shall see. The following lemma is fundamental in this discussion and will be used frequently in what follows.

Lemma 3.10. *Let e and f be local idempotents in a right mininjective ring R. If eR and fR contain isomorphic simple right ideals, then $eR \cong fR$.*

Proof. Let $\alpha : K \to fR$ be monic, where $K \subseteq eR$ is a simple right ideal. By hypothesis $\alpha = a\cdot$ for some $0 \neq a \in R$, and we may assume that $a \in fRe$. Hence $a\cdot : eR \to fR$ is R-linear. We have $S_r \subseteq S_l$ (because R is right mininjective), so $0 \neq \alpha(K) = aK \subseteq aS_r \subseteq aS_l$. This shows that $a \notin J$, so $aeR = aR \nsubseteq fJ$. Hence $a\cdot$ is onto fR because fJ is the unique maximal submodule of fR (as f is local). But then $a\cdot$ is one-to-one because fR is projective and eR is indecomposable. □

If R is a semiperfect ring and $1 = e_1 + \cdots + e_m$, where the e_i are orthogonal local idempotents, assume that $\{e_1R, \ldots, e_nR\}$ is a system of distinct representatives for $\{e_1R, \ldots, e_nR, \ldots, e_mR\}$. Hence $e_iR \cong e_jR$, $1 \leq i, j \leq n$, implies that $i = j$. In this case $\{e_1, \ldots, e_n\}$ is called a *basic set* of idempotents for R, and the idempotent $e = e_1 + \cdots + e_n$ is called a *basic idempotent* for R. The integer n is the number of isomorphism classes of simple right (or left) R-modules. Furthermore, up to isomorphism, the ring eRe is independent of the choice of the e_i, and eRe and R are Morita equivalent rings because $ReR = R$. The ring R is itself called a *basic semiperfect* ring if $m = n$, that is, if $1 = e_1 + \cdots + e_n$, where the e_i are a basic set of local idempotents.

As a first application of Lemma 3.10 we prove

Proposition 3.11. *Let R be a basic, semiperfect, right mininjective ring.*

(1) Let $\{e_1, \ldots, e_n\}$ be a basic set of idempotents in R. If K is any simple right ideal of R then $K \subseteq e_iR$ for some (unique) $i = 1, \ldots, n$.
(2) If $e^2 = e \in R$ is local then $soc(eR)$ is either 0 or a direct sum of homogeneous components of S_r.

Proof. (1). If $K = kR$ is simple, we have $k = e_1k + \cdots + e_nk$. If $e_ik \neq 0 \neq e_jk$ for $i \neq j$, then $e_ikR \cong kR \cong e_jkR$, whence $e_iR \cong e_jR$ by Lemma 3.10, which is a contradiction. Hence $k = e_ik$ for some i, and (1) follows.

(2). Let $e_1, \ldots, e_n$ be as in (1), so that $eR \cong e_iR$ for some $i = 1, \ldots, n$. If $soc(eR) \neq 0$, let $K \subseteq eR$ be simple. If $K' \cong K$, $K' \subseteq R$, then $K' \subseteq e_jR$ for some j by (1). Hence $i = j$ by Lemma 3.10, and (2) follows. □

3.2. Minfull Rings

Call a ring R *right minfull* if it is semiperfect, right mininjective, and $soc(eR) \neq 0$ for each local idempotent $e \in R$. The name derives from the last hypothesis, which enables us to use a counting argument, together with Lemma 3.10, to show that these rings are right and left Kasch and hence that many properties of quasi-Frobenius rings hold in this generality.

Theorem 3.12. *Let R be a right minfull ring. Then:*

(1) R is right and left Kasch.
(2) $soc(eR)$ is homogeneous for each local $e^2 = e \in R$.

(3) $S_r e$ is a simple left ideal for each local $e^2 = e \in R$.

(4) The following conditions are equivalent:

(a) $S_r = S_l$.

(b) $\mathtt{lr}(K) = K$ whenever $e^2 = e \in R$ is local and $K \subseteq Re$ is a simple left ideal.

(c) $soc(Re) = S_r e$ for all local $e^2 = e \in R$.

(d) $soc(Re)$ is simple for all local $e^2 = e \in R$.

Furthermore, if $e_1, \ldots, e_n$ are basic, orthogonal, local idempotents, there exist elements $k_1, \ldots, k_n$ in R and a permutation σ of $\{1, \ldots, n\}$ such that the following hold for each $i = 1, \ldots, n$:

(5) $k_i R \subseteq e_i R$ and $Rk_i \subseteq Re_{\sigma i}$.

(6) $k_i R \cong e_{\sigma i} R / e_{\sigma i} J$ and $Rk_i \cong Re_i / J e_i$.

(7) $Rk_i = S_r e_{\sigma i}$.

(8) $\{k_1 R, \ldots, k_n R\}$ and $\{Rk_1, \ldots, Rk_n\}$ are complete sets of distinct representatives of the simple right and left R-modules, respectively.

Proof. We begin with (5)–(8). Let $e_1, \ldots, e_n$ be basic, orthogonal, local idempotents. For each $i = 1, \ldots, n$, fix a simple right ideal $K_i \subseteq e_i R$. As R is semiperfect, choose $\sigma i \in \{1, \ldots, n\}$ such that $K_i \cong e_{\sigma i} R / e_{\sigma i} J$. This map σ is a permutation of $\{1, \ldots, n\}$ because $\sigma i = \sigma j$ implies that $K_i \cong K_j$, whence $e_i R \cong e_j R$ (by Lemma 3.10), and finally $i = j$ (because the e_i are basic). If $\gamma : e_{\sigma i} R / e_{\sigma i} J \to K_i$ is an isomorphism, write $k_i = \gamma(e_{\sigma i} + e_{\sigma i} J)$. Then $k_i R = K_i \cong e_{\sigma i} R / e_{\sigma i} J$ and $k_i \in e_i R e_{\sigma i}$ proving (5) and half of (6). Because $k_i \in S_r \subseteq S_l$, we obtain $\mathtt{l}(k_i) \supseteq J + R(1 - e_i)$. But $R/[J + R(1 - e_i)] \cong Re_i / J e_i$ is simple (see Lemma 3.6), so it follows that $\mathtt{l}(k_i) = J + R(1 - e_i)$ and hence that $Rk_i \cong Re_i / J e_i$. This proves (6). Now observe that $k_i = k_i e_{\sigma i} \in S_r e_{\sigma i}$. As $k_i \neq 0$, $S_r e_{\sigma i}$ is simple by Theorem 3.2, and (7) follows.

Since R is semiperfect it has exactly n isomorphism classes of simple right and left modules, represented respectively by $e_j R / e_j J$, where $1 \leq j \leq n$, and $Re_i / J e_i$, where $1 \leq i \leq n$. Hence (6) implies both (8) and (1). To prove (2), let $K \subseteq e_i R$ be a simple right ideal. Then $K \cong k_j R$ for some j by (8), so $j = i$ by Lemma 3.10. Hence $soc(e_i R)$ is homogeneous, and (2) follows. Finally, (3) follows from (7), and (4) follows by Theorem 3.7 because R is right Kasch by (1). □

Observe that the Björk example is an artinian, local, right minfull ring in which the conditions in (4) are all satisfied, but which fails to be left minfull (only because it fails to be left mininjective).

Corollary 3.13. *If R is a right minfull, left minannihilator ring then $S_l = S_r$ is finite dimensional as a left R-module.*

Proof. $S_r = S_l$ by Theorem 3.12, and S_r is left artinian by Theorem 3.7. □

Corollary 3.14. *A right minfull ring is left minfull if and only if it is left mininjective.*

Proof. If R is right minfull then $0 \neq S_r e \subseteq soc(Re)$ for each local idempotent e in R by Theorem 3.12. The result follows. □

Corollary 3.15. *If R is a semiperfect, right mininjective ring, the following are equivalent:*

(1) R is right minfull.
(2) R is right Kasch and $soc(eR)$ is homogeneous for each local $e^2 = e \in R$.

Proof. (1) implies (2) by Theorem 3.12. Assume (2). If K is a simple right ideal, then $eK \neq 0$ for some local idempotent e, and so $K \cong eK \subseteq eR$. Hence if $e_1, \ldots, e_n$ are basic idempotents in R, it follows (since R is right Kasch) that every simple right R-module embeds in $e_i R$ for some i. If $soc(e_k R) = 0$ for some k, then some $e_i R$ with $i \neq k$ contains two nonisomorphic simple right ideals, contrary to (2). This proves (1). □

Returning to the basic case, Proposition 3.11 shows that a basic, semiperfect, right mininjective ring is right minfull if (and only if) it is right Kasch. In this case we can improve upon (2) and (4) of Theorem 3.12 and obtain an improved *left* version of Proposition 3.11(1).

Proposition 3.16. *Let R be a basic right minfull ring and let $1 = e_1 + \cdots + e_n$, where the e_i are basic, orthogonal, local idempotents.*

(1) If $k_1, \ldots, k_n$ in R are chosen as in Theorem 3.12, then $soc(e_i R) = soc_{k_i R}(R_R) = Rk_i R$ is a simple ideal.
(2) $S_r = S_l$ if and only if R is a left minannihilator ring; and in this case the only simple left ideals of R are the $S_r e_i = soc(Re_i)$ for $i = 1, \ldots, n$.

Proof. (1). Since $e_i R \ncong e_j R$ when $i \neq j$, this follows from Theorem 2.21 and Corollary 2.22 because each simple right ideal is contained in $e_i R$ for some $i = 1, \ldots, n$.

(2). Assume that $S_r = S_l$. If K is a simple left ideal, then $Ke_i \neq 0$ for some i, so, as $Ke_i \subseteq S_l e_i = S_r e_i$, we have $Ke_i = S_r e_i$ by Theorem 3.2. If $Ke_j \neq 0$ for $j \neq i$, then $Ke_j \cong K \cong Ke_i$, contrary to Theorem 3.12. Since $\Sigma e_i = 1$, it follows that $K = Ke_i = S_r e_i$. This proves the last sentence of (2) and shows that $\mathtt{r}(K) \supseteq J + (1 - e_i)R$. Hence $\mathtt{lr}(K) \subseteq \mathtt{l}(J) \cap Re_i = S_r \cap Re_i = S_r e_i = K$. Thus $\mathtt{lr}(K) = K$, so we have proved that $S_r = S_l$ implies that R is a left minannihilator ring. The converse holds by Proposition 3.7. □

Turning to right and left minfull rings, we have the following immediate consequence of Theorem 3.12.

Proposition 3.17. *Let R be a right and left minfull ring. Then:*

(1) $S_r = S_l$, which we denote as S.
(2) $soc(eR) = eS$ and $soc(Re) = Se$ are simple for all local $e^2 = e \in R$.
(3) S is right and left finite dimensional.
(4) R is a right and left Kasch ring.
(5) R is a right and left minannihilator ring.
(6) If $e_1, \ldots, e_n$ are basic local idempotents, there exists a permutation σ of $\{1, \ldots, n\}$ such that the following hold for all $i = 1, \ldots, n$:
  (a) $Se_{\sigma i} = soc(Re_{\sigma i}) \cong Re_i/Je_i$ and $e_i S = soc(e_i R) \cong e_{\sigma i}R/e_{\sigma i}J$.
  (b) $\{e_1 S, \ldots, e_n S\}$ and $\{Se_1, \ldots, Se_n\}$ are sets of distinct representatives of the simple right and left R-modules, respectively.

Proof. Since R is right and left mininjective, (5) follows by Corollary 2.34. Furthermore $S_r = S_l$, so the rest follows from Theorems 3.12 and 3.7. □

Theorem 3.18. *The following are equivalent for a semiperfect ring R:*

(1) R is right and left minfull.
(2) The dual of every simple R-module is simple.
(3) $S_r e$ and eS_l are simple for every local idempotent $e \in R$.
(4) R is right and left mininjective and right (or left) Kasch.

Proof. (1)⇒(2). This follows from Theorem 2.29 because R is two-sided Kasch (by Proposition 3.17) and two-sided mininjective.

(2)⇒(3). This is because $S_r e \cong (eR/eJ)^*$ and $eS_l \cong (Re/Je)^*$ for all local idempotents e (using Lemma 3.1).

(3)⇒(4). This is by Theorem 3.2.

(4)⇒(1). By symmetry, assume that R is right Kasch. Then $S_r e$ is simple for all local $e = e^2$ by Theorem 3.7, and so $soc(Re) \neq 0$. Since R is also left Kasch (by Theorem 3.12) we similarly get $soc(eR) \neq 0$. □

Note that the Björk example is a local, right and left artinian, right and left Kasch ring that is right minfull but not left minfull.

We conclude this section by proving that being right minfull is a Morita invariant property of rings. The following results (of interest in their own right) will be needed. Recall that a ring R is called semilocal if R/J is semisimple artinian.

Lemma 3.19. *If R is a ring, each of the following is a Morita invariant property:*

(1) $soc(R_R) \subseteq^{ess} R_R$.
(2) $S_r \subseteq S_l$.

Proof. Write $\overline{R} = M_n(R)$ and recall that every right ideal of $\overline{R}$ has the form $[XXX]$ for $X_R \subseteq R_n$. To show that (1) and (2) pass from R to $\overline{R}$, it suffices to show that $soc(\overline{R}_{\overline{R}}) = M_n(S_r)$ and $soc({}_{\overline{R}}\overline{R}) = M_n(S_l)$; we prove the former. Let $soc(\overline{R}_{\overline{R}}) = [SSS]$, $S \subseteq R_n$. Then $S \subseteq (S_r)_n$ because $(S_r)_n = soc(R_n)$, so $soc(\overline{R}_{\overline{R}}) \subseteq M_n(S_r)$. The other inclusion holds because $M_n(S_r) = [(S_r)_n\ (S_r)_n\ (S_r)_n]$ is right $\overline{R}$-semisimple.

Now let $Q = eRe$, where $e^2 = e \in R$ satisfies $ReR = R$. If kQ is simple, $k \in Q$, then kR is simple as in the proof of Proposition 2.11. It follows that $soc(Q_Q) \subseteq eS_re$; the other inclusion is proved similarly. Thus $soc(Q_Q) = eS_re$ and properties (1) and (2) pass from R to Q. □

Theorem 3.20. *Each of the following classes of rings is Morita invariant.*

(1) The right minfull rings.
(2) The right minfull rings R in which $soc(Re)$ is simple for each local idempotent $e^2 = e \in R$.
(3) The right minfull rings in which $S_r \subseteq^{ess} R_R$.

Proof. If R is semiperfect and $P_R \neq 0$ is projective then $P \cong \oplus_i e_i R$ for some local idempotents $e_i \in R$. Hence $soc(eR) \neq 0$ for every local $e^2 = e$ if and only if, for any projective module $P_R \neq 0$, there exists an exact sequence $0 \to S \to P$ where $S_R \neq 0$ is semisimple. Thus (1) follows from Theorem 2.15. Turning to (2), we find that a right minfull ring R satisfies the condition in (2) if and only if $S_r = S_l$ [by Theorem 3.12(4)]. Thus (2) follows from (1) and Lemma 3.19. Similarly, (3) follows from (1) and Lemma 3.19. □

3.3. Nakayama Permutations

If R is a semiperfect ring with basic idempotents $\{e_1, \ldots, e_n\}$, a permutation σ of $\{1, 2, \ldots, n\}$ is called a *Nakayama permutation* for R if

$$soc(Re_{\sigma i}) \cong Re_i/Je_i \qquad \text{and} \qquad soc(e_i R) \cong e_{\sigma i}R/e_{\sigma i}J$$

for each $i = 1, 2, \ldots, n$. Thus every right and left minfull ring admits a Nakayama permutation by Proposition 3.17.

Theorem 3.21. *Let R be a semiperfect ring with a Nakayama permutation in which $S_r \subseteq^{ess} R_R$ (or $S_l \subseteq^{ess} {}_R R$). Then R is right and left minfull.*

Proof. We consider the case when $S_r \subseteq^{ess} R_R$; the other case is analogous. The Nakayama permutation shows that $soc(Re)$ and $soc(eR)$ are both simple for each local $e = e^2$. If $\{e_1, \ldots, e_n\}$ are basic idempotents then R is Kasch because the $e_i R/e_i J$ and Re_i/Je_i are complete systems of representatives of the simple right and left R-modules, respectively. In particular $S_r e_i \neq 0$ and $e_i S_l \neq 0$ for each i by Theorem 3.2. But $soc(e_i R)$ is simple and essential in $e_i R$ for each i, and it follows that $soc(e_i R) \subseteq e_i S_l$. Hence $S_r = \oplus_i soc(e_i R) \subseteq \oplus_i e_i S_l \subseteq S_l$, and so $0 \neq S_r e \subseteq S_l e = soc(Re)$ for each local idempotent e. Hence $S_r e = soc(Re)$ is simple, so R is right mininjective (and right Kasch) by Theorem 3.2. This means that R is right minfull and hence that $S_r = S_l$ [by Theorem 3.12(4) because $soc(Re)$ is simple for all local $e^2 = e$]. But then $eS_l = eS_r = soc(eR)$ is simple for each local $e^2 = e$, so R is left mininjective. Hence R is left minfull. □

Note that the proof of Theorem 3.21 actually shows the following:

Corollary 3.22. *A ring R is right and left minfull if it is a semiperfect, right Kasch ring with $S_r \subseteq^{ess} R_R$, in which $soc(eR)$ and $soc(Re)$ are both simple for each local idempotent e.*

Observe that the essential socle hypothesis in Theorem 3.21 is unnecessary in a local ring R because, in that case, R is right (left) minfull if and only if S_r is simple on the right (S_l is simple on the left). Hence, using Theorem 3.2, R is right and left minfull if and only if it has a Nakayama permutation.

Nontheless, Theorem 3.21 does give the following classical characterization of quasi-Frobenius.

Theorem 3.23 (Nakayama's Theorem). *A ring R is quasi-Frobenius if and only if it is right and left artinian and has a Nakayama permutation.*

Proof. If R satisfies these conditions, Theorem 3.21 shows that R is two-sided minfull and hence two-sided mininjective. Hence R is quasi-Frobenius by Ikeda's Theorem (Theorem 2.30). □

3.4. Min-PF Rings

Recall (Theorem 1.56) that the right PF rings are the semiperfect, right self-injective rings with essential right socle. Accordingly, with an eye on Theorem 3.12, we call a ring R a right *min-PF ring* if R is a semiperfect, right mininjective ring in which $S_r \subseteq^{ess} R_R$ and $\mathtt{lr}(K) = K$ whenever $e^2 = e \in R$ is local and $K \subseteq Re$ is a simple left ideal. Clearly every right min-PF ring is right minfull.

As the name implies, every right PF ring R is right min-PF. To see this it suffices (using Theorem 1.56) to show that R is left minannihilator. But R_R is a cogenerator by Theorem 1.56, so $\mathtt{rl}(T) = T$ for every right ideal T by Lemma 1.40. Hence R is left mininjective and right minannihilator, and so it is left minannihilator by Corollary 2.34 because $S_r = S_l$.

The next result collects the information we have about these right min-PF rings.

Theorem 3.24. *If R is a right min-PF ring, then the following hold:*

(1) R is right and left Kasch.

(2) $J = Z_r$.

(3) $S_r = S_l$ is left finite dimensional.

(4) If $e_1, \dots, e_n$ is a basic set of local idempotents, there exist elements $k_1, \dots, k_n$ in R and a permutation σ of $\{1, \dots, n\}$ such that the following hold for all $i = 1, \dots, n$:

- *(a) $k_i R \subseteq e_i R$ and $Rk_i \subseteq Re_{\sigma i}$.*
- *(b) $k_i R \cong e_{\sigma i} R / e_{\sigma i} J$ and $Rk_i \cong Re_i / Je_i$.*
- *(c) $\{k_1 R, \dots, k_n R\}$ and $\{Rk_1, \dots, Rk_n\}$ are complete sets of distinct representatives of the simple right and left R-modules,respectively.*
- *(d) $soc(Re_{\sigma i}) = Rk_i = S_r e_{\sigma i} \cong Re_i / Je_i$ is simple for each i.*
- *(e) $soc(e_i R)$ is homogeneous with each simple submodule isomorphic to $k_i R \cong e_{\sigma i} R / e_{\sigma i} J$.*

Proof. (1) and (a) through (d) of (4) all follow from Theorem 3.12, (2) is by Proposition 2.27, and (3) is by Corollary 3.13. To prove (e), let $K \subseteq e_i R$ be simple. Then $K \cong k_j R \subseteq e_j R$ for some j by (c), so $j = i$ by Lemma 3.10. This proves (e). □

Corollary 3.25. *Suppose R is a semiperfect, left minannihilator ring in which $S_r \subseteq^{ess} R_R$ and $S_l \subseteq^{ess} {}_RR$ (for example, if R is a semiprimary, left minannihilator ring). Then the following hold:*

(1) R is a right min-PF ring that is left finite dimensional.
(2) if $k \in R$, Rk is simple if and only if kR is simple.
(3) $J = Z_r = Z_l$.

Proof. (1). R is right mininjective by Corollary 2.35. Hence R is right min-PF; it is left finite dimensional by Theorem 3.24 because $S_l \subseteq^{ess} {}_RR$.

(2). If kR is simple then Rk is simple by Theorem 2.21. If Rk is simple, let $\mathtt{r}(k) \subseteq T \subseteq^{max} R_R$. Then $Rk = \mathtt{lr}(k) \supseteq \mathtt{l}(T) \neq 0$ because R is right Kasch (by (1)). Hence $Rk = \mathtt{l}(T)$, so $T \subseteq \mathtt{rl}(T) = \mathtt{r}(k)$. As T is maximal, $T = \mathtt{r}(k)$, whence kR is simple.

(3). We have $J = Z_r$ by Proposition 2.27, and $Z_l \subseteq J$ because R is semipotent by Theorem B.9. But $S_r = S_l \subseteq^{ess} {}_RR$, and it follows that $J \subseteq Z_l$ because $S_r \subseteq \mathtt{l}(J) \subseteq \mathtt{l}(a)$ for all $a \in J$. □

Example 3.26. Every right PF ring is both right and left min-PF.

Proof. By the discussion preceding Theorem 3.24, it remains to prove that $S_l \subseteq^{ess} {}_RR$. To see this, observe first that since R is right self-injective it satisfies $\mathtt{lr}(a) = Ra$ for all $a \in R$ (by Corollary 1.38). Hence let $0 \neq a \in R$; we show that Ra contains a simple left ideal. But if $\mathtt{r}(a) \subseteq T \subseteq^{max} R_R$, then $\mathtt{l}(T) \subseteq \mathtt{lr}(a) = Ra$, and $\mathtt{l}(T)$ is simple by Theorem 2.31 because R is right Kasch (by Theorem 3.24). □

Proposition 3.27. *Suppose that R/A is a right and left min-PF ring for every ideal A of R. Then R is an artinian principal ideal ring.*

Proof. By Theorem 3.24, $soc(R/A)$ is finitely generated and essential in R/A as a left and right module for each ideal A. Hence R/A is artinian by the Vámos Lemma (Lemma 1.52) and so is quasi-Frobenius. Thus R is an artinian principal ideal ring by [53, p. 238]. □

Theorem 3.28. *The right min-PF rings form a Morita invariant class.*

Proof. Let Q denote either $M_n(R)$, where $n \geq 1$, or eRe, where $e^2 = e \in R$ and $ReR = R$. If R is a right min-PF ring then (using Theorem 3.24) R is right minfull with $S_l = S_r \subseteq^{ess} R_R$. Hence Theorem 3.20 and Lemma 3.19 show that Q is a right minfull ring in which $soc(Q_Q) = soc({}_QQ) \subseteq^{ess} Q_Q$. Hence Q is right min-PF by Theorem 3.12. □

3.5. Annihilator Chain Conditions

We begin by applying Theorem 3.7 to give mild chain conditions that guarantee that a right min-PF ring is quasi-Frobenius. If R is a ring, a right ideal of the form $\mathtt{r}(X)$, where $X \subseteq R$, is called a *right annihilator*, and we say that R has the *ACC on right annihilators*, if every ascending chain of these right ideals is eventually constant. We use similar terminology for left annihilators. We need two lemmas about this condition.

Lemma 3.29 (Mewborn–Winton). *If R has ACC on right annihilators, then Z_r is nilpotent.*

Proof. Write $Z = Z_r$. As $Z \supseteq Z^2 \supseteq \cdots$ we get $\mathtt{r}(Z) \subseteq \mathtt{r}(Z^2) \subseteq \cdots$, so let $\mathtt{r}(Z^n) = \mathtt{r}(Z^{n+1})$; we show that $\mathtt{r}(Z^n) = R$. Suppose that $Z^n a \neq 0$ for some $a \in R$, and choose $\mathtt{r}(b)$ maximal in $\{\mathtt{r}(b) \mid Z^n b \neq 0\}$. If $z \in Z$ then $\mathtt{r}(z) \subseteq^{ess} R_R$, so $\mathtt{r}(z) \cap bR \neq 0$, say $0 \neq br$ with $zbr = 0$. Thus $\mathtt{r}(b) \subset \mathtt{r}(zb)$, so, by the choice of b, $Z^n zb = 0$. As $z \in Z$ was arbitrary, this shows that $Z^{n+1} b = 0$, whence $b \in \mathtt{r}(Z^{n+1}) = \mathtt{r}(Z^n)$, which is a contradiction. □

The second lemma we need is a useful condition that a semiprimary ring is left artinian.

Lemma 3.30. *Let R be a semiprimary ring with ACC on right annihilators, in which $S_r = S_l$ is finite dimensional as a left R-module. Then R is left artinian.*

Proof. We use induction on the index of nilpotency n of J, that is, the smallest integer n such that $J^n = 0$. If $n = 1$ then $J = 0$ and R is semisimple artinian. So assume that $n \geq 2$. We have $S_r = \mathtt{l}(J)$ and $S_l = \mathtt{r}(J)$ (because R is semilocal), and we write $\bar{R} = R/A$, where $A = \mathtt{l}(J) = S_r = S_l = \mathtt{r}(J)$. Since ${}_RA$ is artinian, it suffices to show that ${}_R\bar{R} = {}_{\bar{R}}\bar{R}$ is artinian.

Since $A = \mathtt{l}(J)$ is itself a left annihilator, the ring $\bar{R}$ inherits the ACC on right annihilators from R. Moreover, $J(\bar{R}) = (J + A)/A = \bar{J}$, so $\bar{R}/\bar{J} \cong R/(J + A)$ is semisimple and $\bar{J}^{n-1} \subseteq (J^{n-1} + A)/A = 0$ because $J^{n-1} \subseteq \mathtt{r}(J) = A$. Hence $\bar{R}$ is semiprimary and $\bar{J}$ has index of nilpotency at most $n - 1$, so it suffices by induction to show that $soc({}_{\bar{R}}\bar{R}) = soc(\bar{R}_{\bar{R}})$ is finite dimensional as a left $\bar{R}$-module.

If $\bar{x} \in soc(_{\bar{R}}\bar{R}) = \mathrm{r}_{\bar{R}}(\bar{J})$, we have $\bar{J}\bar{x} = \bar{0}$, so $Jx \subseteq A = \mathrm{l}(J)$, whence $JxJ = 0$. Thus $xJ \subseteq \mathrm{r}(J) = A$, so $\bar{x}\bar{J} = \bar{0}$ and $\bar{x} \in \mathrm{l}_{\bar{R}}(\bar{J}) = soc(\bar{R}_{\bar{R}})$. Thus $soc(_{\bar{R}}\bar{R}) \subseteq soc(\bar{R}_{\bar{R}})$, and the other inclusion is similarly proved. Finally, since R has the DCC on left annihilators, we have $\mathrm{l}(J) = \mathrm{l}\{b_1, \ldots, b_m\}$ for some $\{b_1, \ldots, b_m\} \subseteq J$. Then $\theta : \bar{R} \to R^m$ given by $\theta(r + A) = (rb_1, \ldots, rb_m)$ is a well-defined monomorphism of left R-modules. Moreover, $\theta[soc(_{\bar{R}}\bar{R})] = \theta[soc(_R\bar{R})] \subseteq soc[_RR^m] = S_l^m$, so $soc(_{\bar{R}}\bar{R})$ is left finite dimensional by hypothesis. This completes the proof. □

The following theorem is a sharp improvement on Ikeda's theorem (Theorem 2.30) that a right and left artinian, right and left mininjective ring is quasi-Frobenius. Surprisingly, we need only assume that R is semilocal.

Theorem 3.31. *Suppose that R is a semilocal, left and right mininjective ring with ACC on right annihilators in which $S_r \subseteq^{ess} R_R$. Then R is quasi-Frobenius.*

Proof. We have $S_l = S_r$ by the mininjective hypothesis, so $S_r \subseteq \mathrm{r}(J) \subseteq \mathrm{r}(a)$ for all $a \in J$. Since $S_r \subseteq^{ess} R_R$ it follows that $J \subseteq Z_r$. But Z_r is nilpotent by Lemma 3.29, so R is semiprimary (being semilocal). In particular, R is semiperfect and right mininjective, so S_r is finite dimensional as a left R-module by Theorem 3.7. But then R is left artinian by Lemma 3.30, so R has ACC on left annihilators. Since $S_l \subseteq^{ess} {}_RR$ because R is semiprimary, the same argument shows that R is also right artinian, and so it is quasi-Frobenius by Ikeda's theorem. □

We are going to show that every right min-PF ring with ACC on right annihilators is left artinian. We will need the following three lemmas.

Lemma 3.32. *Let R be a ring and let A be an ideal of R such that R/A satisfies ACC on right annihilators. If $Y_1, Y_2, \ldots$ are subsets of $\mathrm{l}(A)$ there exists $n \geq 1$ such that*

$$\mathrm{r}(Y_{n+1} \cdots Y_1) = \mathrm{r}(Y_n \cdots Y_1),$$

where $Y_iY_j = \{xy \mid x \in Y_i \text{ and } y \in Y_j\}$.

Proof. Write $\overline{R} = R/A$ and let $r \mapsto \overline{r}$ denote the coset map $R \to \overline{R}$. Then $\mathrm{r}(\overline{Y}_1) \subseteq \mathrm{r}(\overline{Y}_2\overline{Y}_1) \subseteq \mathrm{r}(\overline{Y}_3\overline{Y}_2\overline{Y}_1) \subseteq \cdots$, so, by hypothesis, there exists $n \geq 2$ such that $\mathrm{r}(\overline{Y}_{n-1} \cdots \overline{Y}_1) = \mathrm{r}(\overline{Y}_n \cdots \overline{Y}_1) = \cdots$. If $b \in \mathrm{r}(Y_{n+1} \cdots Y_1)$, this gives $\overline{Y}_{n+1} \cdots \overline{Y}_1\overline{b} = \overline{0}$, so that $\overline{Y}_{n-1} \cdots \overline{Y}_1\overline{b} = \overline{0}$. Hence $Y_{n-1} \cdots Y_1b \subseteq A \subseteq \mathrm{r}(Y_n)$

and so $b \in \mathbf{r}(Y_n \cdots Y_1)$. Thus $\mathbf{r}(Y_{n+1} \cdots Y_1) \subseteq \mathbf{r}(Y_n \cdots Y_1)$; the other inclusion is clear. □

A one-sided ideal A of a ring R is called right *T-nilpotent* if, given any sequence $a_1, a_2, \dots$ from A, there exists an integer $n \geq 1$ such that $a_n a_{n-1} \cdots a_2 a_1 = 0$.

Lemma 3.33. *If the ring R has ACC on right annihilators then every right T-nilpotent, one-sided ideal is nilpotent.*

Proof. If B is a right T-nilpotent one-sided ideal, we have $\mathbf{r}(B) \subseteq \mathbf{r}(B^2) \subseteq \cdots$, so let $\mathbf{r}(B^n) = \mathbf{r}(B^{n+1}) = \cdots$. If $A = B^n$ then $\mathbf{r}(A) = \mathbf{r}(A^2)$, and we claim that $A^2 = 0$. If not let $Aa_1 \neq 0$ with $a_1 \in A$. Then $A^2 a_1 \neq 0$, so let $Aa_2 a_1 \neq 0$, $a_2 \in A$. This process continues to contradict the right T-nilpotency of B. □

With this we can give a useful condition that a semiperfect, right mininjective ring is semiprimary.

Lemma 3.34. *Suppose R is a semiperfect, right mininjective ring such that $S_r \subseteq^{ess} R_R$. If either R or R/S_r has ACC on right annihilators, then R is semiprimary.*

Proof. We have $J = Z_r$ by Proposition 2.27. Thus J is nilpotent if R has ACC on right annihilators (by Lemma 3.29). Now suppose $\bar{R} = R/S_r$ has ACC on right annihilators. It suffices to show that $\bar{J}$ is right T-nilpotent (then it is nilpotent by Lemma 3.33, so $J^n \subseteq S_r$, whence $J^{n+1} = 0$). So let $a_1, a_2, \dots$ be a sequence from $J = Z_r$. Since $Z_r S_r = 0$ always holds, we have $a_i \in \mathbf{l}(S_r)$ for each i. Hence Lemma 3.32 provides an integer $n \geq 1$ such that $\mathbf{r}(a_{n+1} a_n \cdots a_1) = \mathbf{r}(a_n \cdots a_1)$. But $\mathbf{r}(a_{n+1}) \subseteq^{ess} R_R$, so, if $a_n \cdots a_1 \neq 0$, we have $\mathbf{r}(a_{n+1}) \cap (a_n \cdots a_1 R) \neq 0$. Thus let $0 \neq x = a_n \cdots a_1 r$, $r \in R$, be such that $a_{n+1} x = 0$. Hence $r \in \mathbf{r}(a_{n+1} a_n \cdots a_1) = \mathbf{r}(a_n \cdots a_1)$, so $0 = a_n \cdots a_1 r = x$, which is a contradiction. This proves that $a_n \cdots a_1 = 0$, as required. □

Theorem 3.35. *Let R be a right min-PF ring with ACC on right annihilators. Then R is left artinian.*

Proof. Such a ring R is semiprimary by Lemma 3.34, so $S_r = S_l$ is left finite dimensional by Theorem 3.24. It follows that R is left artinian by Lemma 3.30. □

Recall that the socle series $soc_n(M)$, $n \geq 1$, of a module M is defined by $soc_1(M) = soc(M)$ and

$$\frac{soc_{n+1}(M)}{soc_n(M)} = soc\left(\frac{M}{soc_n(M)}\right)$$

if $n \geq 1$. This is required in the proof of the next result.

Lemma 3.36. *If R is a semilocal ring for which $S_r = S_l$ then*

$$soc_n(R_R) = soc_n({}_RR) = \mathtt{l}(J^n) = \mathtt{r}(J^n) \ \text{ for all } n \geq 1.$$

Proof. We have $S = \mathtt{l}(J) = \mathtt{r}(J)$ because R/J is semisimple, so it holds for $n = 1$. Suppose $soc_k(R_R) = soc_k({}_RR) = \mathtt{l}(J^k) = \mathtt{r}(J^k)$ for some $k \geq 1$. We have $soc_{k+1}(R_R) \subseteq \mathtt{l}(J^{k+1})$ because $soc_{k+1}(R_R)/soc_k(R)$ is right R-semisimple. On the other hand, if $aJ^{k+1} = 0$ then $aJ \subseteq soc_k(R)$, so $[aR + soc_k(R)]/soc_k(R)$ is right R-semisimple (because R/J is semisimple). Hence $aR \subseteq soc_{k+1}(R_R)$ and we have $soc_{k+1}(R_R) = \mathtt{l}(J^{k+1})$. Similarly, $soc_{k+1}({}_RR) = \mathtt{r}(J^{k+1})$. Finally, $\mathtt{l}(J^{k+1}) = \mathtt{r}(J^{k+1})$ follows easily from $\mathtt{l}(J^k) = \mathtt{r}(J^k)$. □

With this we can give an important characterization of when a semiprimary ring is left and right artinian.

Lemma 3.37. *Let R be a semiprimary ring with $S_r = S_l$. If $soc_1(R)$ is right artinian and $soc_2(R)$ is left artinian, then R is left and right artinian.*

Proof. Let n be the index of nilpotency of J. Since R is semiprimary, the theorem is proved when we establish the following statement for $i = 1, 2, \ldots, n$:

$$soc_{n-i+1}(R)/soc_{n-i}(R) \text{is right artinian and} J^{n-i} \text{is left artinian.}$$

We proceed by induction on $i \geq 1$. When $i = 1$, $J^{n-1} \subseteq soc_1(R)$ and so J^{n-1} is left artinian. Moreover, $soc_n(R)/soc_{n-1}(R) = R/soc_{n-1}(R)$ is right artinian because R/J is semisimple and $J \subseteq soc_{n-1}(R)$. Assume by induction that for some $i \geq 1$, $soc_{n-l+1}(R)/soc_{n-l}(R)$ is right artinian and J^{n-l} is left artinian for all $l \leq i$.

Claim. J^{n-i-1} *is left artinian.*

Proof. Let $A = [J^{n-i-1} \cap soc_2(R)] + J^{n-i}$. Observe that A is left artinian, since both $soc_2(R)$ and J^{n-i} are left artinian. Thus, we only need to show that

J^{n-i-1}/A is left artinian as an R-module. Assume not, so that J^{n-i-1}/A is not left artinian as an R/J-module. Let $R/J = \oplus_{i=1}^{r} M_{n_i}(D_i)$, where each D_i is a division ring, n_i is a positive integer, and $i = 1, 2, \ldots, r$. Then there is a homogeneous component H/A of J^{n-i-1}/A corresponding to a simple component, say $M_{n_1}(D_1)$, of R/J such that H/A is an infinite dimensional left vector space over the division ring D_1. Let $\{x_1 + A, x_2 + A, \ldots\}$ be an infinite family of D_1-linearly independent vectors in H/A. Since $R/soc_{n-i}(R)$ is right artinian (by the induction step), it is right noetherian. Thus $[J + soc_{n-i}(R)]/soc_{n-i}(R)$ is right finitely generated, and so $J + soc_{n-i}(R) = y_1R + y_2R + \cdots + y_kR + soc_{n-i}(R)$, for some elements $y_t \in J$, $1 \leq t \leq k$. Let $x \in J^{n-i-1} - A$. If $xy_t = 0$ for all $t = 1, 2, \ldots, k$, then

$$\begin{aligned} xJ &\subseteq x[J + soc_{n-i}(R)] = x[y_1R + y_2R + \cdots + y_kR + soc_{n-i}(R)] \\ &= xy_1R + xy_2R + \cdots + xy_nR + x\, soc_{n-i}(R) = x\, soc_{n-i}(R) \\ &\subseteq J^{n-i-1}soc_{n-i}(R). \end{aligned}$$

Now, $J[J^{n-i-1}soc_{n-i}(R)] \subseteq J^{n-i}soc_{n-i}(R) = 0$ because $soc_{n-i}(R) = \mathbf{r}(J^{n-i})$. And since R/J is semisimple it follows that $J^{n-i-1}soc_{n-i}(R)$ is semisimple as an R/J- module, and so it is semisimple as an R-module. Thus we obtain $J^{n-i-1}soc_{n-i}(R) \subseteq soc_1(R)$. But then $xJ \subseteq J^{n-i-1}soc_{n-i}(R) \subseteq soc_1(R)$, and so $xJ^2 = 0$. Thus $x \in \mathbf{l}(J^2) = soc_2(R)$, and hence $x \in J^{n-i-1} \cap soc_2(R) \subseteq A$, which is impossible. Thus we must have $xy_t \neq 0$, for some t, $1 \leq t \leq k$. With this in mind we are going to arrive at a contradiction by producing an infinite family of D_1-linearly independent elements $\{u_1 + A, u_2 + A, \ldots\}$ in H/A such that $u_p y_q = 0$ for $p = 1, 2, \ldots$ and $q = 1, 2, \ldots, k$.

We begin the process with the element y_1. For each i, $1 \leq i \leq r$, let $M_{n_i}(D_i) = e_i(R/J)$ for some central primitive idempotent e_i of R/J, and denote $m = dim_{D_1}[e_1(J^{n-i}/J^{n-i+1})] + 1$. Observe that $x_p y_q \in J^{n-i}$, which is a left artinian ideal of R by the induction step. Since $m > dim_{D_1}[e_1(J^{n-i}/J^{n-i+1})]$, there exist elements $\alpha_1, \ldots, \alpha_m$ in D_1, not all zero, such that

$$\alpha_1 e_1(x_1y_1 + J^{n-i+1}) + \cdots + \alpha_m e_1(x_my_1 + J^{n-i+1}) = 0 \text{ in } J^{n-i}/J^{n-i+1}.$$

Let $a_i = (\alpha_i, 0, \ldots, 0)$, $1 \leq i \leq m$. Then

$$a_1(x_1y_1 + J^{n-i+1}) + \cdots + a_m(x_my_1 + J^{n-i+1}) = 0 \text{ in } J^{n-i}/J^{n-i+1}.$$

Now, lift $a_1, \ldots, a_m$ to elements $r_1, \ldots, r_m$ in R and define $x_1' = r_1x_1 + r_2x_2 +$

$\cdots + r_m x_m \in H$. Observe that $x_1' y_1 \in J^{n-i+1}$, since

$$\begin{aligned} x_1' y_1 + J^{n-i+1} &= (r_1 x_1 + r_2 x_2 + \cdots + r_m x_m) y_1 + J^{n-i+1} \\ &= r_1(x_1 y_1 + J^{n-i+1}) + \cdots + r_m(x_m y_1 + J^{n-i+1}) \\ &= 0 \text{ in } J^{n-i}/J^{n-i+1}. \end{aligned}$$

Now by using the m elements $x_{m+1}, \ldots, x_{2m}$ and the preceding argument we obtain an element

$$x_2' = r_{m+1} x_{m+1} + r_{m+2} x_{m+2} + \cdots + r_{2m} x_{2m} \in H$$

such that $x_2' y_1 \in J^{n-i+1}$ and $e_1(r_p + J) \neq 0$ in R/J for some p, $m + 1 \leq p \leq 2m$.

In this way we obtain an infinite sequence $x_1', x_2', \ldots \in H$ such that $x_p' y_1 \in J^{n-i+1}$ for all $p \geq 1$ and that $\{x_1' + A, x_2' + A, \ldots\}$ are D_1-linearly independent vectors in H/A. If we apply the same method to the sequence $\{x_1', x_2', \ldots\}$ and to J^{n-i+1}/J^{n-i+2}, we obtain another infinite sequence $x_1'', x_2'', \ldots \in H$ such that $x_p'' y_1 \in J^{n-i+2}$ for all $p \geq 1$ and that $\{x_1'' + A, x_2'' + A, \ldots\}$ are D_1-linearly independent vectors in H/A. If we continue this process, we will obtain an infinite sequence $\{t_1 + A, t_2 + A, \ldots\}$ of D_1-linearly independent vectors of H/A such that $t_p y_1 = 0$ for all $p \geq 1$.

Next, we apply the same process to the element y_2. Note that this process preserves the fact that the previously constructed sequences are annihilated by the element y_1. So, we obtain an infinite sequence $\{t_1' + A, t_2' + A, \ldots\}$ of D_1-linearly independent vectors of H/A such that $t_p' y_1 = 0$ and $t_p' y_2 = 0$ for all $p \geq 1$.

Finally, the same process can be applied to the elements $y_3, \ldots, y_k$ to obtain an infinite sequence $u_1, u_2, \ldots \in H$ such that $u_p y_q = 0$, where $p \geq 1, 1 \leq q \leq k$, and $\{u_1 + A, u_2 + A, \ldots\}$ are D_1-linearly independent vectors in H/A. This is the promised contradiction. Thus J^{n-i-1} is left artinian, and the Claim is proved.

Now we can show that $soc_{n-i}(R)/soc_{n-i-1}(R)$ is right artinian using a similar process and the Claim. The relevant facts in this case are $J^{n-i-1}x \neq 0$ for any $x \in soc_{n-i}(R) - soc_{n-i-1}(R)$, since $soc_{n-i-1}(R) = \mathrm{r}(J^{n-i-1})$, and that $J^{n-i-1} soc_{n-i}(R) \subseteq soc_1(R)$, which is right artinian by hypothesis. This completes the proof. □

Theorem 3.38. *Suppose that R is a two-sided min-PF ring. If either R has ACC on right annihilators or $R/soc(R)$ is right Goldie, then R is quasi-Frobenius* [*where we write* $soc(R) = S_r = S_l$].

Proof. Since R is two-sided minfull, $S_r = S_l$ and both $soc(Re)$ and $soc(eR)$ are simple for every local $e^2 = e \in R$. Since $S_r = S_l$ we are done by Theorem 3.31 if R has the ACC on right annihilators. So assume that $R/soc(R)$ is right Goldie. Since both $soc(Re)$ and $soc(eR)$ are simple for every local $e^2 = e \in R$, it suffices by Theorem 3.4 to prove that R is right and left artinian. But R is semiprimary by Lemma 3.34, whence $soc_2({}_R R) = soc_2(R_R)$. Since R is right and left finite dimensional (by Theorem 3.24), and since $R/soc(R)$ is right Goldie, it follows that R is two-sided artinian by Lemma 3.37. □

The next result is required later and has independent interest.

Lemma 3.39. *Suppose R is semiprimary and $J^2 = 0$. Then R has ACC and DCC on left and right annihilators.*

Proof. Every ascending chain of left annihilators in R has the form $\mathbb{1}(X_1) \subseteq \mathbb{1}(X_2) \subseteq \cdots$, where each X_i is a right ideal and $X_1 \supseteq X_2 \supseteq \cdots$. As $J^2 = 0$, we have

$$J \subseteq \mathbb{1}(X_1 \cap J) \subseteq \mathbb{1}(X_2 \cap J) \subseteq \cdots \quad \text{and} \quad X_1 + J \supseteq X_2 + J \supseteq \cdots \supseteq J.$$

Since R/J is left noetherian and right artinian, there exists $n \geq 1$ such that

$$\mathbb{1}(X_n \cap J) = \mathbb{1}(X_{n+1} \cap J) = \cdots \quad \text{and} \quad X_n + J = X_{n+1} + J = \cdots .$$

Now let $k \geq n$. Since $X_k \subseteq X_n$ and $X_n \subseteq X_k + J$, the modular law gives $X_n = X_n \cap (X_k + J) = X_k + (X_n \cap J)$. Thus

$$\begin{aligned} \mathbb{1}(X_n) &= \mathbb{1}(X_k) \cap \mathbb{1}(X_n \cap J) = \mathbb{1}(X_k) \cap \mathbb{1}(X_k \cap J) \\ &= \mathbb{1}[X_k + (X_k \cap J)] = \mathbb{1}(X_k), \end{aligned}$$

proving the lemma. □

We conclude by settling the $J^2 = 0$ case of the Faith conjecture.

Theorem 3.40. *If R is a semiprimary, right and left mininjective ring with $J^2 = 0$, then R is quasi-Frobenius.*

Proof. Since R is right and left minfull, it is a right and left minannihilator ring by Corollary 2.34, so R is a right and left min-PF ring. Hence R is quasi-Frobenius by Lemma 3.39 and Theorem 3.38. □

If F is a field, the ring $R = \left[\begin{smallmatrix} F & F \\ 0 & F \end{smallmatrix}\right]$ is a right and left artinian ring with $J^2 = 0$, but R is neither right nor left mininjective. Moreover, Camillo's example (Example 2.6) is a commutative, semiprimary, local, mininjective ring with $J^3 = 0$ that is not quasi-Frobenius.

Notes on Chapter 3

The semiperfect rings were introduced by Bass [16] as a generalization of the semiprimary rings and have only finitely many simple right modules represented in the ring using a basic set of local idempotents (see Appendix B). In the right mininjective case this gives combinatorial information about the ring, which in turn leads to the right minfull rings and later to the right min-PF rings. Much of this stems from Rutter [199] in the artinian case.

In Theorem 3.21 the Nakayama permutation refers to the original definition of quasi-Frobenius rings in 1941 by Nakayama [150], where Theorem 3.24 is proved. If e and f are primitive idempotents in a semiperfect ring R, then (eR, Rf) is called an *i-pair* (injective pair) if $soc(eR) \cong fR/fJ$ and $soc(Rf) \cong Re/Je$. These Nakayama permutations are useful in showing that right self-injectivity implies left self-injectivity because of Fuller's theorem [68]: If R is left artinian and e is a local idempotent of R, then eR is injective if and only if there exists a local idempotent f in R such that (eR, Rf) is an i-pair; in this case Rf is also injective. Fuller's theorem was investigated by Baba and Oshiro [13] in the semiprimary case and by Xue [229] in the semiperfect case.

Theorem 3.31 is a descendent of Osofsky's 1966 result [182] that every left artinian, left and right self-injective ring is quasi-Frobenius.

Lemma 3.37 is due to Ara and Park [6, Theorem 2.2].

4
Min-CS Rings

In this chapter, we consider the class of left min-CS rings (for which every minimal left ideal is essential in a direct summand) and show that this weak injectivity property is useful in obtaining semiperfect rings. Indeed, it is proved in Theorem 4.8 that if R is left min-CS, then the dual of every simple right R-module is simple, if and only if R is semiperfect with $S_l = S_r$ and $soc(Re)$ is simple and essential for every local idempotent e of R. The hypotheses of Theorem 4.8 are the weakest known conditions of this type that imply that R is semiperfect.

If we strengthen the left min-CS hypothesis in Theorem 4.8 by requiring that each closed left ideal with simple essential socle be a direct summand of ${}_RR$ (R is left strongly min-CS), we obtain a class of rings that satisfies many of the characteristic properties of left PF rings. If instead of assuming in Theorem 4.8 that the duals of simple right R-modules are simple we suppose, more generally, that R is right Kasch, then we obtain a larger class of rings that still retains many of these properties: It is shown in Theorem 4.10 that R is left CS and right Kasch if and only if it is semiperfect and left continuous with $S_r \subseteq^{ess} {}_RR$.

On the other hand, a new characterization of left PF rings in terms of simple modules alone is obtained by showing in Corollary 4.16 that R is left PF if and only if the ring $S = M_2(R)$ of 2×2-matrices over R is left strongly min-CS and the dual of every simple right S-module is simple. The chapter concludes with some characterizations of the quasi-Frobenius rings among the left continuous, right min-CS rings and some characterizations of the semiperfect, right continuous rings.

In the last section of this chapter, we provide several characterizations of quasi-Frobenius rings in terms of the CS and min-CS conditions.

4.1. Semiperfect Min-CS Rings

We begin with two results about semiperfect rings. The first is a useful sufficient condition that a ring is semiperfect and will be referred to several times in this book.

Lemma 4.1. *Let R be a left Kasch ring in which $\mathtt{r}(L)$ is essential in a summand of R_R for every maximal left ideal L of R. Then R is semiperfect.*

Proof. If $L \subseteq^{max} {}_RR$, we show that the simple left R-module R/L has a projective cover (see Theorem B.21). Since R is left Kasch let $La = 0$, where $0 \neq a \in R$. Then $L = \mathtt{l}(a)$, so $R/L \cong Ra$. By hypothesis, let $\mathtt{rl}(a) = \mathtt{r}(L) \subseteq^{ess} eR$, where $e^2 = e \in R$. Define $\theta : Re \to Ra$ by $x\theta = xa$. Then θ is epic because $a \in eR$, and Ra is simple. Hence we are done if $ker\theta = Re \cap \mathtt{l}(a)$ is small in Re, equivalently (since Re is finitely generated) if $Re \cap \mathtt{l}(a)$ is the only maximal submodule of Re (so e is local). So suppose that $N \subseteq^{max} Re$; we must show that $N \subseteq Re \cap \mathtt{l}(a)$; that is, $N \subseteq \mathtt{l}(a)$. If not, then $N + \mathtt{l}(a) = R$ because $\mathtt{l}(a)$ is maximal. This gives $0 = \mathtt{r}(N) \cap \mathtt{rl}(a)$, whence $0 = [\mathtt{r}(N) \cap eR] \cap \mathtt{rl}(a)$, and so $\mathtt{r}(N) \cap eR = 0$ by the choice of e. Since Re/N is simple the Kasch hypothesis gives an embedding $\sigma : Re/N \to {}_RR$. But if we set $b = \sigma(e + N)$, then $0 \neq b \in eR \cap \mathtt{r}(N)$, which is a contradiction. □

Note that the proof of Lemma 4.1 shows that, in a left Kasch ring, if $L \subseteq_R^{max} R$ and $\mathtt{r}(L) \subseteq^{ess} eR$, where $e^2 = e$, then e is a local idempotent.

Lemma 4.2. *Let R be a semiperfect ring in which $S_l \subseteq^{ess} R_R$. Then:*

(1) $\mathtt{rl}(T)$ is essential in a summand of R_R for each right ideal T of R.
(2) R is left Kasch.

Proof. (2) is Lemma 1.48. To prove (1), write $\mathtt{l}(T) = R(1 - e) + B$, where $e^2 = e$ and $B \subseteq J$ (see Corollary B.19). We show that $\mathtt{rl}(T) \subseteq^{ess} eR$. Since $\mathtt{rl}(T) = eR \cap \mathtt{r}(B)$, it suffices to show that $\mathtt{r}(B) \subseteq^{ess} R_R$. But $B \subseteq J$, so $\mathtt{r}(B) \supseteq \mathtt{r}(J) \supseteq S_l$ and the hypothesis applies. □

A module M is called a *min-CS module* if every simple submodule of M is essential in a direct summand of M. A ring R is called a right *min-CS ring* if R_R is a min-CS module. A module M is called *min-continuous* if M is a min-CS module and every minimal submodule of M that is isomorphic to a direct summand of M is itself a summand of M.

If F is a field, the ring $R = \begin{bmatrix} F & F \\ 0 & F \end{bmatrix}$ is left and right CS by Example 1.20 and hence left and right min-CS, but R is neither right nor left mininjective because $S_r \nsubseteq S_l$ and $S_l \nsubseteq S_r$. Our next result characterizes the right mininjective rings among the *left* min-CS rings.

Proposition 4.3. *The following are equivalent for a left min-CS ring R:*

(1) R is right mininjective.
(2) kR simple, $k \in R$, implies Rk simple (R is right minsymmetric).

In particular, a commutative min-CS ring is mininjective.

Proof. (1) always implies (2) by Theorem 2.21. Conversely, if kR is simple, we must show that $\mathtt{lr}(k) = Rk$. Since Rk is simple by (2), let $Rk \subseteq^{ess} Re$, $e^2 = e$, by the min-CS hypothesis. Then $Rk \subseteq \mathtt{lr}(k) \subseteq \mathtt{lr}(Re) = Re$, so $Rk \subseteq^{ess} \mathtt{lr}(k)$. As Rk is simple, it suffices to show that $\mathtt{lr}(k)$ is semisimple; that is, $\mathtt{lr}(k) \subseteq S_l$. But if $0 \neq a \in \mathtt{lr}(k)$ then $\mathtt{r}(k) \subseteq \mathtt{r}(a) \neq R$, so $\mathtt{r}(k) = \mathtt{r}(a)$ because $\mathtt{r}(k)$ is maximal. Thus $aR \cong kR$ is simple, so Ra is simple by (2) and $a \in S_l$ as required. □

The next two results contain some properties of semiperfect left min-CS rings that will be needed in the sequel. Recall that a ring R is a right min-PF ring if it is semiperfect, right mininjective with $S_r \subseteq^{ess} R_R$ and $\mathtt{lr}(K) = K$ whenever $e^2 = e \in R$ is local and $K \subseteq eR$ is a simple left ideal.

Lemma 4.4. *Let R be a semiperfect left min-CS ring. Then the following hold:*

(1) If $S_l \subseteq S_r$ and Rk is simple, $k \in R$, then kR is simple.
(2) If $S_l = S_r$ the following hold:
  (i) R is a left minannihilator ring.
  (ii) R is right mininjective.
  (iii) Rk is a simple left ideal if and only if kR is a simple right ideal.
(3) If $S_r = S_l \subseteq^{ess} R_R$ then R is a right min-PF ring.

Proof. (1). If Rk is simple let $Rk \subseteq^{ess} Re$, where $e^2 = e \in R$. Then $\mathtt{r}(k) \supseteq J + (1 - e)R$ because $k \in S_l \subseteq S_r$. But Re is indecomposable, so, as R is semiperfect, e is local and $J + (1 - e)R$ is the unique maximal right ideal containing $(1-e)R$ by Lemma 3.6. It follows that $\mathtt{r}(k) = J + (1-e)R$ is maximal, proving (1).

(2). To prove (i), let Rk be a simple left ideal of R. If $0 \neq a \in \mathtt{lr}(k)$ then $\mathtt{r}(k) \subseteq \mathtt{r}(a)$, so aR is simple [$\mathtt{r}(k)$ is maximal by (1)]. It follows that $\mathtt{lr}(k) \subseteq S_r \subseteq S_l$, whence $\mathtt{lr}(k)$ is a semisimple left ideal. But $Rk \subseteq^{ess} \mathtt{lr}(k)$ as in the proof of Proposition 4.3, so $Rk = \mathtt{lr}(k)$, proving (i). Now Proposition 2.33 gives (ii) because $S_r \subseteq S_l$. Finally (ii) implies (iii) by (1).

(3). This follows by (i), (ii), and the definition of a right min-PF ring. □

Lemma 4.5. *Let R be a semiperfect, left Kasch, left min-CS ring. Then the following hold:*

(1) $S_l \subseteq^{ess} {}_RR$ and $soc(Re)$ is simple and essential in Re for all local idempotents $e \in R$.
(2) R is right Kasch if and only if $S_l \subseteq S_r$.
(3) If $\{e_1, \dots, e_n\}$ are basic local idempotents in R then

$$\{soc(Re_1), \dots, soc(Re_n)\}$$

is a complete set of distinct representatives of the simple left R-modules.

Proof. (3). Given $\{e_1, \dots, e_n\}$ as in (3), let $\{e_1, \dots, e_n, \dots, e_m\}$ be a set of orthogonal local idempotents with $1 = e_1 + \cdots + e_m$. As R is left Kasch, let $Re_i/Je_i \cong K_i \subseteq R$ for each $i = 1, 2, \dots, n$, where K_i is a simple left ideal. Since R is a left min-CS ring, each $K_i \subseteq^{ess} Rf_i$ for some $f_i^2 = f_i$, and f_i is local because R is semiperfect. We have $soc(Rf_i) = K_i \cong Re_i/Je_i$ for $i = 1, 2, \dots, n$. But $Re_i/Je_i \cong Re_j/Je_j$ if and only if $i = j$ because $e_1, \dots, e_n$ are basic. Hence $\{f_1, \dots, f_n\}$ is a basic set of local idempotents, and the left ideals $\{K_1, \dots, K_n\}$ are a complete set of representatives of the simple left R-modules. But then $Re_i \cong Rf_{\sigma i}$ for some $\sigma : \{1, \dots, n\} \to \{1, \dots, n\}$, and σ is monic (again because $e_1, \dots, e_n$ are basic). Now (3) follows.

(1). If $e^2 = e \in R$ is local then $Re \cong Rf_i$ for some i where the f_i are as above. Hence $soc(Re)$ is simple and essential in Re. Furthermore, $S_l = \oplus_{i=1}^m soc(Re_i) \subseteq^{ess} \oplus_{i=1}^m Re_i = R$, proving (1).

(2). If $S_l \subseteq S_r$ then R is right Kasch by (1) and Lemma 4.2. Conversely, if R is right Kasch let K be a simple left ideal of R; we must show that $K \subseteq S_r$. We have $K \subseteq^{ess} Re$ for some $e^2 = e \in R$ by hypothesis, so it suffices to show that $S_r e \neq 0$. But e is local (Re is indecomposable), so $S_r e \neq 0$ by Theorem 3.2 □

Corollary 4.6. *The following are equivalent for a ring R:*

(1) R is a left and right min-PF ring.
(2) R is a semiperfect, left and right min-CS ring with $S_r = S_l \subseteq^{ess} R_R$.

Proof. (1)⇒(2). Given (1), R is semiperfect, $S_r \subseteq^{ess} R_R$, and $S_l \subseteq^{ess} {}_RR$. Since R is two-sided mininjective, it is two-sided minannihilator and $S_r = S_l$ by Corollary 2.34. Hence $S_r = S_l$ is essential in both R_R and ${}_RR$, and it remains to show that R is left and right min-CS. If K is a minimal right ideal of R then $K = \mathtt{rl}(K)$ is essential in a direct summand of R by Lemma 4.2, so R is right min-CS. Similarly, R is left min-CS.

(2)⇒(1). Assume R is as in (2). As R is left min-CS it is a right min-PF ring by (3) of Lemma 4.4. But R is a left Kasch ring by Theorem 3.24, so $S_l \subseteq^{ess} {}_RR$ by Lemma 4.5. Hence R is a left min-PF ring, again by (3) of Lemma 4.4. □

In preparation for our next theorem, we need the following lemma about semiperfect rings.

Lemma 4.7. *Let R be a semiperfect ring and assume that $1 = e_1 + \cdots + e_n$, where the e_i are orthogonal local idempotents. If f is a local idempotent, then $R = fR \oplus [\oplus_{i \neq j} e_i R]$ for some $1 \leq j \leq n$.*

Proof. Observe first that $\oplus_{i\neq j} e_i R = (1 - e_j)R$. We have $f = fe_1 f + \cdots + fe_n f$, so, since fRf is a local ring, $fe_j f = a$ is a unit in fRf for some j, say $ab = f = ba$ for some $b \in fRf$. If $fr \in fR \cap (1 - e_j)R$, $r \in R$, then $fr = bar = bfe_j fr = 0$ because $e_j fr = 0$. However, $e_j(fR) \nsubseteq e_j J$ because $e_j f \notin J$ $(f = ab = fe_j fb)$, so $e_j(fR) = e_j R$ by Proposition B.2 because e_j is a local idempotent. In particular, $e_j = e_j fy$ for some $y \in R$, whence $1 - fy \in \mathtt{r}(e_j) = (1 - e_j)R$. It follows that $R = fR + [\oplus_{i\neq j} e_i R]$, completing the proof. □

Lemma 4.7 can be described by saying the decomposition $R = e_1 R \oplus \cdots \oplus e_n R$ in a semiperfect ring R complements local direct summands fR. In fact it complements every direct summand (see [1, Corollary 12.7]).

We can now deduce some important properties of the right mininjective, right Kasch, left min-CS rings. We remind the reader that the dual of every simple right R-module is simple if and only if R is right mininjective and right Kasch (Theorem 2.31).

Theorem 4.8. *Let R be a left min-CS ring such that the dual of every simple right R-module is simple. Then the following statements hold:*

(1) R is semiperfect.
(2) For every $k \in R$, Rk is a minimal left ideal if and only if kR is a minimal right ideal. In particular, $S_l = S_r$.
(3) R is a left minannihilator ring.
(4) $soc(Re)$ is simple and essential in Re, for every local idempotent e of R. In particular, $S_l = S_r \subseteq^{ess} {}_R R$ and R is left finite dimensional.
(5) R is left Kasch if and only if $soc(eR) \neq 0$ for every local idempotent e of R. Moreover, in this case the following assertions hold:
(a) $soc(eR)$ is homogeneous for every local idempotent e of R.
(b) If $\{e_1, \ldots, e_n\}$ is a basic set of local orthogonal idempotents, then there exist elements $k_1, \ldots, k_n$ in R and a (Nakayama) permutation σ of $\{1, \ldots, n\}$ such that the following hold for each $i = 1, \ldots, n$:
(i) $k_i R \subseteq soc(e_i R)$.
(ii) $Rk_i = soc(Re_{\sigma i})$ is simple and essential in $Re_{\sigma i}$.
(iii) $k_i R \cong e_{\sigma i}R/e_{\sigma i}J$, and $Rk_i \cong Re_i/Je_i$.
(iv) $\{k_1 R, \ldots, k_n R\}$ and $\{Rk_1, \ldots, Rk_n\}$ are complete sets of representatives of the isomorphism classes of simple right and left R-modules, respectively.

Conversely, if R is a semiperfect ring with $S_l = S_r$, and $soc(Re)$ is simple for every local idempotent e of R, then the dual of every simple right R-module is simple.

Proof. (1). By Theorem 2.31, R is right Kasch, right mininjective, and $\mathtt{l}(T)$ is a simple left R-module for every maximal right ideal T of R. By Lemma 4.1, R is semiperfect.

(2). If kR is a minimal right ideal of R, then Rk is a minimal left ideal because R is right mininjective (Theorem 2.21). Conversely, suppose that Rk is a minimal left ideal of R. Since R is a left min-CS ring, $Rk \subseteq^{ess} Re$ for some $e^2 = e \in R$. But e is local (R is semiperfect), so $T = J + (1-e)R$ is the unique maximal right ideal containing $(1-e)R$. Since $(1-e)R \subseteq \mathtt{r}(k)$, we have $\mathtt{r}(k) \subseteq T$, whence $\mathtt{lr}(k) \supseteq \mathtt{l}(T)$. By Lemma 2.28, $\mathtt{l}(T) \cong (R/T)^*$, so $\mathtt{l}(T)$ is simple by hypothesis.

However, $Rk \subseteq^{ess} Re$ implies $Rk \subseteq^{ess} \mathtt{lr}(k) \subseteq Re$. Since both Rk and $\mathtt{l}(T)$ are simple, it follows that $\mathtt{l}(T) = Rk$. Thus $\mathtt{r}(k) = \mathtt{rl}(T) = T$ because R is right Kasch, and so kR is a minimal right ideal of R.

(3). $S_l = S_r$ by (2), and so (3) follows by (2) of Lemma 4.4.

(4). As we have already remarked, our hypotheses imply that R is a right Kasch, right mininjective ring. Then it follows from Theorem 3.7 that if e is a local idempotent, then Re has a simple socle. Now let $R = Re_1 \oplus \cdots \oplus Re_n$, where the e_i are local idempotents, and let C_k be the socle of Re_k. Since R is left min-CS, there exists a direct summand Rf of ${}_RR$, $f^2 = f$, such that C_k is essential in Rf. By Lemma 4.7 we have $R = Rf \oplus [\oplus_{i \neq j} Re_i]$ for some $1 \leq j \leq n$. But we must have $j = k$ because $Rf \cap Re_k \supseteq C_k \neq 0$. It follows that $Re_k \cong Rf$, showing that Re_k has simple essential socle.

(5). Every simple left R-module has the form Re/Je, where $e^2 = e$ is local. Since Lemma 3.1 gives $(Re/Je)^* \cong e\,\mathbf{r}(J) = eS_l = eS_r = soc(eR)$, it follows that R is left Kasch if and only if $soc(eR) \neq 0$ for all local $e^2 = e$. In particular, R is right minfull, so (a), (b), and the fact that R is a left Kasch ring follow from Theorem 3.12 and Lemma 4.5.

Finally, for the converse, suppose that R is semiperfect with $S_l = S_r$ and $soc(Re)$ is simple for every local idempotent e of R. Let K be a simple right R-module. Then $K \cong eR/eJ$ for some local idempotent e of R and so $K^* \cong (eR/eJ)^* \cong \mathbf{l}(J) \cdot e = S_l e = soc(Re)$ is simple. □

Corollary 4.9. *A ring R is left and right min-PF if and only if R is left and right min-CS and the dual of every simple R-module is simple.*

Proof. If R is left and right min-PF then R is left and right min-CS by Corollary 4.6. Since left and right min-PF rings are left and right minfull, the dual of every simple R-module is simple by Theorem 3.18. Conversely, if R is left and right min-CS and the dual of every simple R-module is simple, it follows from Theorem 4.8 that R is a semiperfect, left and right mininjective, left and right minannihilator ring with $S_l = S_r$ essential in both ${}_RR$ and R_R. Hence R is a left and right min-PF ring. □

4.2. Continuity

It is an open question whether the left self-injective right Kasch rings are left PF, but a left CS right Kasch ring need not be left PF [the Björk and Camillo examples (Examples 2.5 and 2.6) are not left self-injective, hence not left PF]. However, in the next theorem we show that left CS right Kasch rings are semiperfect.

Theorem 4.10. *A ring R is left CS and right Kasch if and only if R is a semiperfect left continuous ring with $S_r \subseteq^{ess} {}_RR$.*

Proof. Suppose R is left CS and right Kasch and let T be a maximal right ideal of R. Then $\mathbf{l}(T) \subseteq^{ess} Re$, for some $e^2 = e \in R$, so R is semiperfect by Lemma 4.1. Furthermore, since R is right Kasch, R is left continuous because right Kasch rings are left C2 (Proposition 1.46). Moreover, $S_r \neq 0$ (R is right

Kasch), and by the left CS-condition $S_r \subseteq^{ess} Re$ for some idempotent e of R. Thus $(1-e)R \subseteq \mathtt{r}(S_r)$. But $\mathtt{r}(S_r) = J$ because R is right Kasch, which is a contradiction unless $e = 1$. Hence $S_r \subseteq^{ess} {}_RR$. The converse follows from Lemma 4.2. □

In what follows we investigate the class of semiperfect left continuous rings with essential left socle as an interesting generalization of the left PF rings.

Lemma 4.11. *Let R be a semiperfect, left continuous ring, and assume that $S_l \subseteq^{ess} {}_RR$. Then the following hold:*

(1) $Z_r \subseteq J = Z_l$.
(2) $S_l \subseteq S_r$.
(3) $soc(Re)$ is simple and essential in Re for all local $e^2 = e \in R$. (In particular, R is left finitely cogenerated.)
(4) R is a two-sided Kasch ring.
(5) $soc(eR) \neq 0$ for all local $e^2 = e \in R$.

Proof. (1). $Z_r \subseteq J$ in any semiperfect ring and $J = Z_l$ by Utumi's theorem (Theorem 1.26) because R is left continuous.

(2). This follows from $J = Z_l$ because $S_l Z_l = 0$ and $S_r = \mathtt{l}(J)$.

(3). Re is a CS module because R is a left CS ring and summands of CS modules are again CS (by Proposition 1.30). By hypothesis, let S be a simple submodule of Re. Since Re is uniform (it is an indecomposable CS module), it follows that $S = soc(Re) \subseteq^{ess} Re$.

(4). R is right Kasch by (2) and Theorem 4.10. Let $\{e_1, \ldots, e_n\}$ be a basic set of primitive idempotents of R. By (3), $S_i = soc(Re_i)$, $1 \leq i \leq n$, is simple. To show that R is left Kasch, it suffices to show that $\{S_1, \ldots, S_n\}$ is a complete set of distinct representatives of the simple left R-modules. To this end, let $\sigma : S_i \to S_j$ be an isomorphism; we show that $i = j$. If not, then Re_j is Re_i-injective because R is left continuous, so σ can be extended to $\hat{\sigma} : Re_i \to Re_j$, and $\hat{\sigma}$ is monic because $S_i \subseteq^{ess} Re_i$. If $\hat{\sigma}$ is not onto, then $(Re_i)\hat{\sigma} \subseteq Je_j \subseteq Z_l$ because e_j is local and $J = Z_l$. Thus $0 = L(e_i\hat{\sigma}) = (Le_i)\hat{\sigma}$ for some essential left ideal L of R, and so $Le_i = 0$ and $e_i \in Z_l$, which is a contradiction. Thus $i = j$ and R is a left Kasch ring.

(5). R is left Kasch by (4), so, using Lemma 3.1, $0 \neq (Re/Je)^* \cong e\,\mathtt{r}(J) = eS_l \subseteq eS_r = soc(eR)$ by (2). □

Theorem 4.12. *Let R be a semiperfect, left continuous ring with $S_l \subseteq^{ess} {}_RR$. If $\{e_1, \ldots, e_n\}$ is a basic set of local idempotents in R, there exist elements*

$k_1, \ldots, k_n$ of R and a permutation σ of $\{1, 2, \ldots, n\}$ such that the following hold:

(1) $\{Rk_1, \ldots, Rk_n\}$ and $\{k_1R, \ldots, k_nR\}$ are complete sets of distinct representatives of the simple left and right R-modules, respectively.
(2) $k_iR \subseteq soc(e_iR)$ and $Rk_i = soc(Re_{\sigma i})$ for all $i = 1, 2, \ldots, n$.
(3) $k_iR \cong e_{\sigma i}R/e_{\sigma i}J$ and $Rk_i \cong Re_i/J\dot{e}_i$ for all $i = 1, 2, \ldots, n$.

Proof. R is left Kasch and $S_l \subseteq S_r$ by Lemma 4.11. Hence, for any $i = 1, 2, \ldots, n$, Lemma 3.1 gives

$$0 \neq (Re_i/Je_i)^* \cong e_i\,\mathtt{r}(J) = e_iS_l \subseteq e_iS_r = soc(e_iR).$$

Hence choose a simple right ideal $K_i \subseteq e_iS_l$. We have $K_ie_{\sigma i} \neq 0$ for some $\sigma i \in \{1, 2, \ldots, n\}$, so let $0 \neq k_i \in K_ie_{\sigma i}$. Thus $k_iR = K_i$ is simple and $k_i \in e_iRe_{\sigma i}$. Moreover, $k_i \in K_i \subseteq S_l$, so $\mathtt{l}(k_i) \supseteq J + R(1 - e_i)$, a maximal left ideal of R because e_i is local. Hence Rk_i is also simple. But $Rk_i \subseteq Re_{\sigma i}$, so since R is a left min-CS ring, $Rk_i = soc(Re_{\sigma i})$ by Lemma 4.5. Now the maps $re_i \longmapsto rk_i$ and $e_{\sigma i}r \longmapsto k_ir$ are well-defined epimorphisms $Re_i \to Rk_i$ and $e_{\sigma i}R \to k_iR$ respectively, so $Re_i/Je_i \cong Rk_i = soc(Re_{\sigma i})$ and $e_{\sigma i}R/e_{\sigma i}J \cong k_iR$. Since the e_i are basic, these equations imply that both $\{Rk_1, \ldots, Rk_n\}$ and $\{k_1R, \ldots, k_nR\}$ are pairwise nonisomorphic, proving (1). Moreover, $soc(Re_{\sigma i}) \cong Re_i/Je_i$ shows that σ is a permutation. This completes the proof. □

Corollary 4.13. *The following conditions are equivalent for a ring R:*

(1) R is a left CS, left and right Kasch ring.
(2) R is a semiperfect left continuous ring with essential left socle.

Proof. Given (1), R is semiperfect and left continuous by Theorem 4.10. Since R is also left Kasch, it follows from Lemma 4.5 that $S_l \subseteq^{ess} {}_RR$. Hence (1) $\Rightarrow$ (2); (2) $\Rightarrow$ (1) follows from (4) of Lemma 4.11. □

In a left min-CS module M, every simple submodule has a closure in M that is a summand. If *every* closure of a simple submodule of M is a summand, we say that the module M is *strongly min-CS*, and a ring R will be called a *left strongly min-CS* ring if ${}_RR$ is strongly min-CS. The ring $R = \left[\begin{smallmatrix} \mathbb{Z}_4 & \mathbb{Z}_4 \\ 0 & \mathbb{Z}_4 \end{smallmatrix}\right]$ is a right artinian ring that is not right CS, and hence it is not right strongly min-CS by the following lemma. However, it is easily checked that this ring is right min-CS.

Lemma 4.14. *Let M be a module with finitely generated essential socle. Then M is a strongly min-CS module if and only if M is a CS module.*

Proof. Let C be a closed submodule of M; we must show that C is a summand of M. Since C has a finitely generated essential socle, we proceed by induction on the composition length n of $soc(C)$. If $n = 1$ it is our hypothesis. In general, let K denote a simple submodule of C, and let D be a closure of K in C. Then $K \subseteq^{ess} D \subseteq C \subseteq M$, where D is closed in C, and C is closed in M. It follows that D is closed in M by Lemma 1.29, so D is a closure of the simple module K. Hence D is a summand of M by hypothesis, say $M = D \oplus B$. Since $D \subseteq C$ we obtain $C = D \oplus (C \cap B)$, so $soc(C) = soc(D) \oplus soc(C \cap B)$. Since $soc(D) = K$ it follows that $soc(C \cap B)$ has smaller composition length than $soc(C)$. By induction $C \cap B$ is a summand of M and hence of B, say $B = X \oplus (C \cap B)$. But then $M = D \oplus B = D \oplus (C \cap B) \oplus X = C \oplus X$, as required. □

Theorem 4.15. *The following conditions are equivalent for a ring R:*

(1) R is left strongly min-CS and the dual of every simple right R-module is simple.

(2) R is a semiperfect left continuous ring such that $S_l = S_r \subseteq^{ess} {}_RR$.

Moreover, if R satisfies these conditions, then the following hold:

(a) R is left and right Kasch.

(b) $soc(Re)$ is simple and essential in Re and $soc(eR)$ is nonzero and homogeneous, for every local idempotent e of R.

(c) R admits a (Nakayama) permutation of any basic set of primitive idempotents as in (b) of Theorem 4.8.

Proof. (1)⇒(2). It follows from Theorem 4.8 that R is semiperfect, $S_l = S_r \subseteq^{ess} {}_RR$, and S_l is finitely generated, so R is a left CS-ring by Lemma 4.14. Moreover, R is right Kasch by (1) and Theorem 2.31, and so it satisfies the left C2-condition by Proposition 1.46. Therefore R is left continuous, proving (2).

(2)⇒(1). Clearly every left continuous ring is left strongly min-CS. If R is as in (2), then R is a right mininjective ring by Lemma 4.4, and R is a right Kasch ring by Lemma 4.11. Then it follows from Theorem 2.31 that the dual of every simple right R-module is simple.

Finally, the rest of the assertions (a)–(c) follows from Lemma 4.11 and Theorem 4.8. □

In the next corollary we exploit these results to obtain a characterization of left PF rings in terms of simple modules over the 2×2 matrix ring.

Corollary 4.16. *Let R be a ring and write $S = M_2(R)$. Then the following conditions are equivalent:*

(1) R is a left PF ring

(2) S is a left strongly min-CS ring such that the dual of every simple right S-module is simple.

Proof. (1)⇒(2). Given (1), S is left PF by Morita invariance. Hence S is left self-injective and so is left CS. Moreover, S is a left cogenerator that is right Kasch by Theorem 1.56. Hence $\mathtt{lr}(L) = L$ for every left ideal L of S, and so S is right mininjective. It follows that the dual of every simple right S-module is simple by Theorem 2.31. This proves (2).

(2)⇒(1). If (2) holds then S is left continuous by Theorem 4.15, and hence R is left self-injective by Theorem 1.35. Since, by Morita invariance, the dual of every simple right R-module is simple, it follows from Theorem 4.15 that R is a semiperfect ring with essential left socle and hence that R is left PF (again by Theorem 1.56). □

Corollary 4.17. *Let R be a commutative ring. Then the following conditions are equivalent:*

(1) R is a min-CS, Kasch ring.

(2) R is a semiperfect continuous ring with essential socle.

Proof. The implication (2)⇒(1) follows from Theorem 4.15. Conversely, assume that (1) holds. Then R is a mininjective ring by Proposition 4.3, and so the dual of every simple R-module is simple by Theorem 2.31. Then it follows from Theorem 4.8 that R is semiperfect with essential socle, and R satisfies the C2-condition because R is a Kasch ring. Write $R = Re_1 \oplus \cdots \oplus Re_n$, where $\{e_1, \ldots, e_n\}$ is a complete set of orthogonal local idempotents of R. Each Re_i is uniform because $soc(Re_i)$ is simple and essential in Re_i by Theorem 4.8. If A is an ideal of R, then (since R is commutative) $A = \oplus_{i=1}^{n}(A \cap Re_i)$, and so A is essential in a direct summand of R. □

4.3. Quasi-Frobenius Rings

It is a theorem of Utumi that a two-sided artinian, two-sided continuous ring is quasi-Frobenius. Moreover, the Björk example (Example 2.5) is a one-sided continuous, two-sided artinian ring that is not quasi-Frobenius. However, Faith has proved that if R is a one-sided self-injective ring with ACC on left annihilators then R is quasi-Frobenius (see Theorem 1.50). In what follows we will unify and extend the results of Utumi and Faith in one single theorem.

Lemma 4.18. *Let M be a module such that $M/soc(M)$ is finite dimensional. If $\oplus_{i=1}^{\infty} M_i$ is an infinite direct sum of submodules of M, then there exists an integer k such that $M_i \subseteq soc(M)$ for all $i \geq k$.*

Proof. Write $S = soc(M)$, $\overline{M} = M/S$, and $A_k = \oplus_{i=1}^{k} M_i$. Given $n \geq 1$, there exists a submodule $U \subseteq M_{n+1}$ such that $\overline{M}_{n+1} \cap \overline{A}_n = \overline{U}$. Write $S = (A_n \cap S) \oplus T$, so that $U \subseteq A_n + S = A_n \oplus T$ for some $T \subseteq S$. Let $\pi : A_n \oplus T \to T$ be the projection with $ker(\pi) = A_n$. Since $U \subseteq M_{n+1}$, we have $U \cap A_n = 0$, and hence the restriction of the map π to U is monic. Thus U is semisimple, so $U \subseteq S$ and $\overline{M}_{n+1} \cap \overline{A}_n = 0$.

It follows that $\overline{M}_1 \oplus \overline{M}_2 \oplus \overline{M}_3 \oplus \cdots$ is a direct sum, so, since $\overline{M}$ is finite dimensional, there exists k such that $\overline{M}_i = 0$ for all $i \geq k$. Thus $M_i \subseteq S = soc(M)$ for all $i \geq k$, as required. □

Proposition 4.19. *Let M be a CS module. If $M/soc(M)$ is finite dimensional then $M = K \oplus S$, where K is finite dimensional and S is semisimple. In particular, M is a direct sum of uniform submodules.*

Proof. Write $S = soc(M)$, let K be a complement of S in M, and (by the CS hypothesis) write $M = K \oplus T$ for some submodule T of M. Then $K \hookrightarrow M/S$, so K is finite dimensional, and T is a CS module by Proposition 1.30. Thus $soc(T) \subseteq^{ess} A$ for some summand A of T, say $T = A \oplus B$. Hence $M = K \oplus T = K \oplus A \oplus B$. Since $soc(B) = 0$, B is finite dimensional because $B \hookrightarrow M/S$, and we may write $M = (K \oplus B) \oplus A$, where $K \oplus B$ is finite dimensional and A is a CS module with essential socle. So, without loss of generality, we may assume that M has an essential socle.

Suppose that S_1 is a nonclosed simple submodule of M, and let $C(S_1)$ be a closure of S_1 in M. As M is CS, write $M = C(S_1) \oplus M_1$. If S_2 is a nonclosed simple submodule of M_1, write $M = C(S_1) \oplus C(S_2) \oplus M_2$. If this continues indefinitely then Lemma 4.18 shows that some $C(S_m)$ will be in $soc(M)$, which is a contradiction. So we can write $M = C(S_1) \oplus C(S_2) \oplus \cdots \oplus C(S_n) \oplus M_{n+1}$, where each S_i is a nonclosed simple submodule of M, and each simple submodule of M_{n+1} is closed. Since $C = C(S_1) \oplus C(S_2) \oplus \cdots \oplus C(S_n)$ is finite dimensional, we may assume without loss of generality that every simple submodule of M is closed in M.

We can now complete the proof by showing that M is semisimple, and we accomplish this by showing that every finitely generated submodule D of M is semisimple. Suppose D has an infinitely generated socle, and write $soc(D) = \oplus_{j=1}^{\infty} A_i$, where each A_i is infinitely generated. Let $C(A_i)$ be a maximal essential extension of A_i in M. Then $\oplus_{i=1}^{\infty} C(A_i)$ is an infinite direct sum of submodules of M, and by Lemma 4.18 there exists an integer k such that $C(A_i) = A_i$ for all $i \geq k$. By the CS-hypothesis, write $M = A_k \oplus B_k$ for some submodule B_k of M. Since $A_k \subseteq D \subseteq M$, it follows that $D = A_k \oplus (B_k \cap D)$ and hence that A_k is finitely generated, a contradiction. Thus $soc(D)$ is finitely generated. Since every simple submodule of D is a summand, then by splitting off all the simple submodules of D, we can write $D = S \oplus N$, where S is semisimple and N has zero socle. Because $soc(D)$ is essential in D, we infer that $N = 0$ and D is semisimple, whence M is semisimple, as required.

Finally, the last statement follows from the fact that an indecomposable direct summand of a CS module is uniform. □

Lemma 4.20. *Let R be a ring.*

(1) If R has the ACC on left annihilators, then every left T-nilpotent ideal A of R is nilpotent.
(2) If R/S_l has the ACC on left annihilators, then Z_l is nilpotent.
(3) If $J \subseteq Z_l$ and R/S_r has the ACC on left annihilators, then J is nilpotent.

Proof. (1). This follows by Lemma 3.33.

(2). It suffices to show that Z_l is left T-nilpotent [then $(Z_l + S_l)/S_l$ is a nilpotent ideal of R/S_l by (1), and so Z_l is nilpotent because $S_l Z_l = 0$]. Hence let $z_1, z_2, \ldots$ be a sequence from Z_l. Since $Z_l \subseteq \mathtt{r}(S_l)$ we have $\mathtt{l}(z_1 z_2 \cdots z_k) = \mathtt{l}(z_1 z_2 \cdots z_{k+1})$ for some integer $k \geq 1$ by Lemma 3.32. If $z_1 z_2 \cdots z_k \neq 0$, let $0 \neq x \in \mathtt{l}(z_{k+1}) \cap R z_1 z_2 \cdots z_k$, say $x = r z_1 z_2 \cdots z_k$. Then $0 = x z_{k+1} = (r z_1 z_2 \cdots z_k) z_{k+1}$, so $r \in \mathtt{l}(z_1 z_2 \cdots z_{k+1}) = \mathtt{l}(z_1 z_2 \cdots z_k)$, which is a contradiction. So $z_1 z_2 \cdots z_k = 0$, proving (1).

(3). Since $J \subseteq Z_l$ we can repeat the argument in (2) to show that $J^k \subseteq S_r$ for some k. Hence $J^{k+1} \subseteq S_r J = 0$. □

Lemma 4.21. *Suppose R is left continuous. If either R/S_l or R/S_r has the ACC on left annihilators, then R is semiprimary.*

Proof. Suppose first that R/S_l has the ACC on left annihilators. Then Z_l is nilpotent by Lemma 4.20, so J is nilpotent because $Z_l = J$ by Utumi's theorem (Theorem 1.26). For convenience, write $\overline{R} = R/J$ and $\widetilde{R} = R/S_l$, so that $\overline{R}/\overline{S_l} \cong R/(J + S_l) \cong \widetilde{R}/\widetilde{J}$. We must show that $\overline{R}$ is left artinian. The ring $\overline{R} = R/J$ is regular by Utumi's theorem, so its image $R/(J + S_l) \cong \widetilde{R}/\widetilde{J}$ is also regular. But $\widetilde{R}$ is I-finite (it has the ACC on left annihilators), so $\widetilde{R}/\widetilde{J}$ is also I-finite because idempotents lift modulo the nilpotent ideal $\widetilde{J}$. It follows that $\widetilde{R}/\widetilde{J}$, and hence $\overline{R}/\overline{S_l}$ is semisimple artinian. Thus it remains to show that $\overline{S_l}$ is left artinian. But $\overline{S_l} = (S_l + J)/J \cong S_l/(S_l \cap J)$ is semisimple, so we have

$\overline{S_l} \subseteq soc(\overline{R})$. It follows that $\overline{R}/soc(\overline{R})$ is semisimple artinian, being an image of $\overline{R}/\overline{S_l}$. Since $\overline{R}$ is left CS (in fact left continuous by Utumi's theorem), it follows from Proposition 4.19 that $\overline{R}$ is left finite dimensional. Since $\overline{S_l} \subseteq soc(\overline{R})$, it follows that $\overline{S_l}$ is left artinian, as required.

Now suppose that R/S_r has the ACC on left annihilators. Then J is nilpotent by part (3) of Lemma 4.20. In this case write $\widehat{R} = R/S_r$, so again $\overline{R}/\overline{S_r} \cong R/(J + S_r) \cong \widehat{R}/\widehat{J}$. Now $\overline{R}/\overline{S_r}$ is regular (an image of $\overline{R}$) and hence is semiprime. As before, $\widehat{R}/\widehat{J}$ is I-finite, so $\widehat{R}/\widehat{J} \cong \overline{R}/\overline{S_r}$ is regular and I-finite and so is semisimple artinian. Hence $\overline{S_r} = (S_r + J)/J \cong S_r/(S_r \cap J)$ is right semisimple, so $\overline{S_r} \subseteq soc(\overline{R})$. It follows that $\overline{S_r}$ is a sum of minimal left ideals. Hence $\overline{R}/soc(\overline{R})$ is semisimple artinian (an image of $\overline{R}/\overline{S_r}$). Now the proof is completed as in the preceding case. □

Theorem 4.22. *The following conditions on a ring R are equivalent:*

(1) R is quasi-Frobenius.
(2) R is a left continuous, right minannihilator ring with the ACC on left annihilators.
(3) R is a left continuous, right min-CS ring with the ACC on left annihilators.
(4) R is a left continuous, right minannihilator ring and R/S_l is left Goldie.
(5) R is a left continuous, right min-CS ring and R/S_l is left Goldie.
(6) R is a left continuous, right minannihilator ring and R/S_r is left Goldie.
(7) R is a left continuous, right min-CS ring and R/S_r is left Goldie.

Proof. If R satisfies any of the conditions (1) through (7) then R is semiprimary by Lemma 4.21. If R satisfies (2), (4), or (6) then R is a left min-PF ring with $S_r = S_l$ by Corollary 3.25. Then R is a left minannihilator ring by Lemma 4.4, and so R is right min-PF, again by Corollary 3.25. Now Theorem 3.38 shows that R is quasi-Frobenius. However, if R satisfies (3), (5), or (7) then R is a left and right Kasch ring by Lemma 4.11, and hence $S_l = S_r$ by Lemma 4.5 because R is left and right min-CS. Thus R is a left and right min-PF ring by Corollary 4.6, so R is quasi-Frobenius, again by Theorem 3.38. □

Since every left self-injective ring R is left continuous and satisfies the condition $\mathtt{rl}(T) = T$ for all finitely generated right ideals T of R, the following corollary is an immediate consequence of the previous theorem.

Corollary 4.23. *The following conditions on a ring R are equivalent:*

(1) R is quasi-Frobenius.
(2) R is left self-injective and R/S_l is left Goldie.
(3) R is left self-injective and R/S_r is left Goldie.

We remark that the Björk example (Example 2.5) is a left and right artinian, local, left continuous ring R in which the dual of each simple right R-module is simple and every left ideal of R is an annihilator. However, R is not quasi-Frobenius and, in fact, it can be readily seen that R is not right min-CS and the dual of the unique simple left R-module is not simple.

We conclude this section with some characterizations of semiperfect, right continuous rings. The proof requires three lemmas. The first two results are about projective modules. Recall that a module is called *singular* if every element has an essential annihilator.

Lemma 4.24. *If P_R is a finitely generated module that is projective and singular, then $P = 0$.*

Proof. Let $P \cong F/K$, where F_R is free on basis $\{f_1, \ldots, f_n\}$, and put $A_i = \mathbf{r}(f_i + K)$ for each i. Then $A_i \subseteq^{ess} R_R$ by hypothesis, so $f_i A_i \subseteq^{ess} f_i R$, and we obtain $\oplus_{i=1}^n f_i A_i \subseteq^{ess} F$. But $(f_i + K)A_i = 0$ for each i, so $f_i A_i \subseteq K$. It follows that $K \subseteq^{ess} F$. However, $K \subseteq^{\oplus} F$ because $F/K \cong P$ is projective, so $K = F$ and $P = 0$. □

Lemma 4.25. *If R is a semiregular ring in which $Z_r = J$, then R^n satisfies the C2-condition as a right R-module for each $n \geq 1$.*

Proof. Suppose $N \cong M \subseteq^{\oplus} R^n$, $N \subseteq R^n$. By Theorem B.54, R^n is a semiregular module, so let $R^n = P \oplus Q$, where $P \subseteq N$ and $N \cap Q$ is small in R^n by Theorem B.51. Then $N = P \oplus (N \cap Q)$ and $N \cap Q \subseteq rad(R^n) = J^n = Z_r^n$. Thus $N \cap Q$ is singular, and it is finitely generated and projective since this is true of N. Thus $N \cap Q = 0$ by Lemma 4.24, and it follows that $N = P \subseteq^{\oplus} R^n$. □

It is not difficult to prove that an I-finite right C2 ring has the property that monomorphisms $R_R \to R_R$ are epic. This condition arises in the next two results.

Lemma 4.26. *Let M_R be a module and suppose $M = U_1 \oplus U_2 \oplus \cdots \oplus U_n$, where each U_i is uniform. If monomorphisms $M \to M$ are epic then $end(M)$ is semiperfect.*

Proof. By Theorem B.9, it is enough to prove that $end(U_i)$ is local for each i. Given $\alpha \in end(U_i)$ we have $ker(\alpha) \cap ker(1 - \alpha) = 0$ in U_i, so either α or

$1-\alpha$ is monic. But a routine argument shows that monomorphisms in $end(U_i)$ are epic, and the lemma follows. □

We can now prove our characterization of the semiperfect, right continuous rings.

Theorem 4.27. *The following conditions on a ring R are equivalent:*

(1) R is a semiperfect, right continuous ring.
(2) R is a right CS ring, $Z_r = J$, and R has DCC on principal projective right ideals.
(3) R is right quasi-continuous with DCC on principal projective right ideals.
(4) R is a right quasi-continuous, I-finite ring in which monomorphisms $R_R \to R_R$ are epic.

Proof. (1)⇒(2). Let $a_1R \supseteq a_2R \supseteq \cdots$, where each a_kR is projective, $a_k \in R$. Since R satisfies the right C2-condition, $a_kR = e_kR$ for some $e_k^2 = e_k \in R$. Now (2) follows because R is I-finite. We have $Z_r = J$ by Utumi's theorem (Theorem 1.26).

(2)⇒(3). Assume that R satisfies (2), so it is clearly I-finite. Hence $R = e_1R \oplus \cdots \oplus e_nR$, where the e_i are orthogonal primitive idempotents. Moreover, each e_kR is uniform because it is a CS module, so R is semiperfect by Lemma 4.26. Since $J = Z_r$, this implies that R has the right C2-condition by Lemma 4.25, and (3) follows by Lemma 1.21.

(3)⇒(4). The DCC implies that R is I-finite. If $\alpha : R_R \to R_R$ is monic then $\alpha = a\cdot$ is left multiplication by $a = \alpha(1)$, where $\mathtt{r}(a) = 0$, and we must show that $aR = R$. But if $aR \neq R$ then $a^2R \neq aR$, and we obtain $R \supset aR \supset a^2R \supset \cdots$, which is a contradiction because $a^kR \cong aR \cong R$ is projective for each k.

(4)⇒(1). Assume (4). Since R is I-finite, write $1 = e_1 + \cdots + e_n$, where the e_i are orthogonal primitive idempotents. Hence $R = e_1R \oplus \cdots \oplus e_nR$, and each e_kR is uniform because it is a CS module. Thus R is semiperfect by Lemma 4.26. So, to show that R is continuous, it suffices by Lemma 4.25 to prove that $J = Z_r$. We have $Z_r \subseteq J$ because every right ideal $T \nsubseteq J$ in a semiperfect ring contains a nonzero idempotent. Conversely, if $a \in J$ it suffices to prove that each $e_ia \in Z_r$, that is, $\mathtt{r}(e_ia) \subseteq^{ess} R_R$. We do it for e_1a. Since each e_kR is uniform, it is enough to show that $\mathtt{r}(e_1a) \cap e_kR \neq 0$ for each $k = 1, 2, \ldots, n$. If $k = 1$ suppose that $\mathtt{r}(e_1a) \cap e_1R = 0$. Then $x \mapsto e_1ax$ is a monomorphism $e_1R \to e_1R$ and so is epic by (4). This means $e_1R = e_1ae_1R$, which is a contradiction because $a \in J$ and $e_1^2 = e_1 \neq 0$. So assume that

$\mathbf{r}(e_1 a) \cap e_k R = 0$ for some $k \neq 1$. Then the map $\gamma : e_1 a e_k R \to e_k R$ is well defined by $\gamma(e_1 a e_k r) = e_k r$. But R is right quasi-continuous by (4), so $e_k R$ is $e_1 R$-injective by Theorem 1.33. Hence there exists $\hat{\gamma} : e_1 R \to e_k R$ extending γ. If $\hat{\gamma}(e_1) = b \in e_k R$, we have $e_k = \hat{\gamma}(e_1 a e_k) = b e_1 a e_k$. Again this is a

$$\begin{array}{ccc} 0 \to e_1 a e_k R & \hookrightarrow & e_1 R \\ \gamma \downarrow & \swarrow \hat{\gamma} & \\ e_k R & & \end{array}$$

contradiction because $a \in J$. So $\mathbf{r}(e_1 a) \cap e_k R \neq 0$ when $k \neq 1$, and (1) is proved. $\square$

Notes on Chapter 4

Harada [91] calls min-CS modules simple extending modules.

An important source of semiperfect rings is given by Osofsky's theorem (Theorem 1.57), which asserts that a left self-injective, left Kasch ring (left PF ring) is semiperfect and has finitely generated essential left socle. From this theorem (and Lemma 1.41) it follows that a left PF ring is right Kasch, so it is natural to ask whether a left self-injective, right Kasch ring is left PF. This question remains open.

Although it is known (Corollary 7.32) that every left CS, left Kasch ring has finitely generated essential left socle, it is still unknown whether these rings must be semiperfect.

Lemma 4.1, Theorem 4.8, Theorem 4.15, and Corollaries 4.16 and 4.17 are essentially from [85]; Lemma 4.18 and Proposition 4.19 come from [29]. Lemma 4.14 is due to Smith [206]. For Lemma 4.21 see Ara and Park [6], Lee and Tung [134], and Nicholson and Yousif [160]. Theorem 4.27 is motivated by the work of Huynh [99].

Theorem 4.22 extends fundamental work of Armendariz and Park [9], who proved Corollary 4.23. In addition, Theorem 4.22 extends a theorem of Utumi in the case where R is two-sided artinian and two-sided continuous, and it also extends a theorem of Faith [54] where R is one-sided self-injective with ACC on left annihilators. It is an open question whether a left continuous ring R with R/S_l left Goldie is left artinian.

For more detailed information on the continuity condition, the reader is referred to Mohamed and Müller [147] and Dung, Huynh, Smith, and Wisbauer [49].

5

Principally Injective and FP Rings

A ring R is right mininjective if R-linear maps $aR \to R$ extend to $R_R \to R_R$ whenever aR is a simple right ideal. It is natural to enquire about the rings for which this condition is satisfied for *all* principal right ideals aR (for example if R is regular or right self-injective). These rings, called right principally injective (or right P-injective), play a central role in injectivity theory. An example is given of a right P-injective ring that is not left P-injective. If R is right P-injective it is shown that $Z_r = J$, that R is directly finite if and only if monomorphisms $R_R \to R_R$ are epic, and that R has the ACC on right annihilators if and only if it is left artinian. If R is right P-injective and right Kasch then $S_l \subseteq^{ess} {}_RR$. A semiperfect, right P-injective ring R in which $S_r \subseteq^{ess} R_R$ is called a right GPF ring. Hence the right PF rings are precisely the right self-injective, right GPF rings, and these right GPF rings exhibit many of the properties of the right PF rings: They admit a Nakayama permutation, they are right and left Kasch, they are left finitely cogenerated, $S_r = S_l$ is essential on both sides, and $Z_r = J = Z_l$.

Unlike mininjectivity, being right P-injective is not a Morita invariant. In fact, $M_n(R)$ is right P-injective implies that R is right n-injective. Call a ring R right FP-injective if, for every finitely generated submodule K of a free right R-module F, every R-linear map $K \to R_R$ extends to F. Then it is shown that R is right FP-injective if and only if $M_n(R)$ is right P-injective for every $n \geq 1$. These right FP-injective rings form a Morita invariant class.

As in the right P-injective case, the Kasch condition has a profound effect on a right FP-injective ring and leads to an important situation where a right injectivity condition leads to one on the left: If R is right 2-injective and right Kasch then it is left P-injective. This has the remarkable consequence that a right Kasch, right FP-injective ring is left FP-injective. This observation leads to the following result: If R is a semiperfect ring, the following are equivalent: (1) R is right FP-injective and right Kasch, (2) R is right FP-injective, and $J = \mathbf{r}(F)$ for a finite subset $F \subseteq R$, and (3) R is right FP-injective and $S_r \subseteq^{ess} R_R$; remarkably, all three of these conditions are equivalent to their left–right analogues. We call such rings FP rings, and we give several other characterizations. For example, a right FP-injective ring R is an FP ring if and only if it is right finite dimensional and right Kasch; if and only if it is left min-CS and right Kasch; if and only if it is right finitely cogenerated and right min-CS. As an application, we show among other things that a left perfect, right FP-injective ring is quasi-Frobenius if $soc_2(R)$ is finitely generated as a right R-module.

5.1. Principally Injective Rings

If R is a ring, a module M_R is called right *principally injective* (*P-injective*) if every R-homomorphism $\gamma : aR \to M$, $a \in R$, extends to $R \to M$, equivalently if $\gamma = m\cdot$ is multiplication by some element $m \in M$. Every injective module is P-injective, and a ring R is regular if and only if every right R-module is P-injective. In fact, if the map $ar \mapsto r + \mathbf{r}(a)$ from $aR \to R/\mathbf{r}(a)$ is given by left multiplication by $b + \mathbf{r}(a)$, then $aba = a$.

A ring R is called *right principally injective*[1] (or *right P-injective*) if R_R is a P-injective module, that is, if every principal right ideal aR is extensive. Thus every right self-injective ring is right P-injective, and every right P-injective ring is right mininjective. Moreover, neither converse is true: Every regular ring is both right and left P-injective, so there are P-injective rings that are not right self-injective; and the ring $\mathbb{Z}$ of integers is a commutative, noetherian, mininjective ring that is not P-injective.

Lemma 5.1. *The following conditions are equivalent for a ring R:*

(1) R is right P-injective.
(2) $\mathbf{lr}(a) = Ra$ for all $a \in R$.
(3) $\mathbf{r}(a) \subseteq \mathbf{r}(b)$, where $a, b \in R$, implies that $Rb \subseteq Ra$.
(4) $\mathbf{l}[bR \cap \mathbf{r}(a)] = \mathbf{l}(b) + Ra$ for all $a, b \in R$.
(5) If $\gamma : aR \to R, a \in R$, is R-linear, then $\gamma(a) \in Ra$.

Proof. (1)⇒(2). Always $Ra \subseteq \mathbf{lr}(a)$. If $b \in \mathbf{lr}(a)$ then $\mathbf{r}(a) \subseteq \mathbf{r}(b)$, so $\gamma : aR \to R$ is well defined by $\gamma(ar) = br$. Thus $\gamma = c\cdot$ for some $c \in R$ by (1), whence $b = \gamma(a) = ca \in Ra$.

(2)⇒(3). If $\mathbf{r}(a) \subseteq \mathbf{r}(b)$ then $b \in \mathbf{lr}(a)$, so $b \in Ra$ by (2).

(3)⇒(4). Let $x \in \mathbf{l}[bR \cap \mathbf{r}(a)]$. Then $\mathbf{r}(ab) \subseteq \mathbf{r}(xb)$, so $xb = rab$ for some $r \in R$ by (3). Hence $x - ra \in \mathbf{l}(b)$, proving that $\mathbf{l}[bR \cap \mathbf{r}(a)] \subseteq \mathbf{l}(b) + Ra$. The other inclusion always holds.

[1] Right P-injective rings are also referred to as rings with the *right principal extension property*.

(4)⇒(5). Let $\gamma : aR \to R$ be R-linear, and write $\gamma(a) = d$. Then $\mathtt{r}(a) \subseteq \mathtt{r}(d)$, so $d \in \mathtt{lr}(a)$. But $\mathtt{lr}(a) = Ra$ [take $b = 1$ in (4)], so $d \in Ra$.

(5)⇒(1). Let $\gamma : aR \to R_R$. By (5) write $\gamma(a) = ca$, $c \in R$. Then $\gamma = c\cdot$, proving (1). □

Example 5.2. The Björk example (Example 2.5) is a right P-injective ring (it has only one proper left ideal) that is not left P-injective; in fact it is not left mininjective.

Example 5.3. A direct product ΠR_i of rings is right P-injective if and only if each R_i is right P-injective by Lemma 5.1(2).

A multiplicative submonoid U of a ring R is called a *left denominator set* if it satisfies the following conditions: (1) If $u \in U$ and $r \in R$, then $u_1 r = r_1 u$ for some $u_1 \in U$ and $r_1 \in R$ (the left *Ore condition*), and (2) if $ru = r_1 u$ with $r, r_1 \in R$ and $u \in U$, then $u_1 r = u_1 r_1$ for some $u_1 \in U$ (we say U is *left reversible*). When these conditions are satisfied, R can be embedded in a ring of left *quotients* Q in which each element $u \in U$ is a unit and $Q = \{u^{-1}r \mid u \in U, r \in R\}$.

Example 5.4. If R is right P-injective and $U \subseteq R$ is a left denominator set, the ring of quotients $Q = \{u^{-1}r \mid u \in U, r \in R\}$ is also right P-injective. The converse is false.

Proof. If $\mathtt{r}_Q(q) \subseteq \mathtt{r}_Q(p)$, where $q = u^{-1}a$ and $p = v^{-1}b$, then $\mathtt{r}_R(a) \subseteq \mathtt{r}_R(b)$, so $b = ca$, $c \in R$, by hypothesis. It follows that $p = (v^{-1}cu)q$, as required. The converse is false as $\mathbb{Q} \supseteq \mathbb{Z}$ shows. □

Example 5.5. No polynomial ring $R = S[x]$ is right P-injective because $\mathtt{lr}(x) \neq Rx$.

Example 5.6. A domain is right P-injective if and only if it is a division ring. If R is an integral domain in which each finitely generated ideal is principal (a *Bezout domain*), then R/mR is P-injective for all $0 \neq m \in R$.

Proof. The first statement is clear. Write $\bar{r} = r + mR$ in the ring $\bar{R} = R/mR$ for all $r \in R$, and assume that $\mathtt{r}(\bar{a}) \subseteq \mathtt{r}\left(\bar{b}\right)$. By hypothesis, let $mR + aR = dR$, say $m = dx$ and $a = dy$. Then $ax = my$, so $\bar{x} \in \mathtt{r}(\bar{a}) \subseteq \mathtt{r}\left(\bar{b}\right)$. It follows that $bx \in mR = dxR$, so $b \in Rd$ because R is a domain. Hence $\bar{b} \in \bar{R}\bar{d} = \bar{R}\bar{a}$, as required. □

Example 5.7. Let R be right P-injective and assume that, for each $a \in R$, either $Ra \subseteq aR$ or $\mathtt{l}(a) \subseteq \mathtt{r}(a)$. Then the following are equivalent:

(1) R is regular.
(2) $J = 0$.
(3) R is semiprime.
(4) R is *reduced* (that is it has no nonzero nilpotents).

In particular, semiprime, commutative P-injective rings are regular. Furthermore, for reduced rings, being left or right P-injective is equivalent to being (strongly) regular.

Proof. (1)⇒(2) and (2) ⇒(3) are clear.

(3)⇒(4). If $a^2 = 0$ we show that $aRa = 0$ and use (3). If $Ra \subseteq aR$ then $aRa \subseteq a(aR) = 0$; if $\mathtt{l}(a) \subseteq \mathtt{r}(a)$ then $Ra \subseteq \mathtt{l}(a) \subseteq \mathtt{r}(a)$, so again $aRa = 0$.

(4)⇒(1). If $a \in R$ we must show that $a \in aRa$. First we have $\mathtt{r}(a^2) \subseteq \mathtt{r}(a)$ [in fact $a^2r = 0 \Rightarrow (ara)^2 = 0 \Rightarrow ara = 0 \Rightarrow (ar)^2 = 0 \Rightarrow ar = 0$]. This gives $a \in Ra^2$ because R is right P-injective (Lemma 5.1). If $Ra \subseteq aR$ then $a \in Ra^2 = (Ra)a \subseteq (aR)a$; if $\mathtt{l}(a) \subseteq \mathtt{r}(a)$ then, writing $a = ra^2$, we have $(1 - ra) \in \mathtt{l}(a) \subseteq \mathtt{r}(a)$ and so $a = ara$.

The last statements are now routine verifications. □

Example 5.8. A ring is called a right *PP ring* if every principal right ideal is projective. A ring R is a right P-injective, right PP ring if and only if R is regular.

Proof. Let R be a right P-injective, right PP ring. If $a \in R$ then $\mathtt{r}(a) = eR$, where $e^2 = e$ because aR is projective. Hence $Ra = \mathtt{lr}(a) = R(1 - e)$, so R is regular. The converse is clear. □

We now record several results about principal right ideals in a right P-injective ring for reference.

Proposition 5.9. *Let R be a right P-injective ring, and let $a, b \in R$.*

(1) If bR embeds in aR, then Rb is an image of Ra.
(2) If aR is an image of bR, then Ra embeds in Rb.
(3) If $bR \cong aR$, then $Ra \cong Rb$.

Proof. Given $\sigma : bR \to aR$, let $\sigma = v\cdot$, $v \in R$. Then $vb = au$ for some $u \in R$, so define $\theta : Ra \to Rb$ by $(ra)\theta = (ra)u = r(vb)$.

(1). If σ is monic, we have $\mathbf{r}(vb) \subseteq \mathbf{r}(b)$, whence $Rb \subseteq Rvb$ by P-injectivity, say $b = rvb = (ra)\theta$. Thus θ is epic, proving (1).

(2). If σ is epic, write $a = \sigma(bs) = vbs$, $s \in R$. If $(ra)\theta = 0$, $r \in R$, then $0 = (ra)u = r(vb)$, whence $ra = r(vbs) = 0$. Hence θ is monic, proving (2).

(3). If σ is an isomorphism, so also is θ by the proofs of (1) and (2). □

The next result will be referred to several times. Recall that we write $N \subseteq^{\oplus} M$ if the submodule N is a direct summand of the module M.

Proposition 5.10. *Every right P-injective ring is a right C2 ring, but not conversely.*

Proof. If T is a right ideal of R and $T \cong eR$, where $e^2 = e \in R$, then $T = aR$ for some $a \in R$ and T is projective. Hence $\mathbf{r}(a) \subseteq^{\oplus} R_R$, say $\mathbf{r}(a) = fR$, where $f^2 = f \in R$. Hence $Ra = \mathbf{lr}(a) = R(1-f) \subseteq^{\oplus} {}_RR$, and so $T = aR \subseteq^{\oplus} R_R$. The converse is false by Example 5.12 following the next corollary. □

Corollary 5.11. *If R is a right P-injective ring, the following are equivalent for an element $a \in R$:*

(1) aR is projective.
(2) aR is a direct summand of R_R.
(3) aR is a P-injective module.

Proof. (1)⇒(2) by the proof of Proposition 5.10, and (2)⇒(3) is because direct summands of P-injective modules are again P-injective. To prove that (3)⇒(1), write $M = aR$ for clarity and consider $\gamma : aR \to M$ given by $\gamma(ar) = ar$. By (3) we have $\gamma = m\cdot$, $m \in M$. If $m = ab$, $b \in R$, then $a = \gamma(a) = ma = aba$, and $aR = eR$, where $e^2 = e = ab$, proving (1). □

Example 5.12. The trivial extension $R = \left\{\begin{bmatrix} a & v \\ 0 & a \end{bmatrix} \mid a \in F,\ v \in V\right\}$ of the field F by the two-dimensional vector space V is a commutative, local, artinian C2 ring that is not P-injective.

Proof. Let $V = uF \oplus wF$, and write $\bar{u} = \begin{bmatrix} 0 & u \\ 0 & 0 \end{bmatrix}$. Then $\bar{u}R = \begin{bmatrix} 0 & uF \\ 0 & 0 \end{bmatrix}$ and $\begin{bmatrix} 0 & ua \\ 0 & 0 \end{bmatrix} \mapsto \begin{bmatrix} 0 & wa \\ 0 & 0 \end{bmatrix}$ is an R-linear map from $\bar{u}R \to R$ that does not extend to $R \to R$ because $w \notin uF$. Hence R is not P-injective. To see that R is a C2 ring, it suffices (since R is local) to show that if $T \cong R$, with T a right ideal, then $T = R$. But if $\sigma : R \to T$ is an isomorphism, then $T = xR$, where $x = \sigma(1)$, and x is a unit because $\mathbf{r}(x) = 0$ (the monomorphism $x\cdot : R_R \to R_R$ is epic because R is finite dimensional). □

A ring R is called *directly finite* if $ab = 1$ in R implies that $ba = 1$. It is well known that every I-finite ring is directly finite. Whereas the ring of all linear transformations of an infinite dimensional vector space is a right and left P-injective ring that is not directly finite, we do have the following result.

Theorem 5.13. *A right P-injective ring R is directly finite if and only if monomorphisms $R_R \to R_R$ are isomorphisms.*

Proof. Assume that R is directly finite. If $\sigma : R_R \to R_R$ is monic, write $\sigma(1) = a$. Then $\mathtt{r}(a) = 0$ so, by P-injectivity, $Ra = \mathtt{lr}(a) = \mathtt{l}(0) = R$. Hence $ba = 1$ for some $b \in R$, so $ab = 1$ by hypothesis, and σ is onto. Conversely, suppose that $ab = 1$ in R. Then the map $\beta : R \to R$ is monic if we define $\beta(r) = br$ for all $r \in R$. By hypothesis β is epic, so, if $1 = \beta(c)$, then $bc = 1$. Because $ab = 1$, it follows that $c = a$ and hence that $ba = 1$. □

If R is a right self-injective ring then $J = Z_r$ by Theorem 1.26. Even though this fails in a right mininjective ring (consider $\mathbb{Z}_{(p)}$, where p is a prime), it does hold in a right P-injective ring.

Theorem 5.14. *If R is a right P-injective ring then $J = Z_r$.*

Proof. If $a \in Z_r$ then $\mathtt{r}(1-a) = 0$ because $\mathtt{r}(a) \cap \mathtt{r}(1-a) = 0$. Hence $R = \mathtt{lr}(1-a) = R(1-a)$, and it follows that $Z_r \subseteq J$. Conversely, if $a \in J$ we show that $bR \cap \mathtt{r}(a) = 0$, $b \in R$, implies that $b = 0$. In fact Lemma 5.1 gives $\mathtt{l}(b) + Ra = \mathtt{l}[bR \cap \mathtt{r}(a)] = \mathtt{l}(0) = R$, so $\mathtt{l}(b) = R$ because $a \in J$. □

Theorem 5.14 leads to the following result about right P-injective rings.

Proposition 5.15. *Every right P-injective ring with ACC on right annihilators is left artinian.*

Proof. If R is such a ring, observe first that J is nilpotent by Theorem 5.14 and the Mewborn–Winton lemma (Lemma 3.29). Moreover, R has DCC on left annihilators and so has DCC on principal left ideals by Lemma 5.1. Hence R is right perfect by Theorem B.39. Thus R is semiprimary; in particular $S_r \subseteq^{ess} R_R$. Since R is left minannihilator (being right P-injective), it is right min-PF, and so it is left artinian by Theorem 3.35. □

The next result gives one situation where the P-injective hypothesis extends to certain finitely generated right ideals.

Lemma 5.16. *Let R be a right P-injective ring and assume that $Rb_1 \oplus Rb_2 \oplus \cdots \oplus Rb_n$ is a direct sum of left ideals of R.*

(1) Any R-linear map $\gamma : b_1R + b_2R + \cdots + b_nR \to R_R$ extends to $R \to R$.
(2) If $S = b_1R + \cdots + b_kR$ and $T = b_{k+1}R + \cdots + b_nR$, then $\mathtt{l}(S \cap T) = \mathtt{l}(S) + \mathtt{l}(T)$.

Proof. (1). Since R is right P-injective let $\gamma(b_i) = c_ib_i$, where $c_i \in R$ for each i. Similarly, $\gamma(b_1 + b_2 + \cdots + b_n) = c(b_1 + b_2 + \cdots + b_n)$ for some $c \in R$. Hence $c_ib_i = cb_i$ for each i by the direct sum hypothesis, and it follows that $\gamma = c\cdot$ on Σb_iR. Hence γ extends to $c\cdot : R \to R$.

(2) If $x \in \mathtt{l}(S \cap T)$ then $\gamma : S + T \to R$ is well defined by $\gamma(s + t) = xs$. By (1) we have $\gamma = c\cdot$ for some $c \in R$, and it follows that $c \in \mathtt{l}(T)$ and $x - c \in \mathtt{l}(S)$. Hence $x = (x - c) + c \in \mathtt{l}(S) + \mathtt{l}(T)$, which proves that $\mathtt{l}(S \cap T) \subseteq \mathtt{l}(S) + \mathtt{l}(T)$. The other inclusion always holds. □

Proposition 5.17. *Let $\oplus_{i \in I} B_i$ be a direct sum of (two-sided) ideals in a right P-injective ring R. Then we have $A \cap [\oplus_{i \in I} B_i] = \oplus_{i \in I}(A \cap B_i)$ for every left ideal A of R.*

Proof. Let $a \in A \cap [\oplus_{i \in I} B_i]$, write $a = b_1 + b_2 + \cdots + b_n$, $b_i \in B_i$, and let $\pi_k : \oplus_{i=1}^n b_iR \to b_kR$ be the projection. Since $\Sigma_{i=1}^n Rb_i$ is direct, Lemma 5.16 gives $\pi_k = p_k\cdot$, $p_k \in R$. Hence $b_k = p_ka \in A \cap B_k$ for $1 \leq k \leq n$, so $a \in \oplus_{i \in I}(A \cap B_i)$. This shows that $A \cap [\oplus_{i \in I} B_i] \subseteq \oplus_{i \in I}(A \cap B_i)$; the other inclusion always holds. □

We conclude this section with some conditions that an endomorphism ring is P-injective.

Proposition 5.18. *Given a module M_R, write $S = end(M_R)$.*

(1) Assume that M generates $ker(\beta)$ for each $\beta \in S$. Then S is right P-injective if and only if $ker(\beta) \subseteq ker(\gamma)$ implies $\gamma \in S\beta$.
(2) Assume that M cogenerates $M/\beta(M)$ for each $\beta \in S$. Then S is left P-injective if and only if $\gamma(M) \subseteq \beta(M)$ implies $\gamma \in \beta S$.

Proof. (1). If S is right P-injective and $ker(\beta) \subseteq ker(\gamma)$, then $\mathtt{r}_S(\beta) \subseteq \mathtt{r}_S(\gamma)$. Hence $\gamma \in S\beta$ by Lemma 5.1. Conversely, if the condition holds let $\gamma \in \mathtt{lr}(\beta)$; we show that $ker(\beta) \subseteq ker(\gamma)$. Given $x \in ker(\beta)$, use the generation hypothesis to write $x = \sum_i \lambda_i(m_i)$, where, for each i, $m_i \in M$ and $\lambda_i : M \to ker(\beta)$.

Then $\beta\lambda_i = 0$ for each i, so $\gamma\lambda_i = 0$ because $\gamma \in \mathtt{lr}(\beta)$. It follows that $x \in ker(\gamma)$.

(2). Let S be left P-injective. If $\gamma(M) \subseteq \beta(M)$ then $\lambda\beta = 0$ implies $\lambda\gamma = 0$; that is, $l_S(\beta) \subseteq l_S(\gamma)$. Hence $\gamma \in \beta S$ by Lemma 5.1. Conversely, if $\gamma \in \mathtt{rl}(\beta)$, we show that $\gamma(M) \subseteq \beta(M)$. If not and $\gamma(m_0) \notin \beta(M)$, the cogeneration hypothesis yields $\sigma : M/\beta(M) \to M$ such that $\sigma[\gamma(m_0) + \beta(M)] \neq 0$. If $\lambda : M \to M$ is defined by $\lambda(m) = \sigma[m + \beta(M)]$, then $\lambda\gamma \neq 0$ but $\lambda\beta = 0$, which is a contradiction. □

In particular, if M is a generator or a cogenerator, Proposition 5.18 characterizes when the endomorphism ring of M is P-injective.

5.2. Kasch P-Injective Rings

We begin with a result showing that imposing the right Kasch condition on a right P-injective ring makes the *left* socle essential on the left.

Proposition 5.19. *Let R be a right P-injective, right Kasch ring. Then*

(1) S_l is essential in ${}_RR$ and
(2) $\mathtt{l}(J)$ is essential in ${}_RR$.

Proof. (1). If $0 \neq a \in R$, let $\mathtt{r}(a) \subseteq T$, where T is a maximal right ideal. Then $\mathtt{l}(T) \subseteq \mathtt{lr}(a) = Ra$, and (1) follows because $\mathtt{l}(T)$ is simple by Theorem 2.31.

(2). If $0 \neq b \in R$ choose M maximal in bR and let $\sigma : bR/M \to R_R$ be monic. Then define $\gamma : bR \to R_R$ by $\gamma(x) = \sigma(x + M)$ so that $\gamma = c\cdot$ for some $c \in R$ by hypothesis. Hence $cb = \sigma(b + M) \neq 0$ because $b \notin M$ and σ is monic. But $cbJ = \gamma(bJ) = 0$ because $bJ \subseteq M$, so $0 \neq cb \in Rb \cap \mathtt{l}(J)$, as required. □

Proposition 5.20. *The following conditions are equivalent for a semiperfect, right P-injective ring R:*

(1) R is right Kasch.
(2) S_r is essential in ${}_RR$.

Proof. (2)⇒(1) is Lemma 1.48. Given (1), we have $S_r = \mathtt{l}(J)$ because R is semilocal, so (2) follows from Proposition 5.19. □

In proving that a ring is quasi-Frobenius, it is a recurring problem to be able to show that an injectivity condition on the right implies one on the left.

The following result gives one situation where the Kasch condition makes this happen, and it will be used several times throughout the book. The following notion is involved: If $n \geq 1$ is an integer, a module M_R is called *n-injective* if every map from an n-generated right ideal of R to M extends to R. Thus the 1-injective modules are the P-injective modules. A ring R is called *right n-injective* if R_R is an n-injective module.

Lemma 5.21. *If R is right 2-injective and right Kasch, then R is left P-injective.*

Proof. Given $a, b \in R$ such that $\mathtt{l}(a) \subseteq \mathtt{l}(b)$, we must show that $b \in aR$. If $b \notin aR$, let M/aR be a maximal submodule of $(aR+bR)/aR$. By the Kasch hypothesis, let $\sigma : (aR+bR)/M \to R_R$ be monic, and define $\gamma : aR+bR \to R$ by $\gamma(x) = \sigma(x+M)$. Since R is right 2-injective, $\gamma = c\cdot$ for some $c \in R$. Then $cb = \gamma(b) = \sigma(b+M) \neq 0$ because $b \notin M$. However, $ca = 0$ because $a \in M$ and it follows that $cb = 0$ because $\mathtt{l}(a) \subseteq \mathtt{l}(b)$, which is a contradiction. □

In view of Lemma 5.21 we hasten to give

Example 5.22. The Björk example (Example 2.5) is right P-injective but not right 2-injective.

Proof. In the notation of Example 2.5, let $X \neq Y$ be one-dimensional $\bar{F}$-subspaces of F, and consider the (simple) right ideals Xt and Yt. Then $\gamma : Xt \oplus Yt \to Xt$ given by $\gamma(xt+yt) = xt$ is right R-linear, as is easily checked, but if $\gamma = (a+bt)\cdot$, where $a, b \in F$, then $ax = x$ and $ay = 0$ for all $x \in X$ and $y \in Y$, which is a contradiction. □

5.3. Maximal Left Ideals

Recall that a module is called *uniform* if the intersection of any two nonzero submodules is nonzero. (We adopt the convention that uniform modules are nonzero.) We are going to characterize the maximal left ideals of a right P-injective ring R in terms of the uniform principal right ideals uR. In this case write

$$M_u = \{x \in R \mid \mathtt{r}(x) \cap uR \neq 0\}.$$

If x and y are in M_u then $\mathtt{r}(x+y) \cap uR \supseteq [\mathtt{r}(x) \cap uR] \cap [\mathtt{r}(y) \cap uR] \neq 0$ because uR is uniform. Hence M_u is closed under addition and so is a left ideal of R. In fact, we have the following lemma:

Lemma 5.23. *If R is right P-injective and uR is a uniform right ideal, then M_u is the unique maximal left ideal of R containing $\mathtt{l}(u)$.*

Proof. Clearly $\mathtt{l}(u) \subseteq M_u \neq R$. It suffices to show that if $\mathtt{l}(u) \subseteq L \neq R$, where L is a left ideal, then $L \subseteq M_u$. But if $a \in L - M_u$ then $uR \cap \mathtt{r}(a) = 0$, so Lemma 5.1 gives $R = \mathtt{l}[uR \cap \mathtt{r}(a)] = \mathtt{l}(u) + Ra \subseteq L$, which is a contradiction. Hence $L \subseteq M_u$, completing the proof. □

Note that if uR is simple then $M_u = \mathtt{l}(u)$ in Lemma 5.23.

Lemma 5.24. *Let R be right P-injective and let $W = u_1R \oplus \cdots \oplus u_nR$, where each u_iR is a uniform right ideal. Suppose that M is a maximal left ideal of R such that $M \neq M_u$ for all uniform $uR \subseteq R$. Then there exists $m \in M$ such that $\mathtt{r}(1-m) \cap W$ is essential in W.*

Proof. Since $M \neq M_{u_1}$ let $\mathtt{r}(m) \cap u_1R = 0$, $m \in M$. Then $\mathtt{r}(mu_1) \subseteq \mathtt{r}(u_1)$, so $Ru_1 \subseteq Rmu_1$ by Lemma 5.1. Thus $(1-m_1)u_1 = 0$ for some $m_1 \in M$, so $\mathtt{r}(1-m_1) \cap u_1R \neq 0$. If $\mathtt{r}(1-m_1) \cap u_iR \neq 0$ for every $i > 1$ we are done because $\oplus_{i=1}^n[\mathtt{r}(1-m_1) \cap u_iR] \subseteq^{ess} W$. If (say) $\mathtt{r}(1-m_1) \cap u_2R = 0$ then $(1-m_1)u_2R$ is uniform (being isomorphic to u_2R), so $M \neq M_{(1-m_1)u_2}$. Hence, as before, $(1-m)(1-m_1)u_2 = 0$ for some $m \in M$. If we write $m_2 = m + m_1 - mm_1$ this gives $(1-m_2)u_1 = 0$ and $(1-m_2)u_2 = 0$, whence $\mathtt{r}(1-m_2) \cap u_iR \neq 0$ for $i = 1, 2$. Continue in this way to obtain $m \in M$ such that $\mathtt{r}(1-m) \cap u_iR \neq 0$ for each i. The lemma now follows. □

Theorem 5.25. *Let R be right P-injective and right finite dimensional. Then:*

(1) A left ideal M of R is maximal if and only if $M = M_u$ *for some uniform* $uR \subseteq R$.
(2) R *is semilocal.*

Proof. By finite dimensionality, let $W = u_1R \oplus \cdots \oplus u_nR$ be essential in R_R, where each u_iR is uniform.

(1). Each M_u is maximal by Lemma 5.23. If M is a maximal left ideal of R not of the form $M = M_u$, Lemma 5.24 shows that $\mathtt{r}(1-m) \cap W$ is essential in W for some $m \in M$. Hence $\mathtt{r}(1-m) \subseteq^{ess} R_R$, so $1 - m \in Z_r = J$ by Theorem 5.14. Hence m is a unit, which is a contradiction. This proves (1).

(2). Let $x \in M_{u_1} \cap \cdots \cap M_{u_n}$. Then $\mathtt{r}(x) \cap u_iR \neq 0$ for each i, whence $\mathtt{r}(x)$ is essential in R_R (it contains $\oplus_{i=1}^n[\mathtt{r}(x) \cap u_iR]$, which is essential in W). As before, this means that $x \in Z_r = J$. Hence $M_{u_1} \cap \cdots \cap M_{u_n} \subseteq J$. This proves (2) because the reverse inclusion follows from Lemma 5.23. □

Lemma 5.26. *Let R be a right P-injective ring. If $aR \oplus bR$ and $Ra \oplus Rb$ are both direct, where a and b are in R, then $\mathtt{l}(a) + \mathtt{l}(b) = R$.*

Proof. Define $\gamma : (a+b)R \to R$ by $\gamma[(a+b)x] = bx$. This is well defined because $aR \cap bR = 0$, so $\gamma = c\cdot$ is multiplication by an element $c \in R$ by Lemma 5.1. Thus $c(a+b) = b$, whence $ca - (1-c)b = 0$. Since $Ra \cap Rb = 0$ it follows that $c \in \mathtt{l}(a)$ and $(1-c) \in \mathtt{l}(b)$, as required. □

A ring R is called a right *duo ring* if all right ideals of R are two-sided, equivalently if $Ra \subseteq aR$ for all $a \in R$.

Proposition 5.27. *Let R be a right duo, right P-injective ring. The following are equivalent:*

(1) R has a finite number of maximal left ideals.
(2) R is right finite dimensional.

Proof. Given (2), R/J is semisimple artinian by Theorem 5.25. As R/J inherits the right duo condition, it is a finite product of division rings, and (1) follows.

Conversely, let $M_1, \ldots, M_n$ be the maximal left ideals of R; we show that $dim(R_R) \leq n$. If not, assume that $a_1R \oplus \cdots \oplus a_nR \oplus a_{n+1}R$ is direct, where $0 \neq a_i \in R$ for each i. We may assume that $\mathtt{l}(a_1) \subseteq M_1$. Hence assume inductively that $\mathtt{l}(a_i) \subseteq M_i$ for $1 \leq i \leq k$. The right duo hypothesis guarantees that $Ra_i \oplus Ra_{k+1}$ is direct, so $\mathtt{l}(a_{k+1})$ is not contained in M_i for $1 \leq i \leq k$ by Lemma 5.26. Hence we may assume that $\mathtt{l}(a_{k+1}) \subseteq M_{k+1}$. It follows by induction that, after relabeling the M_i, $\mathtt{l}(a_i) \subseteq M_i$ for $1 \leq i \leq n$. But $\mathtt{l}(a_{n+1}) \subseteq M_i$ for some $i = 1, 2, \ldots, n$, which is a contradiction by Lemma 5.26. □

Note that the Björk example is a right P-injective, left duo ring that is not right finite dimensional, so the right duo hypothesis is essential for (2) in Proposition 5.27. Note further that if conditions (1) and (2) are both satisfied in Proposition 5.27, then R/J is a finite product of division rings and so, in addition, R has only finitely many maximal right ideals.

Lemma 5.28. *Let R be a right P-injective, left Kasch ring in which every nonzero right ideal contains a uniform right ideal. Then every maximal left ideal M has the form $M = M_u$ for some $u \in R$ such that uR is uniform.*

Proof. Given a maximal right ideal M, we have $\mathtt{r}(M) \neq 0$ by the Kasch condition, so let $uR \subseteq \mathtt{r}(M)$, where uR is uniform. By Lemma 5.23 it suffices to show that $M_u \subseteq M$. If $x \in M_u$ then $V = uR \cap \mathtt{r}(x) \neq 0$ and $x \in \mathtt{l}(V)$. However, $V \subseteq uR \subseteq \mathtt{r}(M)$, so $M \subseteq \mathtt{l}(V)$. Since M is maximal, it follows that $M = \mathtt{l}(V)$, so $x \in M$ as required. □

Every nonzero right ideal in a ring R has maximal essential extensions in R by Zorn's lemma. The next result identifies one situation when there is uniqueness.

Lemma 5.29. *If R is a right P-injective, right duo ring then every nonzero right ideal has a unique maximal essential extension in R.*

Proof. Let $T \neq 0$ be a right ideal and assume that $T \subseteq C$ and $T \subseteq D$ are maximal essential extensions. It suffices to show that $T \subseteq^{ess} C + D$. If $0 \neq c + d \in C + D$, we must show that $(c + d)R \cap T \neq 0$. Suppose first that $(c + d)\mathtt{r}(c) \neq 0$. Then $(c + d)R \cap dR \neq 0$, so $(c + d)R \cap T \neq 0$ because $T \subseteq^{ess} D$. However, if $(c + d)\mathtt{r}(c) = 0$ then $(c + d) \in \mathtt{lr}(c) = Rc \subseteq C$ by Lemma 5.1 and the right duo hypothesis. Hence $0 \neq (c + d)R \subseteq C$, so $(c + d)R \cap T \neq 0$ because $T \subseteq^{ess} C$. □

If U is a uniform right ideal the maximal essential extensions of U are all uniform and are the maximal members in the set of uniform extensions of U. We call them *maximal uniform right ideals*.

Theorem 5.30. *Let R be a right P-injective, right duo ring in which every nonzero right ideal contains a uniform right ideal. Let $\{U_i \mid i \in I\}$ denote the set of all the maximal uniform right ideals of R. For each $i \in I$ let $M_i = \{m \in R \mid \mathtt{r}(m) \cap U_i \neq 0\}$. Then the following statements hold:*

(1) Each M_i is a maximal left ideal.
(2) $\Sigma_{i \in I} U_i$ is direct and essential in R_R.
(3) The mapping $U_i \mapsto M_i$ is one-to-one.
(4) If R is left Kasch, the map in (3) is a bijection.

Proof. (1). We have $M_i = M_u$ for all $0 \neq u \in U_i$ because U_i is uniform, so this follows from Lemma 5.23.

(2). By Zorn's lemma let $K \subseteq I$ be maximal such that $W = \oplus_{k \in K} U_k$ is direct. It suffices to show that $W \subseteq^{ess} R_R$. Suppose on the contrary that $W \cap T = 0$, where T is a nonzero right ideal. We may assume that T is uniform by hypothesis. By Lemma 5.29, let $T \subseteq U_i$ for a unique $i \in I$. We claim that

$U_i \cap W = 0$, contradicting the maximality of K. But if $U_i \cap W \neq 0$ then, since R is right duo, Proposition 5.17 shows that $\oplus_{k \in K}(U_i \cap U_k) = U_i \cap [\oplus_{k \in K} U_k] = U_i \cap W \neq 0$. Hence $U_i \cap U_k \neq 0$ for some $k \in K$. This implies that $U_i = U_k$ by Lemma 5.29, which is a contradiction.

(3). Suppose that $M_i = M_j$, where $i \neq j$. Then $U_i \neq U_j$, so $U_i \cap U_j = 0$ by Lemma 5.29. Choose $0 \neq u_i \in U_i$ and $0 \neq u_j \in U_j$. Then $u_i R \oplus u_j R$ is direct, whence $Ru_i \oplus Ru_j$ is direct by the duo hypothesis. If $\pi : u_i R \oplus u_j R \to u_i R$ is the projection, we have $\pi = p \cdot$ for some $p \in R$ by Lemma 5.16. Hence $pu_i = u_i$ and $pu_j = 0$, so $0 \neq u_j \in \mathrm{r}(p) \cap U_j$. This proves that $p \in M_j$, and a similar argument shows that $1 - p \in M_i$. This contradicts the assumption that $M_i = M_j$.

(4). If M is a maximal left ideal then (by Lemma 5.28) $M = M_u$ for some $u \in R$ with uR uniform. But $uR \subseteq U_i$ for some $i \in I$, so $M_u = M_i$ because $u \in U_i$. □

5.4. GPF Rings

We now turn to the structure of semiperfect right P-injective rings. Recall that a ring R is called a right pseudo-Frobenius ring (right PF ring) if it is right self-injective and semiperfect and $S_r \subseteq^{ess} R_R$. For convenience we call R a right *generalized pseudo-Frobenius* ring (right *GPF ring)* if it is right P-injective and semiperfect and $S_r \subseteq^{ess} R_R$.

Since right P-injective rings are all left minannihilator rings by Lemma 5.1, the right GPF rings are exactly the right principally injective, right min-PF rings. Hence many of the properties in the next theorem come from Chapter 3, and we collect them here for reference.

Theorem 5.31. *Let R be a right GPF ring. Then the following statements hold:*

(1) R is right and left Kasch.
(2) $S_r = S_l (= S)$ is essential in both R_R and ${}_R R$.
(3) R is left finitely cogenerated.
(4) $\mathtt{l}(S) = J = \mathtt{r}(S)$ and $\mathtt{l}(J) = S = \mathtt{r}(J)$.
(5) $J = Z_r = Z_l$.
(6) $soc(Re) = Se$ is simple and essential in Re for every local idempotent $e \in R$.
(7) $soc(eR)$ is homogeneous and essential in eR for every local idempotent $e \in R$.
(8) The maps $K \mapsto \mathtt{r}(K)$ and $T \mapsto \mathtt{l}(T)$ are mutually inverse lattice isomorphisms between the simple left ideals K and the maximal right ideals T.

(9) If $e_1, \ldots, e_n$ is a basic set of local idempotents, there exist elements $k_1, \ldots, k_n$ in R and a permutation σ of $\{1, \ldots, n\}$ such that the following hold for all $i = 1, \ldots, n$:

(a) $k_i R \subseteq e_i R$ and $Rk_i \subseteq Re_{\sigma i}$.

(b) $k_i R \cong e_{\sigma i} R / e_{\sigma i} J$ and $Rk_i \cong Re_i / Je_i$.

(c) $\{k_1 R, \ldots, k_n R\}$ and $\{Rk_1, \ldots, Rk_n\}$ are complete sets ofdistinct representatives of the simple right and left R-modules, respectively.

(d) $soc(Re_{\sigma i}) = Rk_i = Se_{\sigma i} \cong Re_i / Je_i$ is simple and essential in$Re_{\sigma i}$ for each i.

(e) $soc(e_i R)$ is homogeneous and essential in $e_i R$, with each simplesubmodule isomorphic to $e_{\sigma i} R / e_{\sigma i} J$.

Proof. (1) follows from Theorem 3.24, so $S_r \subseteq^{ess} {}_R R$ by Proposition 5.20. Hence (9) follows from Theorem 3.24, except for $soc(e_i R) \subseteq^{ess} e_i R$ in (d), and this follows because $S_r \subseteq^{ess} R_R$ by hypothesis. Moreover, $S_r = S_l$ by Theorem 3.24, so $S_l \subseteq^{ess} {}_R R$. Now (2) and (3) follow from (d).

(4). We have $\mathtt{l}(S_l) = J$ and $\mathtt{r}(S_r) = J$ because R is left and right Kasch, respectively, and $\mathtt{l}(J) = S_l$ and $\mathtt{r}(J) = S_r$ because R is semilocal.

(5). We have $J = Z_r$ by Theorem 5.14, and $Z_l \subseteq J$ holds in any semiperfect ring because Z_l contains no nonzero idempotent. If $a \in J$ then $S_r a = 0$, so $S \subseteq \mathtt{l}(a)$. Hence $a \in Z_l$ by (2).

(6) and (7). If $e^2 = e$ is local, $Re \cong Re_i$ for some i and $eR \cong e_j R$ for some j, so (d) and (e) apply.

(8). This follows by Theorem 2.32. □

Corollary 5.32. *If a right GPF ring R is actually right 2-injective, then R is a left GPF ring.*

Proof. R is left P-injective by Lemma 5.21; now use Theorem 5.31. □

Corollary 5.33. *The following are equivalent for a right GPF ring R:*

(1) R is left mininjective.

(2) R is a right minannihilator ring.

(3) $soc(eR)$ is simple for all local idempotents $e \in R$.

In this case R is right finite-dimensional.

Proof. (1)⇔(2). Given (1), R is a two-sided mininjective ring and $S_r = S_l$ by hypothesis, so (2) follows by Corollary 2.34. Conversely, given (2), R is a right minannihilator ring in which $S_r = S_l$, so (1) follows from Proposition 2.33.

(1)⇒(3). Given (1), R is a left minfull ring in which $S_r = S_l$, so (3) follows from Theorem 3.12.

(3)⇒(1). We have $S_l = S_r$ by Theorem 5.31, so $eS_l = eS_r = soc(eR)$ for every local $e^2 = e$. Hence (1) follows from Theorem 3.2.

Finally, the last sentence is because S_r is finite dimensional by (3). □

Example 5.34. The Björk example (Example 2.5) is a local, left artinian right GPF ring R with $J^2 = 0$, but

(1) R is not a left GPF ring (indeed R is not left mininjective) and
(2) the conditions in Corollary 5.33 are not satisfied.

5.5. Morita Invariance and FP-Injectivity

Unlike right self-injectivity and right mininjectivity, right P-injectivity is not a Morita invariant. However we do have

Proposition 5.35. *If R is right P-injective, so is eRe for all $e^2 = e \in R$ satisfying $ReR = R$.*

Proof. Write $S = eRe$ and let $\mathtt{r}_S(b) \subseteq \mathtt{r}_S(a)$, where $b, a \in S$. It suffices to show that $\mathtt{r}_R(b) \subseteq \mathtt{r}_R(a)$. (Then $a \in Rb$ by Lemma 5.1, so $a = ea \in eRb = Sb$, as required.) So let $bx = 0$, $x \in R$, and write $1 = \sum_{i=1}^n p_i e q_i$, where $p_i, q_i \in R$. Then $b(exp_ie) = bxp_ie = 0$ for each i and so $a(exp_ie) = 0$ by hypothesis. Hence $ax = \sum_{i=1}^n axp_ieq_i = 0$, as required. □

Thus the reason that Morita invariance fails for right P-injective rings is that this property does not pass from a ring R to the matrix ring $M_n(R)$. The next result will be used to prove this and is of independent interest.

Proposition 5.36. *If $M_n(R)$ is right P-injective then R is right n-injective.*

Proof. Let $\gamma : T \to R_R$ be R-linear, where $T = b_1R + b_2R + \cdots + b_nR$ is an n-generated right ideal of R, and let B and B^γ denote the $n \times n$ matrices with first row $[b_1\ b_2 \cdots b_n]$ and $[\gamma(b_1)\ \gamma(b_2) \cdots \gamma(b_n)]$, respectively, and all other entries zero. If $BX = 0$ with $X \in M_n(R)$ then $B^\gamma X = 0$, so $B^\gamma \in \mathtt{lr}(B) = M_n(R)B$ by hypothesis. If $B^\gamma = CB$, where $C = [c_{ij}] \in M_n(R)$, it follows that $\gamma = c_{11}\cdot$ is multiplication by c_{11}. □

Example 5.37. Being right P-injective and being a right GPF ring are not Morita invariant properties.

Proof. If P-injectivity were a Morita invariant, then every right P-injective ring R would be right n-injective for every $n \geq 1$ by Proposition 5.36. But the Björk example is a right P-injective ring that is not right 2-injective by Lemma 5.21. Hence right P-injectivity is not a Morita invariant. Since the same ring is a right GPF ring that is not right 2-injective, the same argument shows that "right GPF ring" is also not a Morita invariant. □

In view of Example 5.37, the rings R for which every matrix ring $M_n(R)$ is right P-injective become of interest. It turns out that they satisfy a stronger injectivity condition: They are right FP-injective rings. Before studying these rings, we investigate the corresponding condition in an arbitrary module.

A module Q_R is called FP-*injective* (or *absolutely pure*) if, for any finitely generated submodule K of a free right R-module F, every R-linear map $K \to Q_R$ extends to a map $F_R \to Q_R$. Clearly every injective module is FP-injective. If R is a regular ring, every right R-module is FP-injective because every finitely generated submodule of a free module is a summand (by Theorem B.54 and Corollary B.49).

We are going to give several characterizations of these FP-injective modules, and the following matrix notation will be employed. Write M^n and M_n for the direct sum of n copies of the module M, written as row and column matrices, respectively, and write $M_{m\times n}(R)$ for the set of all $m \times n$ matrices over R. As we frequently discuss R-linear maps from R_n or R^n to some module, the following lemma will be used repeatedly. Let $\{\underline{e}_i\}$ and $\{\bar{e}_i\}$ denote the canonical bases of R_n and R^n respectively.

Lemma 5.38. *If R is a ring, then the following statements hold:*

(1) *Every R-linear map $\gamma : R_n \to M_R$ is matrix multiplication $\gamma = \bar{m}\cdot$ by some row $\bar{m} \in M^n$.*
(2) *Every R-linear map $\gamma : R^n \to {}_RM$ is matrix multiplication $\gamma = \cdot\underline{m}$ by some column $\underline{m} \in M_n$.*

Proof. Given γ as in (1), let $\gamma(\underline{e}_i) = m_i$ for $1 \leq i \leq n$. If we write $\underline{r} = [r_1 \cdots r_n]^T$ then $\gamma(\underline{r}) = \gamma(\Sigma \underline{e}_i r_i) = \Sigma m_i r_i = \bar{m}\underline{r}$, where $\bar{m} = [m_1 \cdots m_n]$. Thus $\gamma = \bar{m}\cdot$, proving (1). The proof of (2) is analogous. □

Results about FP-injectivity are scattered throughout the literature, and many have homological proofs. The following theorem gives elementary proofs of several properties of FP-injective modules that will be needed later. Recall that an R-module M_R is said to be *finitely presented* if $M \cong F/K$, where F_R is free and both F and K are finitely generated.

Theorem 5.39. *The following are equivalent for a right module Q_R:*

(1) Q_R *is FP-injective.*
(2) *If* $K_R \subseteq R_n$ *is finitely generated, then every* R*-linear map* $K \to Q_R$ *extends to* R_n.
(3) *If* $\bar{q} \in Q^n$ *and* $A \in M_n(R)$ *satisfy* $\mathrm{r}_{R_n}(A) \subseteq \mathrm{r}_{R_n}(\bar{q})$*, then* $\bar{q} = \bar{x}A$ *for some* $\bar{x} \in Q^n$.
(4) *If* $\bar{q} \in Q^n$ *and* $A \in M_{m\times n}(R)$ *satisfy* $\mathrm{r}_{R_n}(A) \subseteq \mathrm{r}_{R_n}(\bar{q})$*, then* $\bar{q} = \bar{x}A$ *for some* $\bar{x} \in Q^m$.
(5) *If* $L \subseteq M_R$ *and* M/L *is finitely presented, then every* R*-linear map* $L \to Q$ *extends to* M.

Proof. (1)⇔(2). This is clear.

(2)⇒(3). Given $\mathrm{r}_{R_n}(A) \subseteq \mathrm{r}_{R_n}(\bar{q})$, let $\underline{c}_j$ denote column j of A. Then $AR_n = \Sigma_j \underline{c}_j R$ is a finitely generated R-submodule of R_n, and our hypothesis shows that $\alpha : AR_n \to Q_R$ is well defined by $\alpha(A\underline{r}) = \bar{q}\underline{r}$. So (2) gives $\alpha = \bar{x}\cdot$ for some $\bar{x} \in Q^n$. If we write $\bar{q} = [q_1 \cdots q_n] \in Q^n$, we get

$$q_j = \bar{q}\underline{e}_j = \alpha(A\underline{e}_j) = \alpha(\underline{c}_j) = \bar{x}\underline{c}_j.$$

Hence, $\bar{q} = [\bar{x}\underline{c}_1 \cdots \bar{x}\underline{c}_n] = \bar{x}[\underline{c}_1 \cdots \underline{c}_n] = \bar{x}A$, as required.

(3)⇒(4). Let $\mathrm{r}_{R_n}(A) \subseteq \mathrm{r}_{R_n}(\bar{q})$, where A is $m \times n$. If $m = n$ there is nothing to prove. If $m < n$, let $A' = \left[\begin{smallmatrix} A \\ 0 \end{smallmatrix}\right]$ be $n \times n$. Then $\mathrm{r}_{R_n}(A') = \mathrm{r}_{R_n}(A) \subseteq \mathrm{r}_{R_n}(\bar{q})$, so (3) gives $\bar{q} = \bar{y}A' = [\bar{x} \quad \bar{z}]\left[\begin{smallmatrix} A \\ 0 \end{smallmatrix}\right] = \bar{x}A$ for some $\bar{y} = [\bar{x} \quad \bar{z}] \in Q^n$, $\bar{x} \in Q^m$. However, if $m > n$ we let $A' = [A \quad 0]$ be $m \times m$. Then

$$\begin{aligned} \mathrm{r}_{R_m}(A') &= \left\{ \begin{bmatrix} \underline{r} \\ \underline{s} \end{bmatrix} \mid \underline{r} \in R_n,\ \underline{s} \in R_{m-n},\ A\underline{r} = 0 \right\} \\ &= \begin{bmatrix} \mathrm{r}_{R_n}(A) \\ R_{m-n} \end{bmatrix} \subseteq \begin{bmatrix} \mathrm{r}_{R_n}(\bar{q}) \\ R_{m-n} \end{bmatrix} = \mathrm{r}_{R_m}[\bar{q} \quad 0]. \end{aligned}$$

By (3) there exists $\bar{x} \in Q^n$ such that $[\bar{q} \quad 0] = \bar{x}A' = [\bar{x}A \quad 0]$. Thus $\bar{q} = \bar{x}A$.

(4)⇒(2). Let $K = \Sigma_{j=1}^m \underline{c}_j R \subseteq R_n$ and let $\gamma : K \to Q_R$ be R-linear. We must extend γ to R_n. Write $\gamma(\underline{c}_j) = q_j \in Q$, $\bar{q} = [q_1 \cdots q_m] \in Q^m$, and $A = [\underline{c}_1 \cdots \underline{c}_m] \in M_{n\times m}(R)$. Then $\gamma(A\underline{r}) = \bar{q}\underline{r}$ for all $\underline{r} \in R_m$, so $\mathrm{r}_{R_m}(A) \subseteq \mathrm{r}_{R_m}(\bar{q})$. Thus (4) gives $\bar{x} \in Q^n$ such that $\bar{q} = \bar{x}A$. Hence $q_j = \bar{x}\underline{c}_j$ for all j; that is, $\gamma(\underline{c}_j) = \bar{x}\underline{c}_j$ for each j. Hence $\gamma = \bar{x}\cdot$, as required.

(1)⇔(5). Clearly (5)⇒(1), so assume (1) and suppose $\alpha : L \to Q$ is R-linear as in (5). Let $\sigma : F/K \to M/L$ be an isomorphism where F is free and both F and K are finitely generated, and let $\pi : F \to F/K$ and $\theta : M \to M/L$ be the coset maps. As F is projective there exists $\beta : F \to M$ such that $\theta \circ \beta = \sigma \circ \pi$. If $m \in M$ then (since $\sigma \circ \pi$ is onto) we have $\theta(m) = \sigma \circ \pi(f) = \theta \circ \beta(f)$ for some $f \in F$. Thus $m - \beta(f) \in ker\,\theta = L$, so $M = \beta(F) + L$. Moreover,

$$\begin{array}{ccc} & & F \\ & \beta & \downarrow \pi \\ & & \swarrow\ F/K \\ & & \downarrow \sigma \\ & M & \overset{\theta}{\to}\ M/L \to 0 \end{array}$$

$\theta \circ \beta(K) = \sigma \circ \pi(K) = 0$ because $K = ker\,\pi$, so $\beta(K) \subseteq ker\,\theta = L$. It follows that $\alpha \circ \beta : K \to Q$ is defined and so, by (1), $\alpha \circ \beta$ extends to $\delta : F \to Q$. As $M = \beta(F) + L$, define $\hat{\alpha} : M \to Q$ by $\hat{\alpha}[\beta(f) + l] = \delta(f) + \alpha(l)$. It suffices to show that δ is well defined (since then $\hat{\alpha}$ extends α). But if $\beta(f) + l = 0$ then $\beta(f) \in L = ker\,\theta$, so $0 = \theta \circ \beta(f) = \sigma \circ \pi(f)$. Since σ is one-to-one, this gives $\pi(f) = 0$, so $f \in ker\,\pi = K$. Thus $\delta(f) = \alpha \circ \beta(f) = \alpha(-l) = -\alpha(l)$, as required. This proves (5). □

It follows easily from (2) in Theorem 5.39 that a direct sum $Q = \oplus_{i \in I} Q_i$ is FP-injective if and only if each Q_i is FP-injective. It also follows from the definition that the union of an ascending chain of FP-injective submodules of some module is again FP-injective. Since 0 is FP-injective, this means that every module contains maximal FP-injective submodules.

5.6. FP-Injective Rings

Our interest is in the right *FP-injective rings,* that is, the rings R for which R_R is an FP-injective module. Examples include regular and right self-injective rings. The right FP-injective rings turn out to be exactly the rings R that have the property that $M_n(R)$ is right P-injective for every $n \geq 1$. Before proving this, we state a useful annihilator lemma giving conditions that a submodule of a bimodule is an annihilator.

Lemma 5.40. *Given a bimodule ${}_RM_S$, the following conditions are equivalent for ${}_RK \subseteq {}_RM$:*

(1) $\mathtt{r}_S(K) \subseteq \mathtt{r}_S(m)$, $m \in M$, implies that $m \in K$.
(2) $\mathtt{l}_M\mathtt{r}_S(K) = K$.
(3) $K = \mathtt{l}_M(X)$ for some subset $X \subseteq S$.

Proof. (1)⇒(2). If $m \in \mathtt{l}_M\mathtt{r}_S(K)$ then $m \cdot \mathtt{r}_S(K) = 0$, so $\mathtt{r}_S(K) \subseteq \mathtt{r}_S(m)$. Hence (1) implies that $\mathtt{l}_M\mathtt{r}_S(K) \subseteq K$; the reverse inclusion always holds.

(2)⇒(3). This is clear.

(3)⇒(1). If $\mathtt{r}_S(K) \subseteq \mathtt{r}_S(m)$ then $\mathtt{l}_M\mathtt{r}_S(m) \subseteq \mathtt{l}_M\mathtt{r}_S(K)$. This gives (1) because $m \in \mathtt{l}_M\mathtt{r}_S(m)$ and $\mathtt{l}_M\mathtt{r}_S(K) = K$ by (3). □

The right FP-injective rings admit a number of characterizations in addition to those in Theorem 5.39.

Theorem 5.41. *The following are equivalent for a ring R:*

(1) R is right FP-injective.
(2) If $\bar{a}_1, \bar{a}_2, \ldots, \bar{a}_m$ and $\bar{b}$ in R^n satisfy $\cap_i \mathtt{r}_{R_n}(\bar{a}_i) \subseteq \mathtt{r}_{R_n}(\bar{b})$, then $\bar{b} \in \Sigma_i R\bar{a}_i$.
(3) If $n \geq 1$ and ${}_RK \subseteq R^n$ is finitely generated, then $K = \mathtt{l}_{R^n}(X)$ for some set $X \subseteq M_n(R)$.
(4) $M_n(R)$ is a right P-injective ring for each $n \geq 1$.

Proof. (1)⇒(2). Given the situation in (2), let $A \in M_{m\times n}(R)$ be the matrix with the $\bar{a}_i$ as its rows. Then $\mathtt{r}_{R_n}(A) = \cap_i \mathtt{r}_{R_n}(\bar{a}_i) \subseteq \mathtt{r}_{R_n}(\bar{b})$ by hypothesis, so $\bar{b} = \bar{x}A$ for some $\bar{x} \in R^m$ by (1) using Theorem 5.39. If $\bar{x} = [x_1\ x_2 \cdots x_m]$ then $\bar{b} = \Sigma_i x_i \bar{a}_i \in \Sigma_i R\bar{a}_i$, as required.

(2)⇒(3). Let $K = R\bar{a}_1 + \cdots + R\bar{a}_m \subseteq R^n$ as in (3). Then $\mathtt{r}_{M_n(R)}(K) = \cap_i \mathtt{r}_{M_n(R)}(\bar{a}_i)$, so (2) asserts that $\mathtt{r}_{M_n(R)}(K) \subseteq \mathtt{r}_{M_n(R)}(\bar{b})$ implies $\bar{b} \in K$. Now (3) follows from Lemma 5.40 with $M = {}_RR^n{}_{M_n(R)}$.

(3)⇒(4). Given $A \in M_n(R)$, let $\bar{a}_i$ denote row i of A and write $K = R\bar{a}_1 + \cdots + R\bar{a}_n \subseteq R^n$. Then (3) gives $K = \mathtt{l}_{R^n}(X)$ for some set $X \subseteq M_n(R)$. But $M_n(R)\,A = \begin{bmatrix} K \\ \vdots \\ K \end{bmatrix} = \begin{bmatrix} \mathtt{l}_{R^n}(X) \\ \vdots \\ \mathtt{l}_{R^n}(X) \end{bmatrix} = \mathtt{l}_{M_n(R)}(X)$, so (4) follows by Lemma 5.1.

(4)⇒(1). Suppose $\mathtt{r}_{R_n}(A) \subseteq \mathtt{r}_{R_n}(\bar{b})$, where $\bar{b} \in R^n$. If B is the matrix with every row equal to $\bar{b}$, then $\mathtt{r}_{R_n}(A) \subseteq \mathtt{r}_{R_n}(B)$, so (4) gives $B = XA$ for some matrix $X \in M_n(R)$. Hence $\bar{b} = \bar{x}A$, where $\bar{x}$ denotes the first row of X. Thus R_R is FP-injective by Theorem 5.39, proving (1). □

If $A \in M_{m\times n}(R)$ and $\bar{b} \in R^m$, the system of linear equations $\bar{x}A = \bar{b}$ over R is called *consistent* if $A\underline{r} = 0$, $\underline{r} \in R_n$, implies $\bar{b}\underline{r} = 0$; that is, $\mathtt{r}_{R_m}(A) \subseteq \mathtt{r}_{R_m}(\bar{b})$. Hence we can restate (2) in Theorem 5.41 in the following way: R is right FP-injective if and only if every consistent system $\bar{x}A = \bar{b}$ over R has a solution $\bar{x}$.

Corollary 5.42. *Being right FP-injective is a Morita invariant.*

Proof. This follows from condition (4) of Theorem 5.41 and Proposition 5.35. (It also follows from Corollary 5.44 below.) □

Clearly right FP-injective rings are all right P-injective. We record some other properties for reference.

Corollary 5.43. *If R is right FP-injective then:*

(1) R is right F-injective.
(2) $\mathtt{lr}(L) = L$ for all finitely generated left ideals L of R.

Proof. (2) is clear from Theorem 5.41, as is (1) using Proposition 5.36. □

Recall that a module M is called *torsionless* if, given $0 \neq m \in M$, there exists $\lambda \in hom(M, R)$ such that $\lambda(m) \neq 0$, equivalently if M can be embedded in a direct product of copies of R. Using Theorem 5.41 we can now give an elementary proof of the following result.

Corollary 5.44. *A ring R is right FP-injective if and only if every finitely presented left R-module is torsionless.*

Proof. Assume that R is right FP-injective. If ${}_RK$ is a finitely generated submodule of R^n, say $K = R\bar{a}_1 + \cdots + R\bar{a}_m$, we must embed $R^n/K \hookrightarrow ({}_RR)^I$ for some set I. Equivalently, if $\bar{b} \in R^n, \bar{b} \notin K$, we must find $\gamma : R^n \to {}_RR$ such that $K\gamma = 0$ and $\bar{b}\gamma \neq 0$. Suppose no such γ exists so that $K\gamma = 0$ implies $\bar{b}\gamma = 0$. Using Lemma 5.38, this means $\cap_{i=1}^m \mathtt{r}_{R_n}(\bar{a}_i) = \mathtt{r}_{R_n}(K) \subseteq \mathtt{r}_{R_n}(\bar{b})$. But then $\bar{b} \in K$ by Theorem 5.41(2), contrary to hypothesis.

Conversely, suppose $\cap_i \mathtt{r}_{R_n}(\bar{a}_i) \subseteq \mathtt{r}_{R_n}(\bar{b})$, where $\bar{a}_i$ and $\bar{b}$ are in R^n, and write $K = \Sigma_i R\bar{a}_i$. If $\bar{b} \notin K$ then the condition gives $\gamma : R^n \to {}_RR$ such that $K\gamma = 0$ and $\bar{b}\gamma \neq 0$. But $\gamma = \cdot \underline{c}$ for some $\underline{c} \in R_n$ by Lemma 5.38, and so $\underline{c} \in \mathtt{r}_{R_n}(K) = \cap_i \mathtt{r}_{R_n}(\bar{a}_i) \subseteq \mathtt{r}_{R_n}(\bar{b})$. Hence $0 = \bar{b}\underline{c} = \bar{b}\gamma$, which is a contradiction. Thus R is right FP-injective by Theorem 5.41. □

Example 5.45. The Camillo example (Example 2.6) yields a commutative, local, semiprimary FP-injective ring with simple essential socle, that is not self-injective and not noetherian.

Proof. Let $R = span_F\{1, m, x_1, x_2, \ldots\}$ denote the Camillo example, where F is any field and the x_i are commuting indeterminants satisfying the relations $x_i^3 = 0$ for all i, $x_i x_j = 0$ for all $i \neq j$, and $x_i^2 = m$ for all i. We must prove that R is FP-injective but not self-injective. To prove FP-injectivity, let $\gamma : K \to R$ be R-linear, where K is a finitely generated submodule of R_n; we must show that γ extends to $R_n \to R$. Let $K = \Sigma_{j=1}^m \underline{b}_j R$, where $\underline{b}_j = [b_{1j}\ b_{2j} \cdots b_{nj}]^T \in R_n$. Then the elements $\{b_{ij}, \gamma(\underline{b}_j)\}$ are all contained in the subring $S = span_F\{1, m, x_1, \ldots, x_k\}$ of R for some k. This ring S is clearly artinian, and the argument in Example 2.6 shows that S is mininjective. Hence S is quasi-Frobenius by Ikeda's theorem (Theorem 2.30). If we write $K_0 = \Sigma_{i=1}^m \underline{b}_i S$, this means that the restriction $\gamma : K_0 \to S$ extends to $\widehat{\gamma} : S_n \to S$. Thus there exists $\bar{c} \in S^n$ such that $\widehat{\gamma} = \bar{c}\cdot$ is (matrix) multiplication by the row $\bar{c}$. But then $\bar{c}\cdot : R_n \to R$ extends γ, as required. Hence R is FP-injective.

Finally, if $F = \mathbb{Z}_2$ we show that R is not self-injective.[2] Define $\gamma : J \to R$ by $\gamma(a) = a^2$ for all $a \in J$. Then γ is $\mathbb{Z}$-linear because $char(R) = 2$. Moreover, $\gamma(ar) = a^2r^2 = a^2r = \gamma(a)r$ because $J^3 = 0$ and $r^2 - r \in J$ (as $R/J \cong \mathbb{Z}_2$). Hence γ is R-linear and $\gamma(J) = \mathbb{Z}_2 m$ is simple. If R is self-injective, then $\gamma = c\cdot$ for some $c \in R$. It follows that $(c - a)a = 0$ for all $a \in J$. Hence $c \in J$ (otherwise $c - a$ is a unit), say $c = \varepsilon m + \Sigma_{i=1}^n \varepsilon_i x_i$, where $\varepsilon, \varepsilon_i \in \mathbb{Z}_2$. But then $m = x_{n+1}^2 = \gamma(x_{n+1}) = cx_{n+1} = 0$, which is a contradiction. Hence R is not self-injective. □

FP-injectivity is not a left–right symmetric property of rings.

Example 5.46. There exists a left FP-injective ring that is not right FP-injective.

Proof. A ring R is called a right *IF-ring* if every injective right R-module is flat. These rings have been studied by Colby, who showed [41, Theorem 1] that R is a right IF-ring if and only if every finitely presented right R-module is isomorphic to a submodule of a free module. In particular, every right IF-ring is left FP-injective by Corollary 5.44. Colby [41, Example 2] considers the algebra R over a field F with basis $\{1\} \cup \{e_i \mid i \geq 0\} \cup \{x_i \mid i \geq 1\}$ such that 1 is the unity of R and, for all i and j, $e_i e_j = \delta_{ij} e_j$, $x_i e_j = \delta_{i,j+1} x_i$, $e_i x_j = \delta_{ij} x_j$, and $x_i x_j = 0$. He shows that R is a right IF-ring, and so R is left FP-injective. But he also verifies that the R-linear map $x_1 R \to e_0 R$ with $x_1 \mapsto e_0$ cannot be extended to R. Hence R is not right P-injective and so is certainly not right FP-injective. □

We note in passing that in [41, Example 3] Colby also shows that every Bezout domain is FP-injective, and he proves [41, Example 1] that the trivial extension $R = T(\mathbb{Z}, \mathbb{Q}/\mathbb{Z})$ is a commutative FP-injective ring R for which $R/J \cong \mathbb{Z}$ is not regular.

The fact that right FP-injectivity is a Morita invariant (Corollary 5.42) enables us to prove a remarkable symmetry property of this condition. We remind the reader that "right Kasch" is a Morita invariant (a ring R is right Kasch if and only if every semisimple right R-module embeds in a projective right module).

[2] In fact R is not self-injective for any field F. Since R is commutative and semiprimary but not artinian, it is not self-injective by Theorem 6.39, to be proved in Chapter 6.

Theorem 5.47. *Suppose that R is right Kasch and right FP-injective. Then R is left FP-injective.*

Proof. Write $S = M_n(R)$; by Theorem 5.41 it is enough to show that S is left P-injective for any $n \geq 1$. But Corollary 5.42 shows that S is right FP-injective and hence (by Corollary 5.43) is right 2-injective, and that S is right Kasch. Thus S is left P-injective by Lemma 5.21. □

The Kasch hypothesis is essential in Theorem 5.47 by Example 5.46. Note that any nonartinian regular ring (for example an infinite direct product of fields) is left and right FP-injective, but it is neither left nor right Kasch [because a regular ring that is left (right) Kasch is semisimple].

The next result collects some facts about right Kasch, right FP-injective rings that will be needed later.

Proposition 5.48. *Let R be right Kasch and right FP-injective. Then the following hold:*

(1) $\mathtt{rl}(T) = T$ *and* $\mathtt{lr}(L) = L$ *if T and L are finitely generated right, respectively left, ideals of R.*
(2) *A left ideal $K \neq 0$ is minimal if and only if $\mathtt{r}(K)$ is maximal.*
(3) *If a right ideal $T \neq R$ is maximal then $\mathtt{l}(T)$ is minimal.*
(4) $S_l = S_r \subseteq^{ess} {}_RR$, *and* $J = Z_l = Z_r$.
(5) *If $\mathtt{l}(T) = \mathtt{l}(S)$, where T and S are right ideals, with T finitely generated, then $T = S$.*
(6) *If ${}_RL$ is a finitely generated left ideal, then $\mathtt{r}(L)$ is small in R_R if and only if $L \subseteq^{ess} {}_RR$.*
(7) *If T_R is a finitely generated right ideal, and $\mathtt{l}(T)$ is small in ${}_RR$, then $T \subseteq^{ess} R_R$.*

Proof. Observe first that R is left FP-injective by Theorem 5.47.

(1). This follows from Corollary 5.43.

(2) and (3). Because of (1), (3) and the forward implication in (2) follow from Theorem 2.32. If $\mathtt{r}(K)$ is maximal and $0 \neq k \in K$, then $\mathtt{r}(K) \subseteq \mathtt{r}(k)$, whence $\mathtt{r}(K) = \mathtt{r}(k)$. Thus kR is minimal, whence Rk is minimal (because R is right mininjective). But $K \subseteq \mathtt{lr}(K) = \mathtt{lr}(k) = Rk$ by (1), and so $K = Rk$ is a minimal left ideal. This completes the proof of (2).

(4). We have $S_r = S_l$ by Theorem 2.21 because R is left and right mininjective. If $0 \neq a \in R$, let $\mathtt{r}(a) \subseteq T \subseteq^{max} R_R$. Then (1) gives $Ra = \mathtt{lr}(a) \supseteq \mathtt{l}(T)$, so $S_l \subseteq^{ess} {}_RR$ by (3). Since R is left and right P-injective, the rest is by Theorem 5.14.

(5). First $S \subseteq \mathtt{rl}(S) = \mathtt{rl}(T) = T$ by (1). If $S \subset T$ let $S \subseteq M \subseteq^{max} T$ (as T is finitely generated), and let $\sigma : T/M \to R$ be monic. If $\alpha : T \to R$ is defined by $\alpha(t) = \sigma(t + M)$ then $\alpha = a\cdot$ is left multiplication by $a \in R$ by Corollary 5.43. Thus $aS = \alpha(S) = 0$ because $S \subseteq M$, so $a \in \mathtt{l}(S) = \mathtt{l}(T)$. But $aT = \alpha(T) \neq 0$ because $T \nsubseteq M$. This is the desired contradiction.

(6). If $\mathtt{r}(L)$ is small, let $L \cap Rx = 0$, $x \in R$. Since R is right F-injective (by Corollary 5.43) and L is finitely generated, the Ikeda–Nakayama lemma (Lemma 1.37) gives $R = \mathtt{r}(L \cap Rx) = \mathtt{r}(L) + \mathtt{r}(x)$. Hence $\mathtt{r}(x) = R$, and so $x = 0$. This shows that $L \subseteq^{ess} R_R$. Conversely, if $\mathtt{r}(L) + T = R$ then

$$0 = \mathtt{l}[\mathtt{r}(L) + T] = \mathtt{lr}(L) \cap \mathtt{l}(T) \supseteq L \cap \mathtt{l}(T).$$

Hence $\mathtt{l}(T) = 0$ by hypothesis, so $T = R$ by the right Kasch condition.

(7). Let $T \cap xR = 0$, where $x \in R$ and T is a finitely generated right ideal. As in the proof of (6), the Ikeda–Nakayama lemma gives $\mathtt{l}(T) + \mathtt{l}(x) = R$. Thus $\mathtt{l}(x) = R$ by hypothesis, so $x = 0$. □

In connection with (2) and (3) of Proposition 5.48, note that Theorem 2.32 implies that if R is right Kasch and right FP-injective then the maps $K \mapsto \mathtt{r}(K)$ and $T \mapsto \mathtt{l}(T)$ are mutually inverse bijections between the minimal left ideals K of R and the maximal right ideals T of R.

The main application of the symmetry property in Theorem 5.47 is to prove Theorem 5.56 in Section 5.8, which shows that the condition that a ring is semiperfect, right Kasch, and right FP-injective is left–right symmetry. Surprisingly, the ring is only required to be semilocal, so we digress briefly to investigate semilocal rings.

5.7. Semilocal Mininjective Rings

In Chapter 3 we studied semiperfect, right mininjective rings with essential right socle as a natural generalization of the right pseudo-Frobenius rings. In this section we examine the situation when the ring is only assumed to be semilocal, derive several new results about semilocal rings, and apply them to obtain some finite dimensionality conditions for right mininjective rings that will be needed in Section 5.8. Recall that a ring R is called *semilocal* if R/J is semisimple. We begin with the following lemma:

Lemma 5.49. *A semilocal, right and left mininjective ring is right Kasch if and only if it is left Kasch.*

Proof. If R is a semilocal ring, let n denote the number of isomorphism classes of simple right (or left) R-modules. If R is right Kasch, let $\{k_1R, \dots, k_nR\}$, $k_i \in$

R, be a system of distinct representatives of the simple right R-modules. Then Theorem 2.21 shows that each Rk_i is simple (because R is right mininjective), and it suffices to show that $\{Rk_1, \dots, Rk_n\}$ consists of pairwise nonisomorphic right ideals. But $Rk_i \cong Rk_j$ implies $k_iR \cong k_jR$ because R is left mininjective (again by Theorem 2.21), and so $i = j$ as required. □

Note that, in the notation of Lemma 5.49, the proof shows that $\{Rk_1, \dots, Rk_n\}$ is a complete system of representatives of the simple left modules if and only if $\{k_1R, \dots, k_nR\}$ is a complete system of representatives of the simple right modules.

Recall Lemma 1.36, which states that if T and T' are right ideals in a ring R such that R-linear maps $T + T' \to R$ extend to $R \to R$, then

$$\mathtt{l}(T \cap T') = \mathtt{l}(T) + \mathtt{l}(T').$$

We are going to use this several times, often in the case $T + T' = R$. The first application develops a description of S_r in an arbitrary semilocal ring that is of interest in its own right.

Proposition 5.50. *Let R be a semilocal ring. Then either $S_r = 0$ or we have $S_r = \mathtt{lr}(k_1) + \cdots + \mathtt{lr}(k_n)$, where each k_iR is a simple right ideal.*

Proof. Since R is semilocal $S_r = \mathtt{l}(J)$ and $J = T_1 \cap \cdots \cap T_n$, where each T_i is a maximal right ideal. Assume n is minimal, so that no T_i contains the intersection of any of the others. In particular $T_1 + \cap_{i \neq 1} T_i = R$, so Lemma 1.36 gives $S_r = \mathtt{l}(J) = \mathtt{l}(T_1) + \mathtt{l}(\cap_{i \neq 1} T_i)$. But $\cap_{i \neq 1} T_i = T_2 \cap (\cap_{i \neq 1,2} T_i)$, so

$$S_r = \mathtt{l}(T_1) + \mathtt{l}(\cap_{i \neq 1} T_i) = \mathtt{l}(T_1) + [\mathtt{l}(T_2) + \mathtt{l}(\cap_{i \neq 1,2} T_i)]$$

as before. Continuing we obtain $S_r = \mathtt{l}(T_1) + \cdots + \mathtt{l}(T_n)$. If $S_r \neq 0$ we may assume that $\mathtt{l}(T_i) \neq 0$ for each i. In this case, choose $0 \neq k_i \in \mathtt{l}(T_i)$. Then $T_i \subseteq \mathtt{r}(k_i) \neq R$, so $T_i = \mathtt{r}(k_i)$ and $k_iR \cong R/T_i$ is simple. Hence $\mathtt{l}(T_i) = \mathtt{lr}(k_i)$ for each i, as required. □

Note that the ring $\mathbb{Z}_{(p)}$ shows that S_r can be zero even if R is local, commutative, and noetherian.

We can ensure that $S_r \neq 0$ by insisting that R is right Kasch, and the result is a characterization of semilocal rings in terms of simple right ideals.

Theorem 5.51. *Assume that a ring R satisfies $\mathtt{r}(S_r) = J$ (for example if R is right Kasch). Then R is semilocal if and only if $S_r = \mathtt{lr}(k_1) + \cdots + \mathtt{lr}(k_n)$, where k_iR is a simple right ideal of R for each i.*

Proof. Assume that $S_r = \mathtt{lr}(k_1) + \cdots + \mathtt{lr}(k_n)$ as given. Then we obtain

$$J = \mathtt{r}(S_r) = \mathtt{r}\left[\Sigma_{i=1}^n \mathtt{lr}(k_i)\right] = \cap_{i=1}^n \mathtt{rlr}(k_i) = \cap_{i=1}^n \mathtt{r}(k_i).$$

Hence R is semilocal because $\mathtt{r}(k_i)$ is maximal for each i. Since $S_r \neq 0$ [because $\mathtt{r}(S_r) = J$], the converse follows by Proposition 5.50. □

Theorem 5.52. *If R is semilocal and right mininjective then S_r is finitely generated and semisimple (possibly 0) as a left R-module.*

Proof. If $S_r = 0$ there is nothing to prove. Otherwise, Proposition 5.50 gives $S_r = \mathtt{lr}(k_1) + \cdots + \mathtt{lr}(k_n)$, where $k_i R$ is a simple right ideal of R for each i. Hence $S_r = Rk_1 + \cdots + Rk_n$ by right mininjectivity, and each Rk_i is simple by Theorem 2.21, as required. □

Corollary 5.53. *If R is a semilocal, right and left mininjective ring, then $S_r = S_l$ is finitely generated as a left and as a right R-module.*

Proof. $S_r = S_l$ because R is right and left mininjective (Theorem 2.21), so the result follows from Theorem 5.52. □

In Corollary 5.53 the assumption that R is left mininjective gives $S_l \subseteq S_r$. This outcome also follows from the requirement that $S_r \subseteq^{ess} {}_RR$. If we add this to the hypotheses in Theorem 5.52, we obtain

Proposition 5.54. *Suppose R is a semilocal, right mininjective ring in which $S_r \subseteq^{ess} {}_RR$. Then $S_r = S_l$ is finitely generated and essential as a left ideal. In particular, R is left finite dimensional.*

Proof. We have $S_l \subseteq S_r$ because $S_r \subseteq^{ess} {}_RR$, and $S_r \subseteq S_l$ because R is right mininjective. Hence $S_r = S_l$; it is finitely generated on the left by Theorem 5.52, and it is essential in ${}_RR$ by hypothesis. □

5.8. FP Rings

We now return to the right–left symmetry of right Kasch, right FP-injective rings.

Proposition 5.55. *If R is left finite dimensional and right Kasch, then R is semilocal.*

Proof. By the Camps–Dicks theorem (Corollary C.3) it suffices to show that any monomorphism $\sigma : {}_RR \to {}_RR$ is epic. Observe first that R is a left C2 ring by Proposition 1.46, so $\sigma(R) \subseteq^{\oplus} {}_RR$ because σ is monic, say $R = \sigma(R) \oplus K$. Apply σ again to get $R = \sigma^2(R) \oplus \sigma(K) \oplus K$. If $K \neq 0$, this process can continue indefinitely because ${}_RR$ is finite dimensional, so $K = 0$. □

We now come to the main symmetry theorem for right Kasch, right FP-injective rings. Surprisingly, the ring is only assumed to be semilocal.

Theorem 5.56. *The following conditions are equivalent for a ring R:*

(1) R is semilocal, right FP-injective, and right Kasch.
(2) R is semilocal, left FP-injective, and left Kasch.
(3) R is semilocal and right FP-injective and $J = \mathtt{r}\{k_1, \ldots, k_n\}$, where $\{k_1, \ldots, k_n\} \subseteq R$.
(4) R is semilocal and left FP-injective and $J = \mathtt{l}\{m_1, \ldots, m_n\}$, where $\{m_1, \ldots, m_n\} \subseteq R$.
(5) R is right finite dimensional, right FP-injective, and right Kasch.
(6) R is left finite dimensional, left FP-injective, and left Kasch.
(7) R is left finite dimensional, right FP-injective, and right Kasch.
(8) R is right finite dimensional, left FP-injective, and left Kasch.

Moreover, the elements k_i and m_i in (3) and (4) can be chosen so that each of Rk_i, k_iR, Rm_i, and m_iR is simple.

Proof. (1)⇔(2). If R satisfies (1), it is left FP-injective by Theorem 5.47. Thus R is right and left mininjective, right Kasch, and semilocal, so it is left Kasch by Lemma 5.49. Hence (1)⇒(2); the converse is analogous.

(1)⇔(6). Given (1) we have $S_l = S_r \subseteq^{ess} {}_RR$ by Proposition 5.48. Since S_r is finite dimensional as a left R-module by Theorem 5.52, it follows that R is left finite dimensional. This proves (6) because we have proved (1)⇔(2). Conversely, (6)⇒(2) by Theorem 5.25.

(1)⇔(3). Given (1), we have $J = \mathtt{r}(S_r)$ because R is right Kasch. But the proof of Theorem 5.52 shows that $S_r = Rk_1 + \cdots + Rk_n$, $k_i \in R$, with Rk_i simple. Hence $J = \mathtt{r}(S_r) = \mathtt{r}\{k_1, \ldots, k_n\}$, proving (3). Conversely, let K be a simple right R-module; we must show that K embeds in R_R. As R/J is semisimple, K embeds in R/J as a right R/J-module and hence as a right R-module. But (3) implies that R/J is R-embedded in $(R_R)^n$. Thus K embeds in $(R_R)^n$ and hence in R.

(1)⇔(7). We have (1)⇒(7) because we have already proved (1)⇒(6), and (7)⇒(1) by Proposition 5.55.

(2)⇔(4), (1)⇔(5), and (2)⇔(8). These are analogues of (1)⇔(3), (2)⇔(6), and (1)⇔(7), respectively.

Finally, the k_i in the proof of (1)⇒(3) were chosen with $k_i R$ simple, and then Rk_i is simple because R is right mininjective. A similar argument works for Rm_i and $m_i R$. □

Remark. The proof of (3)⇒(1) in Theorem 5.56 shows that if R is semilocal and $J = \mathrm{r}\{k_1, \ldots, k_n\}$ then R is right Kasch.

Since "right Kasch" and "semilocal" are Morita invariants, the rings identified in Theorem 5.56 form a Morita invariant class by Corollary 5.42.

All right PF rings are semiperfect by Theorem 1.56, and this fact is used frequently in the literature. Hence we restrict our attention to the semiperfect rings that satisfy the conditions in Theorem 5.56. Accordingly, a ring R is called an *FP ring* if it is semiperfect and satisfies any of the equivalent conditions in Theorem 5.56. These FP rings admit additional characterizations that show (using Theorem 1.56) that they are simultaneously a natural generalization of both the right and left PF rings.

Theorem 5.57. *The following conditions are equivalent for a ring R:*

(1) R is an FP ring.
(2) R is semiperfect and right FP-injective and $S_r \subseteq^{ess} R_R$.
(3) R is semiperfect and left FP-injective and $S_l \subseteq^{ess} {}_R R$.

Proof. (1)⇒(2) is clear from Part (4) of Proposition 5.48. Given (2), R is right P-injective and so R is right (and left) Kasch by Theorem 5.31. Hence (1)⇔(2); (1)⇔(3) is analogous. □

Since "semiperfect," "right Kasch," and "right FP-injective" are all Morita invariant properties, we have

Theorem 5.58. *The class of FP rings forms a Morita invariant class.*

Recall that if M is a module, the socle series $soc_1(M) \subseteq soc_2(M) \subseteq \cdots$ are submodules defined by setting $soc_1(M) = soc(M)$ and, if $soc_n(M)$ has been specified, by defining $soc_{n+1}(M)$ by

$$soc_{n+1}(M)/soc_n(M) = soc[M/soc_n(M)].$$

Recall further that a module is said to be finitely cogenerated if it has a finitely generated, essential socle.

Proposition 5.59. *If R is any FP ring, write $S = S_r = S_l$. Then the following hold:*

(1) $soc(eR) = eS$ *and* $soc(Re) = Se$ *are simple for all local* $e^2 = e \in R$.
(2) *R is left and right finitely cogenerated.*
(3) *If $\{e_1, \ldots, e_n\}$ is a basic set of local idempotents in R, there exist elements $k_1, \ldots, k_n$ in R and a permutation σ of $\{1, \ldots, n\}$ such that the following hold for all $i = 1, \ldots, n$:*
   (a) $Sk_i = soc(Re_{\sigma i}) \cong Re_i/Je_i$ *and* $k_iS = soc(e_iR) \cong e_{\sigma i}R/e_{\sigma i}J$.
   (b) $\{k_1R, \ldots, k_nR\}$ *and* $\{Rk_1, \ldots, Rk_n\}$ *are sets of distinct representatives of the simple right and left R-modules, respectively.*
(4) $soc_n(R_R) = soc_n({}_RR) = \mathtt{l}(J^n) = \mathtt{r}(J^n)$ *for all* $n \geq 1$.

Proof. Because R is two-sided minfull, (1), (2), and (3) follow from Proposition 3.17 and the fact that S is essential in both R_R and ${}_RR$. Finally, since R is semiperfect with $S_r = S_l$, (4) is Lemma 3.36. □

A module M is called *finitely continuous* if it satisfies the C2-condition and every finitely generated submodule is essential in a direct summand of M; a ring R is called *right finitely continuous* if R_R is a finitely continuous module.

Proposition 5.60. *Every FP ring is left and right finitely continuous.*

Proof. Every finitely generated right (or left) ideal is essential in a direct summand by Corollary 5.43 and Lemma 4.2. Moreover, R satisfies the right and left C2-conditions by Proposition 5.10. Hence R is right and left finitely continuous. □

We can now give more characterizations of the FP rings in terms of the min-CS condition introduced in Chapter 4. Recall that a ring R is called right finitely cogenerated if S_r is finitely generated and essential in R_R.

Theorem 5.61. *The following conditions are equivalent for a ring R:*

(1) *R is an FP ring.*
(2) *R is right Kasch, right FP-injective, and left min-CS.*

Lemma 5.64. *Let A be an ideal of R that is right or left T-nilpotent and is finitely generated as a right or left ideal. Then A is nilpotent.*

Proof. Let $X = \{x_1, x_2, \ldots, x_n\}$ generate A_R or ${}_RA$. By Lemma 5.63 there exists an integer m such that the product of any m elements of X is zero. If $A = x_1R + \cdots + x_nR$, and if $a_1, a_2, \ldots, a_m$ are in A, the product $a_1a_2 \cdots a_m$ takes the form $a_1a_2 \cdots a_m = \Sigma_{i_j} x_{i_1}x_{i_2} \cdots x_{i_m} r_{(i_1,i_2,\ldots,i_m)}$ for elements $r_{(i_1,i_2,\ldots,i_m)} \in R$ because A is a left ideal. Hence $A^m = 0$. A similar argument works if $A = Rx_1 + \cdots + Rx_n$. □

Lemma 5.65. *Let A be an ideal of R such that A/A^2 is right finitely generated, and let M_R be a module.*

(1) If M/MA is finitely generated then MA/MA^2 is finitely generated.
(2) In particular A^k/A^{k+1} is right finitely generated for all $k \geq 1$.

Proof. Write $A = \Sigma_{i=1}^n a_iR + A^2$, $a_i \in A$, and assume that $M = \Sigma_{j=1}^k m_jR + MA$, $m_j \in M$. Then

$$MA = \Sigma_{j=1}^k m_jA + MA^2 = \Sigma_{j=1}^k m_j(\Sigma_{i=1}^n a_iR + A^2) + MA^2$$
$$= \Sigma_{j=1}^k \Sigma_{i=1}^n m_ja_iR + MA^2,$$

which shows that MA/MA^2 is finitely generated. This proves (1), and (2) follows from (1) by induction. □

Theorem 5.66. *Let R be a left perfect, right FP-injective ring. Then:*

(1) R is quasi-Frobenius if and only if $soc_2(R)$ is finitely generated as a right R-module.
(2) R is quasi-Frobenius if and only if $R/soc(R)$ is finitely cogenerated as a left R-module [where $soc(R)$ means $S_r = S_l$].
(3) If R is also right perfect, then R is quasi-Frobenius if and only if $soc_2(R)$ is finitely generated as a left R-module.

Proof. Note first that R is an FP ring by Theorem 5.57 because left perfect rings have essential right socle. Write $S = S_r = S_l$.

(1). Since R is left perfect, R/S has an essential right socle. But then R/S is finitely cogenerated as a right R-module because $soc_2(R)$ is right finitely generated. Hence ${}_RJ$ is finitely generated by Proposition 5.62, so J is nilpotent by Lemma 5.64. Thus R is semiprimary, so each J^k/J^{k+1} is semisimple as a left module. But then each J^k/J^{k+1} is artinian on the left by Lemma 5.65, so R is left artinian. Hence R is quasi-Frobenius by Theorem 3.31.

(2). Proposition 5.62 shows that J_R is finitely generated, so R is right artinian as in (1). Since R is two-sided mininjective, it is quasi-Frobenius by Theorem 3.31.

(3). Now R is right perfect, and so R/S has an essential left socle. Hence the hypothesis shows that R/S is left finitely cogenerated, and the result follows from (2). □

5.9. Group Rings

If G is a group and R is a ring, the *group ring* RG is the free R-module on G as basis, with multiplication determined by the ring and group axioms and the requirement that $rg = gr$ for all $r \in R$ and $g \in G$. Hence

$$RG = \{\Sigma_{g\in G} a_g g \mid a_g \in R\}$$

where $a_g = 0$ except for finitely many g. If $h \in G$, the projection $\pi_h : RG \to R$ given by $\pi_h(\Sigma_{g\in G} a_g g) = a_h$ is right and left R-linear. We write $\pi_1 = \pi$ and note that, if $x \in RG$, then $\pi_h(x) = \pi(xh^{-1})$ and

$$x = \Sigma_{g\in G}\pi_g(x)g = \Sigma_{g\in G}\pi(xg^{-1})g. \qquad (*)$$

We use this formula to prove the following important theorem.

Theorem 5.67 (Connell's Theorem). *If G is a finite group the group ring RG is right self-injective if and only if R is right self-injective.*

Proof. Assume that R is right self-injective, and let $\gamma : Y \to RG$ be RG-linear, where $Y \subseteq RG$ is a right ideal. Then $\pi \circ \gamma : Y \to R$ is right R-linear, so (as R_R is injective) extend $\pi \circ \gamma$ to $\lambda : RG \to R$. Now use λ to define $\alpha : RG \to RG$ by $\alpha(x) = \Sigma_{g\in G}\lambda(xg)g^{-1}$. Then α is RG-linear because

$$\alpha(xh) = \Sigma_{g\in G}\lambda(xhg)g^{-1} = \Sigma_{k\in G}\lambda(xk)k^{-1}h = \alpha(x)h$$

for all $h \in G$. Finally, α extends γ. If $y \in Y$ then (*) gives

$$\begin{aligned}\alpha(y) = \Sigma_{g\in G}\lambda(yg)g^{-1} &= \Sigma_{g\in G}\pi[\gamma(yg)]g^{-1}\\ &= \Sigma_{g\in G}\pi[\gamma(y)g]g^{-1}\\ &= \Sigma_{m\in G}\pi[\gamma(y)m^{-1}]m\\ &= \gamma(y).\end{aligned}$$

Hence RG is right self-injective. Conversely, if $\gamma : T \to R$ is R-linear where $T \subseteq R$, define $\bar{\gamma} : TG \to RG$ by $\bar{\gamma}(\Sigma t_g g) = \Sigma\gamma(t_g)g$. This is RG-linear, so $\bar{\gamma} = x\cdot$ for some $x \in RG$. Hence if $t \in T$, we have $\gamma(t) = \bar{\gamma}(t) = xt = \pi(x)t$, so $\gamma = \pi(x)\cdot$. Thus R is right self-injective. □

The next result explores the situation for P-injective rings. A group is called *locally finite* if every finitely generated subgroup is finite.

Theorem 5.68. *Let G be a group, let R be a ring, and let RG denote the group ring.*

(1) If RG is right P-injective then R is right P-injective and G is locally finite.
(2) If R is right self-injective and G is locally finite, then RG is right P-injective.

Proof. (1). Suppose that $\mathrm{r}_R(a) \subseteq \mathrm{r}_R(b)$, where $a, b \in R$. Then $\mathrm{r}_{RG}(a) \subseteq \mathrm{r}_{RG}(b)$, so $a \in RGb$ by hypothesis. It follows that $a \in Rb$, proving that R is right P-injective. Recall that the augmentation ideal of RG is defined by $\omega(RG) = \{\Sigma_{g\in G}a_g g \mid \Sigma_{g\in G}a_g = 0\}$. To show that G is locally finite, it suffices to show that the subgroup $\langle H, g\rangle$ generated by H and g is a finite group whenever $g \in G$ and H is a finite subgroup of G. We have $\omega(RH) = \Sigma_{h\in H}(1-h)RH = \mathrm{l}_{RH}(y)$, where y is the sum of the elements of the finite group H. Define $x = (1-g)y$ so that $\mathrm{r}_{RG}(y) \subseteq \mathrm{r}_{RG}(x)$.

Claim. $\mathrm{r}_{RG}(y) \neq \mathrm{r}_{RG}(x)$.

Proof. Otherwise $RGx = RGy$ by hypothesis, so write $y = zx = z(1-g)y$, where $z \in RG$. It follows that $[1 - z(1-g)] \in \mathrm{l}_{RG}(y) = RG\,\omega(RH)$. But then $1 \in RG(1-g) + \omega(RH) \subseteq \omega(RG)$, which is a contradiction. This proves the Claim.

So choose $z \in \mathrm{r}_{RG}(x)$ such that $z \notin \mathrm{r}_{RG}(y)$. Then $yz \neq 0$ and $(1-g)yz = 0$. Since $(1-h)y = 0$ for all $h \in H$, this gives $\langle H, g\rangle \subseteq \{h \in G \mid (1-h)yz = 0\}$, a finite group by a result of Passman [190, p. 105].

(2). Let $\mathbf{r}_{RG}(y) \subseteq \mathbf{r}_{RG}(x)$, where $y, x \in RG$. Because G is locally finite, there is a finite subgroup H of G such that $y, x \in RH$. Hence $\mathbf{r}_{RH}(y) = \mathbf{r}_{RG}(y) \cap RH \subseteq \mathbf{r}_{RG}(x) \cap RH = \mathbf{r}_{RH}(x)$. Since RH is right self-injective by Connell's theorem, this implies that $x \in RHy \subseteq RGy$. This proves (2). □

Corollary 5.69. *If F is a field and G is a group, the following are equivalent:*

(1) FG is right P-injective.
(2) G is locally finite.
(3) FG is left P-injective.

Example 5.70. If R is right P-injective and G is finite, RG need not be right P-injective.

Proof. The Björk example R is a right P-injective ring, and we may assume that 6 is a unit in R and that R contains a primitive cube root of unity. If $G = S_3$ is the symmetric group of order 6, we have $RG \cong R \times R \times M_2(R)$. [Indeed, if $h \in G$ has order 3 and $e = \frac{1}{3}(1 + h + h^2)$, then e is a central idempotent in RG, $RGe \cong R \times R$, and $RG(1 - e) \cong M_2(R)$.] Hence if RG were right P-injective then $M_2(R)$ would be right P-injective, whence R would be right 2-injective by Theorem 5.36, contrary to Example 5.22. □

Notes on Chapter 5

The right P-injective rings were first discussed in 1952 by Ikeda [102]. The equivalence of the first two conditions in Lemma 5.1 was noted by Ikeda and Nakayama [104].

The theorem that $J = Z_r$ in any right principally injective ring (Theorem 5.14) has been extended by Page and Zhou [188, Theorem 2.2]. Proposition 5.15 is due to Rutter [199].

The commutative case of Proposition 5.17 is due to Camillo [26] who, in addition, shows that if R is a commutative, finite dimensional, P-injective ring, then R has finitely many maximal ideals. In Proposition 5.27, we show that finite dimensionality is in fact equivalent to having finitely many maximal ideals, assuming only that R is right duo. In fact, many results in Section 5.3 stem from [25].

The fact that every right module over a regular ring is FP-injective comes from two theorems of Zelmanowitz in 1972: The direct sum of regular modules is regular [236, Theorem 2.8], and every finitely generated submodule of a free module over a regular ring is a direct summand [236, Theorem 1.6].

The equivalence of (1) and (4) in Theorem 5.41 was privately communicated to the authors by Puninski. The result in Corollary 5.44 is due to Jain [105], and the usual proof is homological. For detailed information on FP-injectivity we refer the reader to the book by Stenström [208]. The argument in Lemma 5.64 is adapted from Lam [132]. Statement (1) in Theorem 5.66 is an extension of a result of Clark and Huynh [36].

The theorem that, for a finite group, the group ring RG is right self-injective if and only if R is right self-injective was proved in 1963 by Connell [43]. In 1971 Renault [195] showed that G must be finite if RG is right self-injective and, in 1975, Farkas [69] showed that, if F is a field, FG is right P-injective if and only if G is locally finite.

6
Simple Injective and Dual Rings

We now introduce another class of right mininjective rings that seems to effectively mirror the properties of right self-injective rings. A ring R is called right simple injective if every R-linear map with simple image from a right ideal to R extends to R. Examples include right mininjective rings and right self-injective rings, but there are simple injective rings that are not right P-injective, and there exist right simple injective rings that are not left simple injective. If R is right simple injective, so also is eRe, where e is an idempotent satisfying $ReR = R$, but simple injectivity is not a Morita invariant [we characterize when $M_n(R)$ is right simple injective].

We show that a simple injective ring R is right Kasch if and only if every right ideal is an annihilator and, in this case, that R is left P-injective, $soc(eR)$ is simple and essential in eR for every local idempotent e in R, and R is left finite dimensional if and only if it is semilocal. In fact, if R is semiperfect, right simple injective and $soc(eR) \neq 0$ for every local idempotent e (the analogue of the right minfull rings), then R is a right and left Kasch, right and left finitely cogenerated, right continuous ring in which $S_r = S_l (= S)$ and both $soc(eR) = eS$ and $soc(Re) = Se$ are simple and essential in eR and Re, respectively, for every local idempotent e in R. Furthermore, it is shown that a semiprimary ring is right self-injective if and only if it is right simple injective, and we conjecture that a left perfect, right simple injective ring is right self-injective. Finally, we show that a ring R is quasi-Frobenius if and only if R is a left perfect, left and right simple injective ring; if and only if R is a right simple injective, right Goldie ring with $S_r \subseteq^{ess} R_R$; if and only if R is a left perfect, right simple injective ring in which $soc_2(R)$ is countably generated as a left R-module.

Kaplansky called a ring a dual ring if every one-sided ideal is an annihilator. We show that the simple injective rings are closely related to these dual rings. After giving a short proof that every dual ring is semiperfect, we show that R

is a dual ring if and only if R is a two-sided Kasch, two-sided simple injective ring; if and only if R is a semiperfect, two-sided simple injective ring with $soc(eR) \neq 0$ for every local idempotent e in R. Furthermore, after developing some properties of $AB5^*$ modules and rings, we show that every dual ring is right and left quotient finite dimensional. We then show that R is a left and right PF ring if and only if $M_2(R)$ is a dual ring, and that R is quasi-Frobenius if and only if it is a right semiartinian dual ring.

Finally, we call a ring R a right Ikeda–Nakayama ring (a right IN ring) if $\mathbf{l}(T \cap T') = \mathbf{l}(T) + \mathbf{l}(T')$ for all right ideals T and T' of R. These rings are closely related to the dual rings: A ring R is a dual ring if and only if R is a right and left IN ring and the dual of every simple right R-module is simple. Moreover, we show that every right IN ring is right quasi-continuous and that a ring R is right self-injective if and only if $M_2(R)$ is a right IN ring. In fact, R is quasi-Frobenius if and only if R is a left perfect, right and left IN ring.

6.1. Examples

A ring R is called right *simple injective* if, for every right ideal T of R, every R-linear map $\gamma : T \to R$ with $\gamma(T)$ simple extends to R, that is, if $\gamma = c\cdot$ is left multiplication by an element c of R. Here is another characterization that will be needed.

Lemma 6.1. *A ring R is right simple injective if and only if an R-linear map $\gamma : T \to R_R$ extends to $R_R \to R_R$ whenever T is a right ideal of R and $\gamma(T)$ is semisimple and finitely generated.*

Proof. Assume that R is right simple injective. If $\gamma(T) = 0$ then $\gamma = 0\cdot$. Otherwise, let $\gamma(T) = K_1 \oplus \cdots \oplus K_n$, where the K_i are simple right ideals. If $\pi_i : \gamma(T) \to K_i$ is the projection, let $\pi_i \circ \gamma = c_i\cdot$, where $c_i \in R$ by hypothesis. It is routine to verify that $\gamma = (c_1 + \cdots + c_n)\cdot$, as required. □

In particular, if R is right simple injective and T is a finitely generated right ideal of R, then every R-linear map $T \to S_r$ extends to R.

Every ring with zero right socle is right simple injective, so every domain is right simple injective (if the socle is not zero it is a division ring). Thus $\mathbb{Z}$ is a commutative, noetherian, simple injective ring that is not self-injective. Since every right simple injective ring is clearly right mininjective, Example 6.2 shows that the following inclusions are strict:

$$\{\text{right self-injective}\} \subset \{\text{right simple injective}\} \subset \{\text{right mininjective}\}.$$

Example 6.2. If the field is taken to be $\mathbb{Z}_2$, the Camillo example R (Example 2.6) is a commutative, local ring with $J^3 = 0$ that is mininjective with simple essential socle but is not simple injective.

Proof. Recall that $R = \mathbb{Z}_2[1, m, x_1, x_2, \dots]$, where the x_i are commuting indeterminants satisfying the relations $x_i^3 = 0$ for all i, $x_i x_j = 0$ for all $i \neq j$, and $x_i^2 = m$ for all i. Hence $J = span_{\mathbb{Z}_2}\{m, x_1, x_2, \dots\}$, $R/J \cong \mathbb{Z}_2$ and $J^3 = 0$. Moreover, $soc(R)$ is simple and essential in R by Example 2.6. Finally, R is not simple injective by the argument in Example 5.45 because, in the notation of that example, $\gamma(J) = \mathbb{Z}_2 m$ is simple. □

We give an example later of a left simple injective ring that is not right simple injective.

Proposition 6.3. *A direct product $\Pi_{i \in I} R_i$ of rings is right simple injective if and only if R_i is right simple injective for each $i \in I$.*

Proof. If $R \times S$ is right simple injective, it is routine to show that the same is true of R and S. Conversely, write $R = \Pi_{i \in I} R_i$, assume that each R_i is right simple injective, and let $\gamma : T \to R$ be R-linear, where $\gamma(T)$ is simple. Let σ_i and π_i be the canonical inclusion and projection maps for R. For each $k \in I$, write $\bar{e}_k = \sigma_k(1)$ and define $T_k = \{x \in R_k \mid \sigma_k(x) \in T\}$, a right ideal of R_k. Then $\sigma_k(T_k) \subseteq T$ because, if $x = \pi_k(\bar{t}) \in T_k, \bar{t} \in T$, then $\sigma_k(x) = \bar{t}\bar{e}_k \in T$. Hence the map $\gamma_k = \pi_k \gamma \sigma_k : T_k \to R_k$ is defined and R_k-linear, and $\gamma_k(T_k) \subseteq \pi_k[\gamma(T)]$ is zero or simple as a right ideal of R_k because π_k is a ring homomorphism. By hypothesis $\gamma_k = c_k \cdot$ for some $c_k \in R_k$. Now let $\bar{t} = \langle t_i \rangle \in T$ and write $\gamma(\bar{t}) = \bar{s} = \langle s_i \rangle$. Then each $t_k \in T_k$ because $\sigma_k(t_k) = \bar{t}\bar{e}_k$, so $\gamma\sigma_k(t_k) = \gamma(\bar{t}\bar{e}_k) = \bar{s}\bar{e}_k = \sigma_k(s_k)$. Hence $c_k t_k = \gamma_k(t_k) = \pi_k \gamma \sigma_k(t_k) = \pi_k \sigma_k(s_k) = s_k$ for each $k \in I$. If $\bar{c} = \langle c_i \rangle \in R$, we obtain $\bar{c}\bar{t} = \bar{s} = \gamma(\bar{t})$. As this holds for all $\bar{t} \in T$, we have $\gamma = \bar{c} \cdot$, as required. □

Example 6.4. If R is a ring for which $S_r = eR$ for a central idempotent e, then R is right simple injective by Proposition 6.3 because eR is semisimple and $soc[(1 - e)R] = 0$.

Recall that a ring is called right universally mininjective if every right module is mininjective, equivalently if $K^2 \neq 0$ for every simple right ideal K of R (see Theorem 2.36).

Example 6.5. If R is a right universally mininjective ring in which every idempotent is central, then R is right simple injective. In particular, every semiprime ring with all idempotents central is right and left simple injective.

Proof. If $\gamma : T \to R_R$ has $\gamma(T)$ simple then γ splits because $\gamma(T)$ is projective (by hypothesis), say $T = K \oplus P$, where $K = ker(\gamma)$ and $P \cong \gamma(T)$. Then $P = eR$, $e^2 = e$, again by hypothesis. If $c = \gamma(e)$, then $cp = \gamma(p)$ for all $p \in P$, and $cK = 0$ because e is central. It follows that $\gamma = c\cdot$, as required. □

Since $\mathbb{Z}$ is simple injective but not P-injective, Example 6.2 shows that there is no inclusion relationship between the right simple injective rings and the right P-injective rings. The following example presents a non-self-injective commutative ring with both properties.

Example 6.6 (Clark Example). We construct a commutative ring R with ideal lattice

$$0 = Rv_0 \subset Rv_1 \subset Rv_2 \subset \cdots \subset V \subset \cdots \subset Rp^2 \subset Rp \subset R,$$

where p and v_i, $i \geq 0$, satisfy $pv_k = v_{k-1}$ for all $k \geq 1$, and where V is the only nonprincipal ideal. Thus R is local with simple, essential socle. Furthermore, R is P-injective and simple injective, but not self-injective.

Proof. Let D be a discrete valuation ring, that is, a commutative integral domain with ideal lattice $0 \subset \cdots \subset Dp^n \subset \cdots \subset Dp^2 \subset Dp \subset D$. [For example $D = \mathbb{Z}_{(p)}$, with p a prime, or $D = F[[x]]$, with F any field (and we take $p = x$).] If we write U for the group of units of D, then $Dp^{n+1} - Dp^n = Up^n$ and the field of quotients is $Q = \{up^k \mid k \in \mathbb{Z}, u \in U\}$.

Claim 1. The lattice of D-submodules of ${}_DQ$ that contain D is

$$D \subset Dp^{-1} \subset Dp^{-2} \subset \cdots \subset Q.$$

Proof. Clearly $D \subset Dp^{-1} \subset Dp^{-2} \subset \cdots \subset Q$. If $D \subseteq {}_DA \neq Q$ then there exists $m \geq 0$ such that $p^{-m} \in A$ and $p^{-(m+1)} \notin A$. If $x \in A - Dp^{-m}$ write $x = up^{-k}$, $k \geq 1$, $u \in U$. Then $p^{-k} \in A$, so $k \leq m$, whence $x = (up^{m-k})p^{-m} \in Dp^{-m}$. This proves Claim 1.

Now define ${}_DV = Q/D$ and $v_m = p^{-m} + D \in V$, $m \geq 0$ (so $v_0 = 1 + D = 0$). Then $pv_k = v_{k-1}$ for each $k \geq 1$ and we have

Claim 2. $0 = Dv_0 \subset Dv_1 \subset Dv_2 \subset \cdots \subset V$ is the D-submodule lattice of ${}_DV$; and ${}_DV$ is a divisible module (and hence is injective).

Proof. We have $Dv_m = Dp^{-m}/D$, so $Dv_0 \subseteq Dv_1 \subseteq Dv_2 \subseteq \cdots \subseteq V$ by Claim 1. It is routine to verify that $v_{m+1} \notin Dv_m$. If $0 \neq d \in D$ we must show that $dV = V$. Let $d = up^n$, $u \in U$, $n \geq 0$. If $v = d_1v_m$, $d_1 \in D$, then $v = dw$, where $w = u^{-1}d_1v_{m+n}$. This proves Claim 2.

Now let R be the trivial extension of D by V; that is, $R = D \oplus V$, where the multiplication is defined by $(d + v)(d' + v') = dd' + (dv' + d'v)$. Then $Rv_m = R(0 + v_m) = 0 \oplus Dv_m$ for all $m \geq 0$, and $Rp^n = R(p^n + 0) = Dp^n \oplus V$ for all $n \geq 0$ because $V = p^nV$ by Claim 2.

Claim 3. $0 = Rv_0 \subset Rv_1 \subset Rv_2 \subset \cdots \subset V \subset \cdots \subset Rp^2 \subset Rp \subset R$ *is the ideal lattice of* R.

Proof. If A is an ideal of R, $A \neq 0, R$, assume first that $A \subseteq V$. If $A_0 = \{v \in V \mid 0 + v \in A\}$ then $A_0 = V$ or $A_0 = Dv_m$ for some m by Claim 2, so $A = V$ or $A = Rv_m$. However, if $A \nsubseteq 0 + V$ consider the ideal A, of D defined by $A_1 = \{d \in D \mid d + v \in A \text{ for some } v \in V\}$. Since $A \neq R$, we have $A_1 = Dp^n$, $n \geq 1$. Then $A = Dp^n \oplus V$ because V is divisible; that is, $A = Rp^n$. This proves Claim 3.

Hence R is "almost" a principal ideal ring. But V is not finitely generated by Claim 2 because $V = \Sigma_m Rv_m = \cup_m Rv_m$. However, R is P-injective; indeed every ideal is an annihilator. In fact one verifies that

$$Rv_m = \mathtt{r}(Rp^m) \text{ and } Rp^m = \mathtt{r}(Rv_m) \text{ for all } m \geq 0, \quad \text{and } \mathtt{r}(V) = V.$$

Since every ideal except V is principal, and since V has no simple image (no maximal submodule), this shows that R is simple injective. However, R is not self-injective. Indeed $\gamma : V \to R$ is well defined by $\gamma(0 + dv_m) = 0 + dv_{m-1}$ because $v_{m-1} = pv_m$. A routine computation shows that γ is R-linear and that $\gamma = c\cdot$ is impossible for $c \in R$. So R is not self-injective. □

Observe that in Example 6.6 we have $R/V \cong D$, $R/Rv_m \cong R$ for all $m \geq 0$, and $R/Rp^n \cong D/Dp^n$ for each $n \geq 0$.

6.2. Matrix Rings

We begin with a characterization of when the ring $T_n(R)$ of upper triangular matrices over R is right simple injective. Surprisingly it is equivalent to right min-injectivity.

Proposition 6.7. *The following are equivalent for a ring R:*

(1) *$T_n(R)$ is right simple injective for every $n \geq 2$.*
(2) *$T_n(R)$ is right simple injective for some $n \geq 2$.*
(3) *$T_n(R)$ is right mininjective for some $n \geq 2$.*
(4) *$S_r = 0$.*
(5) *$soc[T_n(R)_{T_n(R)}] = 0$ for every $n \geq 1$.*

Proof. The implications (5)⇒(1)⇒(2)⇒(3) are clear.

(3)⇒(4). Let e_{ij} denote the standard matrix unit. If kR is simple, $k \in R$, then $K = e_{1n}kR$ is a simple right ideal of $T_n(R)$ and we define $\gamma : K \to T_n(R)$ by $\gamma(e_{1n}x) = e_{nn}x$. Then γ is $T_n(R)$-linear, so $\gamma = C\cdot$, $C \in T_n(R)$, by (3). Hence $e_{nn}k = C\,e_{1n}k$, so $k = 0$ because $n \geq 2$, which is a contradiction. This proves (4).

(4)⇒(5). Write $T_n = T_n(R)$ and proceed by induction on $n \geq 1$. The case $n = 1$ is (4). In general, if $\tilde{a}T_n$ is simple, where $\tilde{a} = [a_{ij}] \in T_n$, we show $\tilde{a} = 0$, a contradiction. Write $\tilde{a} = \left[\begin{smallmatrix} a_{11} & \bar{y} \\ 0 & \tilde{z} \end{smallmatrix}\right]$, where $\tilde{z} \in T_{n-1}$ and $\bar{y} \in R^{n-1}$. If $a_{11} \neq 0$ it follows easily that $a_{11}R$ is simple (since $\tilde{a}T_n$ is simple), contrary to (4). Hence $a_{11} = 0$. It remains to show that $\bar{y} = 0$ since then $\tilde{a} = \left[\begin{smallmatrix} 0 & 0 \\ 0 & \tilde{z} \end{smallmatrix}\right]$ and so $\tilde{a}T_n = \left[\begin{smallmatrix} 0 & 0 \\ 0 & \tilde{z}T_{n-1} \end{smallmatrix}\right]$. It follows that $\tilde{z}T_{n-1}$ is simple, so $\tilde{z} = 0$ by induction, as required. So we show that $\bar{y} = 0$.

We have $\bar{y} = [a_{12}\ a_{13} \cdots a_{1n}]$. If $a_{1j} \neq 0$ we show that $a_{1j}R$ is simple, contrary to (4). Let $0 \neq a_{1j}r \in a_{1j}R$. If we write $\tilde{b}_j = \tilde{a}(e_{jj}r)$ then $0 \neq \tilde{b}_j \in \tilde{a}T_n$ and so, as before, $\tilde{a}T_n = \tilde{b}_jT_n$. This implies that $\tilde{a} \in \tilde{b}_jT_n$, whence $a_{1j} \in a_{1j}rR$, as required. □

Although the right simple injective rings do not form a Morita invariant class, they do have the following property.

Proposition 6.8. *Let R be right simple injective. If $e^2 = e \in R$ satisfies $ReR = R$, then eRe is right simple injective.*

Proof. Write $S = eRe$ and let $\gamma : T \to S$ be S-linear, where T is a right ideal of S and $\gamma(T)$ is S-simple. Define $\bar{\gamma} : TR \to R_R$ by $\bar{\gamma}(\Sigma t_i r_i) = \Sigma\gamma(t_i)r_i$.

Claim. *$\bar{\gamma}$ is well defined and $\bar{\gamma}(TR)$ is simple as a right R-module.*

Proof. Let $\Sigma t_i r_i = 0$. If $r \in R$ we get $0 = \Sigma t_i r_i r e = \Sigma t_i(er_i re)$, whence $0 = \Sigma\gamma(t_i)(er_i re) = [\Sigma\gamma(t_i)r_i]re$. It follows that $\Sigma\gamma(t_i)r_i = 0$ because $ReR = R$, so $\bar{\gamma}$ is well defined. But then $\bar{\gamma}(TR) = \gamma(T)R = KR$, where we write

$K = \gamma(T)$. Choose $0 \neq k \in K$ so that $K = kS$ because K_S is simple. Then $KR = kSR \subseteq kR \subseteq KR$, so $KR = kR$ and we must show that kR is simple. Suppose that $0 \neq kr \in kR$. Then $krReR \neq 0$, so there exists $b \in R$ such that $0 \neq krbe = (ke)rbe \in kS$. Since $kS = K$ is S-simple, we get $krbeS = kS$, so $k \in krbeS \subseteq krR$. This proves the Claim.

Because R is right simple injective, the Claim gives $\bar{\gamma} = c\cdot$, where $c \in R$. If $t \in T$ then

$$\gamma(t) = e\gamma(t) = e\bar{\gamma}(t) = e(ct) = (ec)t = (ec)et = (ece)t.$$

It follows that $\gamma = (ece)\cdot$, as required. □

We will show later (Example 6.21) that right simple injectivity is *not* a Morita invariant.

If $\mathcal{C}$ is a class of rings closed under isomorphisms and under $R \mapsto eRe$, where $e^2 = e \in R$ and $ReR = R$, it is not difficult to show that $\mathcal{C}_0 = \{R \mid M_n(R) \in \mathcal{C}$ for all $n \geq 1\}$ is the largest Morita invariant class contained in $\mathcal{C}$. If $\mathcal{C}$ is the class of all right P-injective rings, then $\mathcal{C}_0$ is the important class of all right FP-injective rings by Theorem 5.41. By analogy, the class $\mathcal{C}_0$ is of interest when $\mathcal{C}$ is the class of right simple injective rings. The next result characterizes when the matrix ring $M_n(R)$ is right simple injective for a particular value of n. As before, let R_n denote the set of $n \times 1$ column matrices over R.

Theorem 6.9. *The following conditions are equivalent for a ring R and an integer $n \geq 1$:*

(1) $M_n(R)$ is right simple injective.
(2) For each right R-submodule T of R_n, every R-linear map $\gamma : T \to R$ with $\gamma(T)$ simple can be extended to $R_n \to R$.
(3) For each right R-submodule T of R_n, every R-linear map $\gamma : T \to R_n$ with $\gamma(T)$ simple can be extended to $R_n \to R_n$.

Proof. We prove it for $n = 2$; the general argument is analogous.

(1)⇒(2). Given $\gamma : T \to R_R$, where $T \subseteq R_2$ and $\gamma(T) = K$ is simple, consider the right ideal $\bar{T} = [T\ T]$ of $M_2(R)$. The map $\bar{\gamma} : \bar{T} \to M_2(R)$ defined by $\bar{\gamma}[\underline{x}_1\ \underline{x}_2] = \begin{bmatrix} \gamma(\underline{x}_1) & \gamma(\underline{x}_2) \\ 0 & 0 \end{bmatrix}$ is $M_2(R)$-linear and $\bar{\gamma}(\bar{T}) = \begin{bmatrix} K & K \\ 0 & 0 \end{bmatrix}$ is a simple right ideal of $M_2(R)$. By (1) we have $\bar{\gamma} = C\cdot$, where $C \in M_2(R)$, so $\gamma = \bar{c}\cdot$, where $\bar{c}$ is the first row of C.

(2)⇒(3). Given (2), consider $\gamma : T \to R_2$, where $T \subseteq R_2$ is a right R-submodule and $\gamma(T)$ is R-simple. If $\pi_i : R_2 \to R$ is the ith projection, then

$\pi_i\gamma(T)$ is zero or simple, so (2) provides an R-linear map $\gamma_i : R_2 \to R$ extending $\pi_i \circ \gamma$. Then $\hat{\gamma} : R_2 \to R_2$ extends γ, where $\hat{\gamma}(\underline{x}) = [\gamma_1(\underline{x})\ \gamma_2(\underline{x})]^T$ for all $\underline{x} \in R_2$.

(3)⇒(1). Write $S = M_2(R)$, and consider $\gamma : T \to S_S$, where T is a right ideal of S and $\gamma(T)$ is S-simple. Then $T = [T_0\ T_0]$, where $T_0 = \{\underline{x} \in R_2 \mid [\underline{x}\ 0] \in T\}$ is a right R-submodule of R_2. Moreover, the S-linearity of γ shows that $\gamma[\underline{x}\ 0] = [\underline{y}\ 0]$ for some $\underline{y} \in R_2$, and writing $\underline{y} = \gamma_0(\underline{x})$ yields an R-linear map $\gamma_0 : T_0 \to R_2$ such that $\gamma[\underline{x}\ 0] = [\gamma_0(\underline{x})\ 0]$ for all $\underline{x} \in T_0$.

Claim. $\gamma_0(T_0)$ *is* R-*simple.*

Proof. Let $0 \neq \underline{y} \in \gamma_0(T_0)$, say $\underline{y} = \gamma_0(\underline{x})$. Then $0 \neq [\underline{y}\ 0] = \gamma[\underline{x}\ 0] \in \gamma(T)$. If $\underline{y}_1 \in \gamma_0(T_0)$ then $[\underline{y}_1\ 0] \in \gamma(T)$, so, since $\gamma(T)$ is S-simple, there exists $B = \left[\begin{smallmatrix} a & b \\ d & c \end{smallmatrix}\right] \in S$ such that $[\underline{y}_1\ 0] = [\underline{y}\ 0]B = [\underline{y}a\ \underline{y}b]$. Hence $\underline{y}_1 = \underline{y}a \in \underline{y}R$, which proves the Claim.

Given the Claim, γ_0 extends to an R-linear map $\hat{\gamma} : R_2 \to R_2$ by (3). Hence $\hat{\gamma} = C\cdot$ for some matrix $C \in S$. If $[\underline{x}\ \underline{y}] \in T$ it follows that

$$\gamma[\underline{x}\ \underline{y}] = [\gamma_0(\underline{x})\ 0] + [\gamma_0(\underline{y})\ 0]\begin{bmatrix} 0 & 1 \\ 0 & 0 \end{bmatrix} = [C\underline{x}\ C\underline{y}] = C[\underline{x}\ \underline{y}].$$

In other words, $\gamma = C\cdot$, and (1) is proved. □

Motivated by Theorem 6.9, we call an R-module M_R *simple quasi-injective* if, for any submodule $X \subseteq M$, any R-morphism $X \to M$ with simple image extends to an R-morphism $M \to M$. Thus a ring R is right simple injective if and only if R_R is a simple quasi-injective module. More generally, Theorem 6.9 gives the following corollary:

Corollary 6.10. *The following are equivalent for a ring R:*

(1) $M_n(R)$ is right simple injective for every $n \geq 1$.
(2) Every finitely generated free right R-module is simple quasi-injective.

The rings satisfying these conditions are a Morita invariant class.

The following property of simple quasi-injective modules will be used later.

Lemma 6.11. *If M_R is a simple quasi-injective module and $end(M)$ is local, then $soc(M)$ is either 0 or simple and essential in M.*

Proof. Suppose $soc(M) \neq 0$, and let $K \subseteq M$ be simple. If $0 \neq T \subseteq M$ is a submodule, it suffices to show that $K \subseteq T$. If not, define $\gamma : K \oplus T \to M$ by $\gamma(k+t) = k$. By hypothesis, let $\hat{\gamma} \in end(M)$ extend γ. Then $(1_M - \hat{\gamma})(k) = 0$ for each $k \in K$, so $\hat{\gamma} \notin J[end(M)]$. Hence $\hat{\gamma}$ is invertible in the local ring $end(M)$, so the fact that $\hat{\gamma}(T) = 0$ means $T = 0$, contrary to assumption. □

If M_R is simple quasi-injective and $N \subseteq^{\oplus} M$ then N_R is also simple quasi-injective. Indeed, if $X \subseteq N$ and $\gamma : X \to N$ has simple image then γ extends to $\widehat{\gamma} : M \to M$, so $\pi \circ (\widehat{\gamma}_{|N}) : N \to N$ extends γ, where $\pi : M \to N$ is any projection.

Proposition 6.12. *Let R be a right simple injective ring.*

(1) If e is a local idempotent in R then $soc(eR)$ is either zero or simple and essential in eR.
(2) If R is semiperfect, the following conditions are equivalent:
  (a) $soc(eR) \neq 0$ for each local idempotent e.
  (b) S_r is finitely generated and essential in R_R.

Proof. The module eR is simple quasi-injective by the preceding remark, and $end(eR) \cong eRe$ is a local ring. Hence (1) follows from Lemma 6.11.

If $1 = e_1 + \cdots + e_n$, where the e_i are orthogonal local idempotents, then $S_r = \oplus_{i=1}^{n} soc(e_i R)$ and (a)⇒(b) follows from (1). The converse is clear. □

6.3. The Kasch Condition

Note first that the Kasch condition is independent of simple injectivity. Indeed, the ring $\mathbb{Z}$ of integers is simple injective (it has zero socle) but not Kasch, whereas the ring in Example 6.2 is a commutative, local Kasch ring that is not simple injective.

If R is a semiperfect, right simple injective ring with $soc(eR) \neq 0$ for each local idempotent e, then R is right minfull, and hence it is right (and left) Kasch by Theorem 3.12. However, the Kasch condition is much stronger in a right simple injective ring R, as the next result shows.

Lemma 6.13. *The following conditions are equivalent for a right simple injective ring R:*

(1) R is right Kasch.
(2) If $X \subseteq^{max} T$, where $T \subseteq R$ is a right ideal, then $X = T \cap \mathbf{r}(c)$ for some $c \in R$.
(3) $\mathbf{rl}(T) = T$ for every right ideal T of R.

(4) $\mathbf{r}\mathbf{l}(T) = T$ for every essential right ideal T of R.
(5) T is a direct summand of $\mathbf{r}\mathbf{l}(T)$ for every right ideal T of R.
(6) Every right ideal of R is a direct summand of a right annihilator.

Proof. (1)⇔(2). If X and T are as in (2), let $\sigma : T/X \to R$ be monic by (1), and define $\gamma : T \to R$ by $\gamma(t) = \sigma(t + X)$. Then $\gamma = c\cdot$ for $c \in R$ by simple injectivity, so $X = \{t \in T \mid ct = 0\} = T \cap \mathbf{r}(c)$. For the converse, take $T = R$ in (2) to show that $\mathbf{r}\mathbf{l}(X) = X$ for every maximal right ideal X of R.

(1)⇒(3). Always $T \subseteq \mathbf{r}\mathbf{l}(T)$. If $b \in \mathbf{r}\mathbf{l}(T) - T$ let $T \subseteq M \subseteq^{max} (bR + T)$. By the Kasch hypothesis let $\sigma : (bR + T)/M \to R$ be monic, and then define $\gamma : bR + T \to R$ by $\gamma(x) = \sigma(x + M)$. Since $im(\gamma) = im(\sigma)$ is simple, right simple injectivity gives $\gamma = c\cdot$, where $c \in R$. Hence $cb = \sigma(b + M) \neq 0$ because $b \notin M$. But if $t \in T$ then $ct = \sigma(t + M) = 0$ because $T \subseteq M$, so $c \in \mathbf{l}(T)$. Since $b \in \mathbf{r}[\mathbf{l}(T)]$ this gives $cb = 0$, which is a contradiction.

(3)⇒(4). This is obvious.

(4)⇒(5). If T is a right ideal of R, there exists a right ideal C such that $T \oplus C \subseteq^{ess} R_R$ by Lemma 1.7. Hence $T \subseteq \mathbf{r}\mathbf{l}(T) \subseteq \mathbf{r}\mathbf{l}(T \oplus C) = T \oplus C$ by (4), so $\mathbf{r}\mathbf{l}(T) = T \oplus [C \cap \mathbf{r}\mathbf{l}(T)]$ by the modular law.

(5)⇒(6). This is obvious.

(6)⇒(1). If $T \subseteq^{max} R_R$, we show that T is itself a right annihilator [which gives (1)]. Let $T \oplus K = \mathbf{r}(X)$ by (6), where K is a right ideal. If $K = 0$ we are done; if $K \neq 0$ then $T \oplus K = R$, so there exists $e^2 = e \in R$ such that $T = eR = \mathbf{r}(1 - e)$. □

Using Lemma 6.13, we list a number of properties of the right Kasch, right simple injective rings for reference later.

Proposition 6.14. *Let R be a right simple injective, right Kasch ring. Then R has the following properties:*

(1) $\mathbf{r}\mathbf{l}(T) = T$ *for every right ideal T of R.*
(2) *R is left P-injective, and hence right and left mininjective.*
(3) $S_r = S_l$.
(4) $J = Z_l = \mathbf{r}(S_r)$.
(5) $\mathbf{l}(J) \subseteq^{ess} {}_RR$.
(6) *If $e^2 = e$ is local then $soc(Re)$ is simple and essential in Re.*
(7) *For the following conditions, we have (c)⇒(a)⇔(b):*
  (a) *$soc(eR) \neq 0$ for each local $e^2 = e \in R$.*
  (b) *$soc(eR)$ is simple and essential in eR for each local $e^2 = e \in R$.*
  (c) *R is left Kasch.*

Proof. (1) is part of Lemma 6.13, and shows that R is left P-injective because $\mathtt{rl}(aR) = aR$ for each $a \in R$ (see Lemma 5.1). This gives (2), which in turn implies (3) by Theorem 2.21. As to (4), we have $J = Z_l$ because R is left P-injective (by Theorem 5.14), and $\mathtt{r}(S_r) \subseteq J$ because R is right Kasch (the other inclusion always holds).

(5). Let $0 \neq b \in R$. Since bR has maximal submodules, let $\gamma : bR \to R$ have simple image by the Kasch condition. Then $\gamma = c\cdot$ by simple injectivity, so $cb = \gamma(b) \neq 0$ whereas $cbJ \subseteq \gamma(bR)J \subseteq S_r J = 0$. Thus $0 \neq cb \in Rb \cap \mathtt{l}(J)$.

(6). We have $\mathtt{l}(J)e \cong [eR/eJ]^*$ by Lemma 3.1. Since eR/eJ is simple (because e is local), and since R is right mininjective and right Kasch, it follows by Theorem 2.31 that $\mathtt{l}(J)e$ is a simple submodule of $soc(Re)$. Hence (3) gives $\mathtt{l}(J)e \subseteq soc(Re) = S_l \cap Re = S_l e = S_r e \subseteq \mathtt{l}(J)e$. It follows that $soc(Re) = \mathtt{l}(J)e$ is simple; it is essential in Re by (5).

(7). Clearly (b)⇒(a), and the converse follows from Proposition 6.12. To show that (c)⇒(a), observe that $e \cdot \mathtt{r}(J) \cong [Re/Je]^*$, so $e \cdot \mathtt{r}(J)$ is simple because e is local, and R is left Kasch by (c) and left mininjective by (2). This proves (a). □

We know (Proposition 5.55) that any left finite dimensional, right Kasch ring is semilocal. The next theorem provides a converse in the presence of right simple injectivity.

Theorem 6.15. *Let R be right simple injective and right Kasch. Then*

R is left finite dimensional if and only if R is semilocal.

In this case R has the following properties:

(1) $S_r = S_l$ is finitely generated and essential both as a right and a left R-module.
(2) R is left Kasch.
(3) $soc(eR)$ is simple and essential in eR for each local $e^2 = e \in R$.
(4) $soc(Re)$ is simple and essential in Re for each local $e^2 = e \in R$.
(5) $dim({}_R R) = length[(R/J)_R]$.

Proof. The forward implication holds without the simple injective hypothesis by Proposition 5.55. Conversely, if R is semilocal then $S_r = \mathtt{l}(J)$. By Proposition 6.14, $S_r = S_l = \mathtt{l}(J) \subseteq^{ess} {}_R R$ and R is right and left mininjective. But then S_r is finitely generated as a left R-module by Theorem 5.52. In particular, R is left finite dimensional.

(1). We have already seen that $S_r = S_l \subseteq^{ess} {}_RR$. The fact that $S_r = S_l$ is finitely generated on both sides follows from Theorem 5.52 because R is semilocal and two-sided mininjective. That $S_r \subseteq^{ess} R_R$ is shown in (3).

(2). Since R is semilocal, right and left mininjective, and right Kasch, it is left Kasch by Lemma 5.49.

(3) and (4). These follow from (2) and Proposition 6.14.

(5). Observe that $\mathtt{l}(J) \cong [\bar{R}_R]^*$, where we write $\bar{R} = R/J$. But $\bar{R}_R$ is semisimple by hypothesis, say $\bar{R}_R = \oplus_{i=1}^n K_i$, where each K_i is a simple right R-module. Hence

$$S_l = S_r = \mathtt{l}(J) \cong [\bar{R}_R]^* = [\oplus_{i=1}^n K_i]^* \cong \oplus_{i=1}^n K_i^*.$$

Since each K_i^* is simple (because R is right mininjective and right Kasch), and since $S_l \subseteq^{ess} {}_RR$, this shows that $dim({}_RR) = dim({}_RS_l) = n = length(\bar{R}_R)$. □

We now turn to the semiperfect right simple injective rings and study the analogue of the right minfull rings. This is a class of right simple injective rings in which both Kasch conditions must hold. Recall that a module M is called finitely cogenerated if $soc(M)$ is finitely generated and essential in M.

Theorem 6.16. *Assume that R is a semiperfect, right simple injective ring in which $soc(eR) \neq 0$ for every local idempotent e of R. Then the following hold:*

(1) R is right and left Kasch.
(2) $S_r = S_l$, which we denote as S.
(3) $\mathtt{rl}(T) = T$ for all right ideals T of R, so R is left P-injective.
(4) R is right continuous.
(5) $soc(eR) = eS$ and $soc(Re) = Se$ are simple and essential in eR and Re, respectively, for every local idempotent $e \in R$.
(6) If $e_1, \dots, e_n$ are basic local idempotents then $\{e_1S, \dots, e_nS\}$ and $\{Se_1, \dots, Se_n\}$ are systems of distinct representatives of the simple right and left R-modules, respectively.
(7) R is right and left finitely cogenerated.
(8) $Z_r = Z_l = J$.

Proof. As R is right minfull, it is a right and left Kasch ring by Theorem 3.12, proving (1). Then (2) and (3) follow by Proposition 6.14. Moreover, $S_r \subseteq^{ess} R_R$ by Proposition 6.12, so (2) and (3) imply that R is a right CS ring by Lemma 4.2. This proves (4) because R is a right C2 ring by Proposition 1.46. Since R is left mininjective by (3), Theorem 6.15 and (1) show that R is also left minfull. Hence Proposition 3.17 proves (6) and shows that $soc(eR) = eS$ and $soc(Re) = Se$ are simple for every local idempotent e. With this, Proposition 6.14 completes the proof of (5).

Since R is semiperfect, 1 is a finite sum of orthogonal, local idempotents, so (5) implies that S is finitely generated and essential in both R_R and ${}_RR$. This is (7). Finally, $Z_r \subseteq J$ because R is semiperfect, and $J \subseteq Z_r$ because $S \subseteq^{ess} R_R$ (since $JS = 0$). Hence $Z_r = J$; similarly $Z_l = J$. □

6.4. Dual Rings

Following Kaplansky a ring R is called a *dual ring* if $\mathtt{rl}(T) = T$ for all right ideals T, and $\mathtt{lr}(L) = L$ for all left ideals L. The name comes from the fact that, in such a ring, the maps $T \longmapsto \mathtt{l}(T)$ and $L \longmapsto \mathtt{r}(L)$ give a duality between the right ideals T and the left ideals L of R. The following proposition contains some of the most important features of dual rings. Surprisingly, they are all semiperfect.

Proposition 6.17. *Let R be a dual ring. Then the following hold:*

(1) If $\{T_i \mid i \in I\}$ and $\{L_i \mid i \in I\}$ are sets of right and left ideals of R, respectively, then

$$\mathtt{l}[\cap_{i\in I} T_i] = \Sigma_{i\in I}\mathtt{l}(T_i) \quad \text{and} \quad \mathtt{r}[\cap_{i\in I} L_i] = \Sigma_{i\in I}\mathtt{r}(L_i).$$

(2) R is semiperfect.

(3) Every R-linear map from a right (respectively left) ideal of R to R with finitely generated image is given by left (respectively right) multiplication by an element of R. In particular, R is right and left F-injective and right and left simple injective.

Proof. (1). $\mathtt{l}(\cap_{i\in I} T_i) = \mathtt{l}[\cap_{i\in I}\mathtt{rl}(T_i)] = \mathtt{lr}[\Sigma_{i\in I}\mathtt{l}(T_i)] = \Sigma_{i\in I}\mathtt{l}(T_i)$. The proof of the other equation is similar.

(2). We show that every right ideal T of R has a supplement in R (that is, a right ideal T' minimal with respect to $T + T' = R$) and apply Theorem B.28. Given T, let L be a left ideal maximal with respect to $\mathtt{l}(T) \cap L = 0$. Then $R = \mathtt{r}(0) = \mathtt{r}[\mathtt{l}(T) \cap L] = T + \mathtt{r}(L)$ using (1); we claim that $\mathtt{r}(L)$ is a supplement for T. To see this, let $T + X = R$, where $X \subseteq \mathtt{r}(L)$ is a right ideal. Then $L \subseteq \mathtt{l}(X)$ so, since $\mathtt{l}(T) \cap \mathtt{l}(X) = \mathtt{l}(T + X) = 0$, the maximality of L implies that $L = \mathtt{l}(X)$. Hence $\mathtt{r}(L) = \mathtt{rl}(X) = X$, as required.

(3). If $\alpha : T \to R_R$ is R-linear, where T is a right ideal and $\alpha(T)$ is finitely generated, then $T = a_1R + \cdots + a_nR + ker(\alpha)$ for some $a_i \in T$. If $n = 1$ then the restrictions of α to a_1R and to $ker(\alpha)$ both extend to R (because R is right P-injective), so T extends using (1) and Lemma 1.36. The general result follows by induction on n. □

In connection with (1) of Proposition 6.17, we remark that a ring R satisfies $\mathtt{rl}(T) = T$ for all right ideals T if and only if $\mathtt{r}[\mathtt{l}(T) \cap Rb] = T + \mathtt{r}(b)$ for all right ideals T and all $b \in R$.

Call a ring R a right *dual ring* if $\mathtt{rl}(T) = T$ for all right ideals T (and define left *dual rings* analogously). Each right dual ring is clearly left P-injective and right Kasch, and a right simple injective ring is right dual if and only if it is right Kasch (by Lemma 6.13). Hence every dual ring is right and left Kasch and right and left simple injective. In fact these properties characterize the dual rings.

Theorem 6.18. *The following conditions on a ring R are equivalent:*

(1) R is a dual ring.
(2) R is a two-sided Kasch, two-sided simple injective ring.
(3) R is semiperfect and two-sided simple injective with $soc(eR) \neq 0$ [respectively $soc(Re) \neq 0$] for every local idempotent $e \in R$.

Proof. We have already observed that (1)⇒(2), and (2)⇒(3) follows from Theorem 6.16 and Proposition 6.14. If R is as in (3) with $soc(eR) \neq 0$, then R is left Kasch and right dual by Theorem 6.16. Since R is left simple injective, it follows from Theorem 6.15 that $soc(Re) \neq 0$ for each local idempotent e. Hence R is left dual by Theorem 6.16 again. The case where $soc(Re) \neq 0$ is analogous. □

The following theorem contains more important properties of dual rings. They are contained in Theorem 6.16 and Proposition 6.14 and are collected here for reference.

Theorem 6.19. *If R is a dual ring, then the following statements hold:*

(1) R is right and left Kasch.
(2) $S_r = S_l$, which we denote as S.
(3) R is right and left continuous.
(4) $soc(eR) = eS$ and $soc(Re) = Se$ are simple and essential in eR and Re, respectively, for every local idempotent $e \in R$.

(5) If $e_1, \ldots, e_n$ are basic local idempotents then $\{e_1S, \ldots, e_nS\}$ and $\{Se_1, \ldots, Se_n\}$ are systems of distinct representatives of the simple right and left R-modules, respectively.
(6) R is right and left finitely cogenerated.
(7) $Z_r = Z_l = J$.

Recall that a ring R is called a right PF ring if it is semiperfect, right self-injective, and $S_r \subseteq^{ess} R_R$. Theorem 6.19 gives the following characterization of the left and right PF rings.

Corollary 6.20. *The following are equivalent for a ring R:*

(1) R is a left and right PF ring.
(2) $M_2(R)$ is a dual ring.
(3) $M_n(R)$ is a dual ring for each $n \geq 1$.

Proof. (1)⇒(3). Each $M_n(R)$ is a left and right PF ring because being a right PF ring is a Morita invariant. Hence (3) follows by Proposition 6.14 because right PF rings are right simple injective and right Kasch.

(3)⇒(2). This is obvious.

(2)⇒(1). If $M_2(R)$ is a dual ring then it is right and left continuous by Theorem 6.19 and so is right and left self-injective by Theorem 1.35. Since R is semiperfect by Proposition 6.17, (1) follows from Theorem 6.15 because "right Kasch" and "right self-injective" both pass from $M_2(R)$ to R. □

As an application of our results on dual rings, we construct a right simple injective ring R such that the matrix ring $M_2(R)$ is not right simple injective. This shows that simple injectivity is not a Morita invariant.

Example 6.21. Right simple injectivity is not a Morita invariant.

Proof. Let R denote the Clark example (Example 6.6). Then R is a commutative, local ring with simple essential socle that is simple injective but not self-injective. Write $S = M_2(R)$. If right simple injectivity were a Morita invariant, then S would be right simple injective. As S is right Kasch (right Kasch being a Morita invariant) it follows by Proposition 6.14 that $\mathtt{rl}(T) = T$ for each right ideal T of S. Similarly, $\mathtt{lr}(L) = L$ for each left ideal L; that is, S is a dual ring. But every dual ring is left and right continuous by Theorem 6.19, and it follows that R is self-injective by Theorem 1.35, which is a contradiction. □

6.5. The AB5* Condition

We are going to prove that dual rings exhibit another important property: They are right and left quotient finite dimensional. Here a module M_R is called *quotient finite dimensional* if every image of M is (Goldie) finite dimensional. We accomplish this by first showing that dual rings satisfy the right and left AB5*-condition.

To motivate the condition, consider the following notion. A family $\{M_i \mid i \in I\}$ of submodules of a module M is called *direct* if, for any M_i and M_j, there is a submodule M_k in the family such that $M_i + M_j \subseteq M_k$. The following lemma is an immediate consequence of the definition.

Lemma 6.22. *Let M be a module, let K be a submodule of M, and let $\{M_i \mid i \in I\}$ be a direct family of submodules of M. Then*

$$K \cap (\Sigma_{i \in I} M_i) = \Sigma_{i \in I} (K \cap M_i).$$

The lattice property in Lemma 6.22 is known as the $AB5$-condition; our interest here is in the modules for which the dual condition holds.

A family $\{M_i \mid i \in I\}$ of submodules of a module M is called *inverse* if, for any M_i and M_j, there is a submodule M_k such that $M_k \subseteq M_i \cap M_j$ (for example if the M_i are linearly ordered). A module M is said to satisfy the *$AB5^*$-condition*[1] (and is called an *$AB5^*$ module*) if, for any submodule K of M and any inverse family $\{M_i \mid i \in I\}$ of submodules of M, we have

$$K + (\cap_{i \in I} M_i) = \cap_{i \in I} (K + M_i).$$

Note that $K + (\cap_{i \in I} M_i) \subseteq \cap_{i \in I} (K + M_i)$ always holds.

Unlike the $AB5$-condition, not every module satisfies the $AB5^*$-condition. For example, if $M_{\mathbb{Z}} = \mathbb{Z}$, take $K = 2\mathbb{Z}$ and, if p is an odd prime, take $M_i = p^i\mathbb{Z}$ for $i \geq 1$. Then $K + (\cap_{i \in I} M_i) = K$ whereas $\cap_{i \in I} (K + M_i) = \mathbb{Z}$. However, we have

Example 6.23. If the submodules of M are linearly ordered by inclusion, then M is an $AB5^*$ module.

Proof. Let K and M_i be submodules of M, $i \in I$. If $K \subseteq M_i$ for all i, then $K \subseteq \cap_{i \in I} M_i$, so $K + (\cap_{i \in I} M_i) = \cap_{i \in I} M_i = \cap_{i \in I} (K + M_i)$. Otherwise, suppose $K \nsubseteq M_t$ for some $t \in I$ so, by hypothesis, $M_t \subseteq K$. Then $\cap_{i \in I} M_i \subseteq K$, so $K = K + (\cap_{i \in I} M_i) \subseteq \cap_{i \in I} (K + M_i) \subseteq K + M_t = K$. □

[1] The term originated in Grothendieck's treatment of category theory.

The next example will be needed in the following.

Example 6.24. Every artinian module M is an $AB5^*$ module.

Proof. Let $\mathcal{F} = \{M_i \mid i \in I\}$ be an inverse family of submodules of M, and let M_0 be a minimal member of $\mathcal{F}$. If $i \in I$, there exist $t \in I$ such that $M_t \subseteq M_0 \cap M_i$. Hence $M_t \subseteq M_0$, so $M_t = M_0$ and it follows that $M_0 \subseteq M_i$ for each $i \in I$. Hence, if K is any submodule of M, we have

$$K + M_0 \subseteq K + \cap_{i \in I} M_i \subseteq \cap_{i \in I}(K + M_i) \subseteq K + M_0,$$

and the proof is complete. □

In the next proposition we collect some properties of $AB5^*$ modules for reference.

Proposition 6.25. *Let M denote an $AB5^*$ module.*

(1) Every submodule and image of M is an $AB5^*$ module.
(2) $E^{(\mathbb{N})}$ is not an $AB5^*$ module for any $E_R \neq 0$.
(3) Let $\mathcal{S}(M)$ be the set of nonisomorphic simple images of submodules of M. If $\mathcal{S}(M)$ is finite, then M is quotient finite dimensional.

Proof. (1). The proof is a straightforward consequence of the definition.

(2). Write $L = \oplus_{i \in \mathbb{N}} E_i$, where $E_i = E$ for each $i \in \mathbb{N}$. Let $\pi_i : L \to E$ denote the ith projection map, and define submodules K and L_i, $i \in \mathbb{N}$, of L by $K = \{k \in L \mid \Sigma_{i \in \mathbb{N}} \pi_i(k) = 0\}$ and $L_i = \oplus_{j \geq i} E_j$. Then $\{L_i \mid i \in \mathbb{N}\}$ is an inverse family, and $K + L_i = L$ for each $i \in \mathbb{N}$ because $E_i = E$ for each i. Hence $\cap_{i \in \mathbb{N}}(K + L_i) = L \neq K = K + 0 = K + \cap_{i \in \mathbb{N}} L_i$. Thus L is not $AB5^*$, as required.

(3). Suppose N is a quotient of M such that N is not finite dimensional. Then N contains an infinite direct sum $K = \oplus_{i \in \mathbb{N}} K_i$ of nonzero principal submodules. For each $i \in \mathbb{N}$, let T_i be a maximal submodule of K_i, and write $S_i = K_i/T_i$. If $T = \oplus_{i \in \mathbb{N}} T_i$ then $B = N/T$ contains a copy of the infinite direct sum $\oplus_{i \in \mathbb{N}} S_i$. Since $\mathcal{S}(B)$ is finite by hypothesis (every simple submodule of a quotient of B bears the same relationship to M), it follows that B contains an infinite direct sum of copies of S_j for some $j \in \mathbb{N}$. This contradicts (2) because B is $AB5^*$ by (1). Thus N is finite dimensional. □

Since semilocal rings have only finitely many nonisomorphic simple modules, we obtain

Corollary 6.26. *If R is semilocal, every left and right $AB5^*$ module is finite dimensional.*

Corollary 6.27. *A ring R is right artinian if and only if R is a left perfect, right $AB5^*$ ring.*

Proof. If R is right artinian, then R is left perfect, and it is a right $AB5^*$ ring by Example 6.24. Conversely, if $M \neq 0$ is a principal right R-module then $soc(M) \subseteq^{ess} M$ by Theorem B.32 and Lemma B.31 because R is left perfect, and $soc(M)$ is finitely generated by Proposition 6.25 because R is a right $AB5^*$ ring. In other words, M is finitely cogenerated by Lemma 1.51, and so R is right artinian by Vámos' lemma (Lemma 1.52). □

The following lemma, interesting in its own right, will be required in the proof of our main theorem (Theorem 6.29).

Lemma 6.28. *Let R be a ring that satisfies the following two conditions:*

(1) $\mathtt{r}(L_1 \cap L_2) = \mathtt{r}(L_1) + \mathtt{r}(L_2)$ for all left ideals L_1 and L_2 of R.
(2) $\mathtt{rl}(T) = T$ for all right ideals T of R.

Then R is a right $AB5^*$ ring.

Proof. Let K be a right ideal of R and let $\{T_i \mid i \in I\}$ be an inverse family of right ideals of R. Then (1) and (2) give

$$\begin{aligned}\cap_{i\in I}(K + T_i) &= \cap_{i\in I}\{\mathtt{rl}(K) + \mathtt{rl}(T_i)\}\\ &= \cap_{i\in I}\mathtt{r}\{\mathtt{l}(K) \cap \mathtt{l}(T_i)\}\\ &= \mathtt{r}[\Sigma_{i\in I}\{\mathtt{l}(K) \cap \mathtt{l}(T_i)\}].\end{aligned}$$

But $\{\mathtt{l}(T_i) \mid i \in I\}$ is a direct family of left ideals of R. Indeed, given T_i and T_j choose $k \in I$ such that $T_k \subseteq T_i \cap T_j$; then $\mathtt{l}(T_k) \supseteq \mathtt{l}(T_i \cap T_j) \supseteq \mathtt{l}(T_i) + \mathtt{l}(T_j)$. Hence, using Lemma 6.22, we continue, obtaining

$$\begin{aligned}\cap_{i\in I}(K + T_i) &= \mathtt{r}[\mathtt{l}(K) \cap \Sigma_{i\in I}\mathtt{l}(T_i)]\\ &= \mathtt{rl}(K) + \mathtt{r}[\Sigma_{i\in I}\mathtt{l}(T_i)]\\ &= \mathtt{rl}(K) + \cap_{i\in I}[\mathtt{rl}(T_i)]\\ &= K + (\cap_{i\in I}T_i).\end{aligned}$$

This proves the $AB5^*$-condition. □

We can now deduce the promised properties of dual rings.

Theorem 6.29. *Let R be a dual ring. Then the following hold:*

(1) R is a right and left AB5 ring.*
(2) R is right and left quotient finite dimensional.

Proof. (1) follows from Lemma 6.28 and Proposition 6.17. To prove (2), we must show that every principal R-module M is finite dimensional. But M is an image of R, so this follows from (1) and Corollary 6.26 because the dual ring R is semiperfect (by Proposition 6.17). □

6.6. Ikeda–Nakayama Rings

A ring R is called a right *Ikeda–Nakayama ring* (right *IN ring*) if the left annihilator of the intersection of any two right ideals is the sum of the two left annihilators, that is, if

$$\mathtt{l}(T \cap T') = \mathtt{l}(T) + \mathtt{l}(T') \quad \text{for all right ideals } T \text{ and } T' \text{ of } R.$$

Thus the Ikeda–Nakayama lemma (Lemma 1.37) asserts that every right P-injective, right IN ring is right F-injective.

Examples of right IN rings include the right self-injective rings, the right uniserial rings, and the right uniform domains. Every dual ring is a right and left IN ring by Proposition 6.17, and $\mathbb{Z}$ is a commutative, noetherian IN ring that is not self-injective and is not dual. In this section we show, among other things, that R is a dual ring if and only if R is a left and right IN ring and the dual of every simple right R-module is simple.

Lemma 6.30. *Let R be a right IN ring. Then the following hold:*

(1) If T is a right ideal of R then $T \subseteq^{ess} \mathtt{rl}(T)$.
(2) If T is a right ideal of R and $\mathtt{l}(T) \subseteq J$ then $T \subseteq^{ess} R_R$.
(3) If C is a closed right ideal of R then $C = \mathtt{rl}(C)$.

Proof. (1). Suppose that $T \cap xR = 0$, where $x \in \mathtt{rl}(T)$. Then $R = \mathtt{l}(0) = \mathtt{l}(T) + \mathtt{l}(x)$ by the IN hypothesis. But $\mathtt{l}(T) \subseteq \mathtt{l}(x)$ so $R = \mathtt{l}(x)$ and $x = 0$.

(2). Now assume that $T \cap xR = 0$, where $x \in R$. Then $R = \mathtt{l}(T) + \mathtt{l}(x)$ as before, so $R = J + \mathtt{l}(x)$ by hypothesis. Again $\mathtt{l}(x) = R$ so that $x = 0$.

(3). This follows from (1). □

We are going to prove that the right IN rings are all right quasi-continuous; that is, they satisfy the right CS- and C3-conditions. The next result implies this and actually characterizes the right quasi-continuous rings. Recall (Theorem 1.31) that a module M is quasi-continuous if and only if $M = C \oplus D$ whenever C and D are submodules that are complements of each other.

Theorem 6.31. *The following are equivalent for a ring R:*

(1) R is right quasi-continuous.
(2) If $P \cap Q = 0$, where P and Q are right ideals of R, then $P \subseteq^{ess} eR$ and $Q \subseteq (1-e)R$ for some $e^2 = e \in R$.
(3) If $P \cap Q = 0$, where P and Q are right ideals of R, then $P \subseteq eR$ and $Q \subseteq (1-e)R$ for some $e^2 = e \in R$.
(4) If $P \cap Q = 0$, where P and Q are right ideals of R, then $R = \mathtt{l}(P) + \mathtt{l}(Q)$.

Proof. (1)⇒(2). If $P \cap Q = 0$, let $C \supseteq Q$ be a complement of P, and then let $D \supseteq P$ be a complement of C. Then C is also a complement of D by Proposition 1.27, so $R = C \oplus D$ by (1) and there exists $e^2 = e$ such that $D = eR$ and $C = (1-e)R$. It remains to show that $P \subseteq^{ess} D$. If $0 \neq X \subseteq D$ then $X \oplus C \supset C$, so $(X \oplus C) \cap P \neq 0$ by the maximality of C. If $0 \neq p = x + c$ with the obvious notation, then $c = p - x \in C \cap D = 0$ and so $0 \neq p \in X \cap P$, as required.

(2)⇒(3). This is clear.

(3)⇒(4). If $P \cap Q = 0$ and e is as in (3), then (4) follows because $\mathtt{l}(P) \supseteq R(1-e)$ and $\mathtt{l}(Q) \supseteq Re$.

(4)⇒(1). Let C and D be right ideals of R that are complements of each other; we must show that $R = C \oplus D$. Since $C \cap D = 0$ we have $R = \mathtt{l}(C) + \mathtt{l}(D)$ by (4), say $1 = f + e$, where $f \in \mathtt{l}(C)$ and $e \in \mathtt{l}(D)$. Then $1 - e = f \in \mathtt{l}(C)$, so $ec = c$ for all $c \in C$. Similarly, $fd = d$ for all $d \in D$, so

$$C = eC, \qquad D = fD, \qquad \text{and} \qquad eD = 0 = fC.$$

Clearly, $D \subseteq \mathtt{r}(e) \subseteq \mathtt{r}(e^2)$ and $\mathtt{r}(e^2) \cap C = 0$ and so, since D is a complement of C, $D = \mathtt{r}(e) = \mathtt{r}(e^2)$. Similarly, $C = \mathtt{r}(f) = \mathtt{r}(f^2)$. Since $f = 1 - e$, it suffices to show that f (and hence e) is an idempotent. As $C \oplus D \subseteq^{ess} R_R$ by Lemma 1.7, it is enough to show that $efR \cap (C \oplus D) = 0$. Suppose $efr \in C \oplus D$, where $r \in R$. Then we have $e^2 f^2 r = 0$ because $ef = fe$ and $eD = 0 = fC$. Since $\mathtt{r}(e) = \mathtt{r}(e^2)$ and $\mathtt{r}(f) = \mathtt{r}(f^2)$, this gives $efr = 0$, as required. □

Note that the property in (2) of Theorem 6.31 is a strengthened version of the CS-condition.

The next theorem is an immediate consequence of condition (4) in Theorem 6.31.

Theorem 6.32. *Every right IN ring is right quasi-continuous.*

Proof. If $P \cap Q = 0$, where P and Q are right ideals of R, then $R = \mathtt{l}(P \cap Q) = \mathtt{l}(P) + \mathtt{l}(Q)$ because R is a right IN ring. Hence R is right quasi-continuous by Theorem 6.31. □

With this we can improve upon the characterization of right self-injective rings in Theorem 1.35.

Corollary 6.33. *The following are equivalent for a ring R:*

(1) $M_2(R)$ is a right IN ring.
(2) $M_n(R)$ is a right IN ring for all $n \geq 2$.
(3) R is right self-injective.

Proof. (2)⇒(1) is clear, and (3)⇒(2) because self-injectivity is a Morita invariant. Given (1), $M_2(R)$ is right quasi-continuous by Theorem 6.32, so (3) follows by Theorem 1.35. □

The converse to Theorem 6.32 is false by the following example.

Example 6.34. There exists a commutative, local continuous ring that is not an IN ring.

Proof. Consider the following special case of the Camillo example (Example 2.6). Let $R = \mathbb{Z}_2[x_1, x_2, \ldots]$, where $x_i^3 = 0$ for all i, $x_i x_j = 0$ for all $i \neq j$, and $x_i^2 = x_j^2 = m \neq 0$ for all i and j. We have $J = span\{m, x_1, x_2, \ldots\}$, and R has simple essential socle $J^2 = \mathbb{Z}_2 m$. In particular, R is uniform and so is a CS ring; C2 also holds because $\mathtt{r}(a) = 0$, $a \in R$, implies that a is a unit. Hence R is continuous. To see that R is not an IN ring, let S be the ideal generated by $\{x_1 + x_2, x_1 + x_4, \ldots, x_1 + x_{2k}, \ldots\}$ and let T be the ideal generated by $\{x_1 + x_3, x_1 + x_5, \ldots, x_1 + x_{2k+1}, \ldots\}$.

Claim. $\mathtt{r}(S) = x_3 R + x_5 R + \cdots$ and $\mathtt{r}(T) = x_2 R + x_4 R + \cdots$.

Proof. Write $A = x_3 R + x_5 R + \cdots$. Clearly $\mathtt{r}(S) \supseteq A$. Suppose that $q \in \mathtt{r}(S)$; we must show that $q \in A$. If $q \notin A$ then, since A contains m and all x_k with $k \geq 3$ odd, we may assume that q has the form $q = x_1 + p$ or $q = p$, where p denotes a sum of terms x_{2i} with $i \geq 1$. Thus $x_1 p = 0$. Let x_{2n}, $n \geq 1$, be the largest term in p, so that $x_{2n+2} \cdot p = 0$. If $q = x_1 + p$ then $0 = (x_1 + x_{2n+2})(x_1 + p) = x_1^2 = m$, a contradiction. So $q = p$ and $0 =$

$(x_1 + x_{2n}) \cdot p = x_{2n}^2 = m$, again a contradiction. Hence $q \in A$ and we have shown that $\mathtt{r}(S) = A$. A similar argument works for $\mathtt{r}(T)$, proving the Claim.

It follows from the Claim that $\mathtt{r}(S) + \mathtt{r}(T) = \Sigma_{i=2}^{\infty} x_i R$. However, $S \cap T = socR = \mathbb{Z}_2 m$ because $\{x_1, x_2, \ldots\}$ is $\mathbb{Z}_2$-independent. Hence $\mathtt{r}(S \cap T) = \mathtt{r}(socR) = J = \Sigma_{i=1}^{\infty} x_i R$. Hence $\mathtt{r}(S \cap T) \neq \mathtt{r}(S) + \mathtt{r}(T)$, so R is not a right IN ring. □

Note that a right IN ring need not satisfy C2 as the ring $\mathbb{Z}$ of integers shows. Moreover, if F is a field the ring $R = \left[\begin{smallmatrix} F & F \\ 0 & F \end{smallmatrix}\right]$ is a right and left artinian, right and left CS ring that is not a right IN ring. Indeed, if $x = \left[\begin{smallmatrix} 0 & 0 \\ 0 & 1 \end{smallmatrix}\right]$ and $y = \left[\begin{smallmatrix} 0 & 1 \\ 0 & 1 \end{smallmatrix}\right]$ then $xR \cap yR = 0$ but $\mathtt{l}(x) + \mathtt{l}(y) \subseteq \left[\begin{smallmatrix} F & F \\ 0 & 0 \end{smallmatrix}\right] \neq R$.

We now consider the effect of the Kasch condition on a right IN ring. We begin with

Proposition 6.35. *Every left Kasch, right IN ring R is a semiperfect, right continuous ring with $S_l \subseteq^{ess} R_R$.*

Proof. Indeed, R is a right CS ring by Theorem 6.32, so R is a semiperfect, right continuous ring with $S_l \subseteq^{ess} R_R$ by Theorem 4.10. □

The next result uses Theorem 6.31 to improve upon Theorem 6.15.

Proposition 6.36. *The following are equivalent for a ring R:*

(1) R is a left and right Kasch, right simple injective ring.
(2) R is a semiperfect, right simple injective ring with $S_r \subseteq^{ess} R_R$.

Proof. (2)⇒(1) follows from Theorem 6.16.

(1)⇒(2). Let P and Q be right ideals of R with $P \cap Q = 0$. Since R is right Kasch, $\mathtt{rl}(P) = P$ and $\mathtt{rl}(Q) = Q$ by Proposition 6.14, so $\mathtt{r}[\mathtt{l}(P) + \mathtt{l}(Q)] = \mathtt{rl}(P) \cap \mathtt{rl}(Q) = 0$. Since R is left Kasch, it follows that $\mathtt{l}(P) + \mathtt{l}(Q) = R$ and hence that R is right quasi-continuous by Theorem 6.31. In particular, R is a right CS ring so, since R is left Kasch, R is semiperfect by Lemma 4.1. Now $S_r \subseteq^{ess} R_R$ by Theorem 6.15. □

Every dual ring is clearly right and left Kasch. We are going to characterize the dual rings in terms of the IN rings, and the next lemma will be needed.

Lemma 6.37. *Assume that R is right Kasch and that, for all $x \in R$ and all right ideals T of R, $Rx \subseteq^{ess} \mathtt{lr}(x)$ and $\mathtt{l}(T \cap xR) = \mathtt{l}(T) + \mathtt{l}(x)$. Then $\mathtt{rl}(T) = T$ for all right ideals T of R.*

Proof. Suppose that $x \in \mathtt{rl}(T)$ so that $\mathtt{l}(T) \subseteq \mathtt{l}(x)$; we must show that $x \in T$. Consider the right ideal $K = \{k \in R \mid xk \in T\}$. Then $xK \subseteq T$, so $T \cap xR = xK$, and our hypothesis yields $\mathtt{l}(xK) = \mathtt{l}(T \cap xR) = \mathtt{l}(T) + \mathtt{l}(x) = \mathtt{l}(x)$. This in turn implies that $Rx \cap \mathtt{l}(K) = 0$. However, the fact that $\mathtt{r}(x) \subseteq K$ gives $\mathtt{l}(K) \subseteq \mathtt{lr}(x)$, so $\mathtt{l}(K) = 0$ because $Rx \subseteq^{ess} \mathtt{lr}(x)$ by hypothesis. But then $K = R$ by the Kasch hypothesis, so $x \in T$. □

We can use the preceding results to characterize dual rings in terms of IN rings. Recall that, by Theorem 2.31, the dual of every simple right R-module is simple if and only if R is right mininjective and right Kasch.

Theorem 6.38. *A ring R is a dual ring if and only if R is a left and right IN ring and the dual of every simple right R-module is simple*

Proof. Every dual ring is a two-sided IN ring (by Proposition 6.17) that is right mininjective and right Kasch (Theorem 6.18), so the dual of every simple right module is simple by Theorem 2.31.

Conversely, assume that the conditions are satisfied, so that R is a right mininjective, right Kasch, two-sided IN ring. Hence R is a two-sided CS ring by Theorem 6.32. But then Lemma 6.37 shows that R is right dual, and that it remains to show that R is left Kasch. But this follows by Lemma 4.5 because R is semiperfect by (the left–right analogue of) Lemma 4.1, and $S_r \subseteq S_l$ because R is right mininjective. □

6.7. Applications to Quasi-Frobenius Rings

We are going to give some characterizations of quasi-Frobenius rings in terms of dual and IN rings. Recall that a ring R is called right *semiartinian* if every nonzero (principal) right module has a nonzero socle, and that R is left perfect if and only if it is right semiartinian and I-finite.

Theorem 6.39. *The following conditions on a ring R are equivalent:*

(1) R is quasi-Frobenius.
(2) R is a left perfect, right and left self-injective ring.
(3) R is a left perfect, right and left simple injective ring.

(4) R is a right semiartinian dual ring.
(5) R is a left perfect, right and left IN ring.

Proof. (1)⇒(2)⇒(3) and (1)⇒(5) are clear.

(3)⇒(4). Given (3), R is left and right Kasch by Theorem 3.12, and so (4) follows by Theorem 6.18 because left perfect rings are right semiartinian.

(4)⇒(1). First, R is right quotient finite dimensional by Theorem 6.29. But then every quotient of R_R is finitely cogenerated because R is right semiartinian, so R is right artinian by Vámos' lemma (Lemma 1.52). Being right and left min-injective, R is quasi-Frobenius by Ikeda's theorem (Theorem 2.30).

(5)⇒(4). Given (5), R is right semiartinian (being left perfect), and it is left and right quasi-continuous by Theorem 6.32. Since R is left perfect, it has DCC on principal right ideals and so is right continuous by Theorem 4.27. Moreover, $S_r \subseteq^{ess} R_R$ and so R is left and right Kasch by Lemma 4.11. Finally, $Rx \subseteq^{ess} \mathtt{lr}(x)$ and $xR \subseteq^{ess} \mathtt{rl}(x)$ for all $x \in R$ (by the CS-conditions), so it follows that R is a dual ring by Lemma 6.37. This proves (4). □

There exist commutative noetherian IN rings that are not quasi-Frobenius (for example $\mathbb{Z}$). However, we do have the following result.

Corollary 6.40. *Suppose that R is a left and right IN ring with ACC on right annihilators in which monomorphisms $R_R \to R_R$ are epic. Then R is quasi-Frobenius.*

Proof. R is right quasi-continuous by Theorem 6.32, so R is a semiperfect, right continuous ring by Theorem 4.27. Hence $J = Z_r$ by Utumi's theorem (Theorem 1.26), so J is nilpotent by the Mewborn–Winton lemma (Lemma 3.29). Thus R is semiprimary and the result follows from Theorem 6.39. □

The Clark example (Example 6.6) is a commutative, local simple injective ring that is not quasi-Frobenius, so the right semiartinian hypothesis is essential in (4) of Theorem 6.39. Similarly, the following example shows that the simple injectivity hypothesis is essential in (3) of Theorem 6.39, and the two-sided dual condition is essential in (4).

Example 6.41. The Björk example R (Example 2.5) is a local, left artinian, two-sided Kasch ring with simple essential left socle. Moreover, R need not be right artinian (indeed R need not be right finite dimensional).

(1) R is a left IN ring in which $\mathtt{lr}(L) = L$ for all left ideals L. However, R need not be a right IN ring.

(2) If the dimension of F over $\bar{F}$ is finite and greater than 1, then R is a right and left artinian, left IN ring that is not a right IN ring. In addition, R is neither right nor left simple injective.

Proof. (1). The first two assertions follow because the only left ideals of R are 0, J, and R. If K and M are simple right ideals and $K \cap M = 0$ then $R = \mathbf{l}(K \cap M)$, whereas $\mathbf{l}(K) + \mathbf{l}(M) = J + J = J$. Hence R is not a right IN ring.

(2) It remains to show that R is neither right nor left simple injective. R is not left simple injective because it is not left mininjective. If R were right simple injective then it would be left P-injective by Proposition 6.14, which is a contradiction. □

Example 6.41 exhibits a left and right artinian, left IN ring that is not quasi-Frobenius. In contrast, Example 6.34 gives a commutative, local, continuous, principally injective ring with $J^3 = 0$ that is not artinian. The next example exhibits a commutative, local IN ring with simple essential socle that is not principally injective.

If R is a ring and ${}_RV_R$ is a bimodule, recall that the *trivial extension* $T(R, V)$ of R by V is the additive group $R \oplus V$ with multiplication given by $(r + v)(s + w) = rs + (rw + vs)$.

Example 6.42. We construct a commutative, local IN ring R with simple essential socle that is not Kasch and not principally injective.

Proof. Let $\mathbb{Z}_{2^\infty}$ denote the Prüfer group whose only subgroups are $0 \subset \mathbb{Z}x_1 \subset \mathbb{Z}x_2 \subset \cdots \subset \mathbb{Z}_{2^\infty}$, and where $2x_{i+1} = x_i$ for each $i \geq 1$. Writing $V = \mathbb{Z}_{2^\infty}$ for convenience, we have that the trivial extension $R = T(\mathbb{Z}, V)$ is a commutative, local ring with $J = 0 + V$ and with simple essential socle $0 + \mathbb{Z}x_1$. However, R is not principally injective because $3r \mapsto r$ from $3R \to R$ does not extend to R; R is not Kasch since $R/3R \cong \mathbb{Z}_3$ does not embed in R; and R does not satisfy the right C2-condition because $3R \cong R$ but $3R \neq R$. However, since $V = \mathbb{Z}_{2^\infty}$ is a divisible group, it can be verified that the only ideals of R are $nR = (n + 0)R$, where $0 < n \in \mathbb{Z}$, and $0 + H$, where $H \subseteq V$ is a subgroup. We claim that R is an IN ring.

Since $0 + H \subseteq nR$ for all subgroups H of V and all $0 < n \in \mathbb{Z}$, and since $\mathbf{l}(T \cap T') = \mathbf{l}(T) + \mathbf{l}(T')$ holds whenever the right ideals T and T' are comparable, we must show that $\mathbf{l}(nR \cap n'R) = \mathbf{l}(nR) + \mathbf{l}(n'R)$ for all positive integers n and n'. One verifies that $\mathbf{l}(nR) = 0 + \mathbf{l}_V(n)$ and that $\mathbf{l}_V(m2^t) = x_t$ whenever m is odd and $t \geq 0$ (we take $x_0 = 0$). If we write $n = m2^t$ and

$n' = m'2^{t'}$, then $nR \cap n'R = n''R$, where $n'' = lcm(n, n') = m''2^{t''}$, $m'' = lcm(m, m')$, and $t'' = max(t, t')$. But then

$$\begin{aligned} \mathtt{l}(nR \cap n'R) &= 0 + \mathtt{l}_V(n'') = 0 + \mathbb{Z}x_{t''} \\ &= (0 + \mathbb{Z}x_t) + (0 + \mathbb{Z}x_{t'}) \\ &= \mathtt{l}(nR) + \mathtt{l}(n'R), \end{aligned}$$

as required. □

The next theorem gives weak finiteness conditions that force a right simple injective ring to be quasi-Frobenius. To prove it we need the following facts about a right mininjective ring.

Lemma 6.43. *Let R be a right mininjective ring in which S_r is right finitely generated. Then the following hold:*

(1) $R/\mathtt{l}(S_r)$ is semisimple artinian.
(2) If $S_r \subseteq^{ess} R_R$ then $J \subseteq Z_r$ and R/Z_r is semisimple artinian.

Proof. If $S_r = k_1R \oplus \cdots \oplus k_nR$, where each k_iR is simple, then $\mathtt{l}(S_r) = \cap_{i=1}^{n}\mathtt{l}(k_i)$ and (1) follows because each Rk_i is simple (as R is right mininjective). If, in addition, $S_r \subseteq^{ess} R_R$ then $\mathtt{l}(S_r) \subseteq Z_r$. As the other inclusion always holds, we have $\mathtt{l}(S_r) = Z_r$ so R/Z_r is semisimple by (1). But then $J \subseteq Z_r$, proving (2). □

Recall that a ring R is called a right *Goldie ring* if it has the ACC on right annihilators and R_R is finite dimensional. Since right artinian rings are right noetherian and have $S_r \subseteq^{ess} R_R$, the following result is a simple injective version of Theorem 1.50.

Theorem 6.44. *The following conditions are equivalent for a ring R:*

(1) R is quasi-Frobenius.
(2) R is a right simple injective, right noetherian ring with $S_r \subseteq^{ess} R_R$.
(3) R is a right simple injective, right Goldie ring with $S_r \subseteq^{ess} R_R$.

Proof. Clearly (1)⇒(2)⇒(3). Assume that (3) holds. Since R has ACC on right annihilators, Z_r is nilpotent by the Mewborn–Winton lemma (Lemma 3.29). Hence $Z_r \subseteq J$. Moreover, S_r is finitely generated because R_R is finite dimensional, so R is semiprimary with $J = Z_r$ by Lemma 6.43. Since $S_r \subseteq^{ess} R_R$, Proposition 6.12 shows that $soc(eR) \neq 0$ for each local idempotent $e \in R$.

Hence Theorem 6.16 shows that R is left finite dimensional with $S_r = S_l$, whence R is left artinian by Lemma 3.30. Since Theorem 6.16 also shows that R is right Kasch, it is left and right mininjective (by Proposition 6.14). Hence R is quasi-Frobenius by Theorem 2.30. □

Note that the hypothesis $S_r \subseteq^{ess} R_R$ is needed in (2) of Theorem 6.44 because $\mathbb{Z}$ is a commutative, noetherian simple injective ring that is not quasi-Frobenius. The Clark example (Example 6.6) shows that the ACC on right annihilators is essential in (3) of Theorem 6.44.

We now return to our study of when a left perfect ring is quasi-Frobenius. We need the following result about relative injectivity.

Lemma 6.45. *Let R be a semiperfect ring. A module E_R is injective if and only if E is eR-injective for every local idempotent $e \in R$.*

Proof. If $1 = f_1 + \cdots + f_n$, where the f_i are orthogonal local idempotents, then E is injective if and only if E is $R = \oplus_{i=1}^n f_i R$-injective (by the Baer criterion), and this holds if and only if E is $f_i R$-injective for each i (by Azumaya's lemma – Lemma 1.13). □

We require the next basic result which uses the following language. If E_R and M_R are R-modules, E is said to be a *simple M-injective* module if, for any submodule $X \subseteq M$ and any R-morphism $\gamma : X \to E$ such that $im(\gamma)$ is simple, there exists an R-morphism $\widehat{\gamma} : M \to E$ such that $\widehat{\gamma}_{|X} = \gamma$. Thus M is simple quasi-injective (see Corollary 6.10) if it is simple M-injective; in particular, a ring R is right simple injective if R_R is simple R-injective.

Lemma 6.46 (Baba–Oshiro). *Let e be a local idempotent in a semiprimary ring R. If eR is simple fR-injective for every local idempotent $f \in R$, then eR is injective.*

Proof. By Lemma 6.45, we need only show that eR is fR-injective for each local idempotent $f \in R$. So let $\gamma : T \to eR$ be R-linear, where $T \subseteq fR$ is a right ideal. We must show that γ extends to $fR \to eR$. Suppose not. Then $\gamma \neq 0$, so $T \nsubseteq ker(\gamma)$. Since J is nilpotent, choose $n_1 \geq 0$ such that

$$TJ^{n_1} \nsubseteq ker(\gamma) \quad \text{but} \quad TJ^{n_1+1} \subseteq ker(\gamma).$$

Since R is semilocal and $\gamma(TJ^{n_1})J = 0$, it follows that $\gamma(TJ^{n_1}) \neq 0$ is a semisimple submodule of eR. Hence Lemma 6.11 shows that $\gamma(TJ^{n_1}) =$

$soc(eR)$ is simple and so, by hypothesis, there exists $\theta_1 : fR \to eR$ such that $(\theta_1)_{|TJ^{n_1}} = \gamma_{|TJ^{n_1}}$. Write $\gamma_1 = \gamma - (\theta_1)_{|T} : T \to R$ so that $TJ^{n_1} \subseteq ker(\gamma_1)$. Then γ_1 does not extend to $fR \to eR$ (if $\widehat{\gamma_1}$ extends γ_1 then $\widehat{\gamma_1} + \theta_1$ extends γ). In particular $T \nsubseteq ker(\gamma_1)$, so choose $n_2 \geq 0$ such that

$$TJ^{n_2} \nsubseteq ker(\gamma_1) \quad \text{but} \quad TJ^{n_2+1} \subseteq ker(\gamma_1).$$

Then $n_2 < n_1$ (for if $n_2 \geq n_1$ then $TJ^{n_2} \subseteq TJ^{n_1} \subseteq ker(\gamma_1)$, which is a contradiction). Continue the process to obtain integers $n_1 > n_2 > \cdots > n_k > \cdots \geq 0$ and maps $\gamma_k : T \to eR$ that do not extend to $fR \to eR$, such that

$$TJ^{n_k} \nsubseteq ker(\gamma_{k-1}) \quad \text{but} \quad TJ^{n_k+1} \subseteq ker(\gamma_{k-1}).$$

This is a contradiction because some $n_k = 0$, whence $T \subseteq ker(\gamma_{k-1})$ and $\gamma_{k-1} = 0$ can be extended. □

Using the Baba–Oshiro lemma, we can prove the following basic result.

Theorem 6.47. *Let R be a semiprimary ring. Then R is right self-injective if and only if R is right simple injective.*

Proof. If R is right simple injective, it suffices to show that eR is injective for all local $e^2 = e \in R$. By the Baba–Oshiro lemma it suffices to show that eR is simple gR-injective for all local idempotents $g \in R$. To this end, let $X \subseteq gR$ be a right ideal and let $\gamma : X \to eR$ be R-linear with simple image. Then γ extends to $\hat{\gamma} : R \to R$ because R is right simple injective. If $\pi : R \to eR$ is the projection then $(\pi \circ \hat{\gamma})_{|gR} : gR \to eR$ extends γ. This completes the proof. □

The Faith conjecture asserts that every semiprimary, right self-injective ring is quasi-Frobenius. Theorem 6.47 makes it easier to find a counterexample to the Faith conjecture: Find a semiprimary, right simple injective ring that is not quasi-Frobenius.

The following construction will be used in Example 6.49 at the end of this section. If M is a module and α is an ordinal, the *α-socle* of M is a submodule $soc_\alpha(M)$ of M defined inductively as follows:

(1) $soc_0(M) = 0$;

(2) if $soc_\alpha(M)$ is defined, $soc_{\alpha+1}(M)$ is given by

$$\frac{soc_{\alpha+1}(M)}{soc_\alpha(M)} = soc\left[\frac{M}{soc_\alpha(M)}\right];$$

(3) if α is a limit ordinal then $soc_\alpha(M) = \cup_{\beta<\alpha} soc_\beta(M)$.

Note that $soc_1(M) = soc(M)$ for every module M. The series $soc_1(M) \subseteq soc_2(M) \subseteq \cdots$ is called the *Loewy series* of the module M. Since M is a set, we have $soc_\gamma(M) = soc_{\gamma+1}(M) = \cdots$ for some ordinal γ. A module is called *semiartinian* if every nonzero factor module has nonzero socle. The higher socles provide a description of the semiartinian modules that we need.

Lemma 6.48. *Let M be a module.*

(1) $soc_\alpha(M)$ is semiartinian for every ordinal α.
(2) M is semiartinian if and only if $soc_\alpha(M) = M$ for some ordinal α.

Proof. For convenience write $S_\alpha = soc_\alpha(M)$.

(1). It is clear for $\alpha = 0$, so assume that $\alpha > 0$ and S_β is semiartinian for each $\beta < \alpha$. If $N \subset S_\alpha$, we must show that $soc(S_\alpha/N) \neq 0$. Clearly $S_\alpha \nsubseteq N$, so let $\mu \leq \alpha$ be the smallest ordinal such that $S_\mu \nsubseteq N$. Thus $S_\beta \subseteq N$ for all $\beta < \mu$, so μ is not a limit ordinal (for otherwise $S_\mu = \cup_{\beta<\mu} S_\beta \subseteq N$). If $\mu = \gamma + 1$, define

$$\theta : soc[M/S_\gamma] = S_{\gamma+1}/S_\gamma \to S_{\gamma+1}/(N \cap S_{\gamma+1})$$

by $\theta(x + S_\gamma) = x + (N \cap S_{\gamma+1})$. Then θ is well defined because $S_\gamma \subseteq N \cap S_{\gamma+1}$, and so it is an R-epimorphism. Hence $(S_{\gamma+1} + N)/N \cong S_{\gamma+1}/(N \cap S_{\gamma+1})$ is a semisimple submodule of S_α/N, and it is nonzero because $S_{\gamma+1} = S_\mu \nsubseteq N$. This completes the induction.

(2). If M is semiartinian, assume that $S_\alpha \neq M$ for every ordinal α. Since M is a set, $S_\gamma = S_{\gamma+1}$ for some ordinal γ. This implies that $soc(M/S_\gamma) = 0$, which, contradicts our hypothesis because $M/S_\gamma \neq 0$. Hence $S_\alpha = M$ for some ordinal α. The converse is by (1). $\square$

We conclude by using Lemma 6.48 and Theorem 6.39 to construct an example of a left simple injective ring that is not right simple injective.

Example 6.49. There exists a left perfect, left simple injective ring that is not right simple injective.

Proof. Let R be a left perfect ring that is not right perfect (see Example B.36). Since R is a set, there exists an ordinal β such that $soc_\beta({}_R R) = soc_{\beta+1}({}_R R)$. Since R is not right perfect, it is not left semiartinian by Theorem B.32, so $soc_\beta({}_R R) \neq R$ by Lemma 6.48. Now observe that $soc_\alpha({}_R R)$ is an ideal of R for every ordinal α (by induction on α). It follows that $S = R/soc_\beta({}_R R)$ is a left perfect ring and that $soc({}_S S) = soc({}_R S) = soc_{\beta+1}({}_R R)/soc_\beta({}_R R) = 0$.

Hence S is left simple injective but not quasi-Frobenius. But then S is not right simple injective by Theorem 6.39.

A specific example is as follows: Let F be a field and let R be the ring of all lower triangular, countably infinite square matrices over F with only finitely many nonzero off-diagonal entries. Let S be the F-subalgebra of R generated by 1 and $J(R)$. Then S is a left perfect, left simple injective ring that is neither right perfect nor right simple injective. Moreover, S is not left self-injective because it is not left finite dimensional. □

6.8. The Second Socle

The Faith conjecture asserts that every left perfect, right self-injective ring is quasi-Frobenius, and this is open even for semiprimary rings. In this section we show that the conjecture is true if the second right socle is countably generated. We begin with the following important result.

Lemma 6.50 (Osofsky's Lemma). *If R is a left perfect ring in which J/J^2 is right finitely generated, then R is right artinian.*

Proof. Let $J = F_1 + J^2$, where F_1 is a finitely generated right ideal. Then $J^2 = F_1 J + J^3$. But J^2/J^3 is also right finitely generated by Lemma 5.65, say $J^2 = \sum_{i=1}^n x_i R + J^3$. Thus $x_i = y_i + z_i$, $y_i \in F_1 J$, $z_i \in J^3$, and so

$$J^2 = \sum_{i=1}^{n} (y_i + z_i)R + J^3 = \sum_{i=1}^{n} y_i R + J^3 = F_2 + J^3,$$

where $F_2 = \sum_{i=1}^n y_i R \subseteq F_1 J \subseteq F_1$. If we continue this process, we get a chain of finitely generated right ideals $F_1 \supseteq F_2 \supseteq \cdots$ such that $J^k = F_k + J^{k+1}$ and $F_k J \supseteq F_{k+1}$ for each $k \geq 1$. Since R is left perfect, we have $F_n = F_{n+1}$ for some $n \geq 1$ by Björk's theorem (see Theorem B.39). This means $F_n J = F_n$, and hence $F_n = 0$ by Nakayama's lemma (Lemma 1.47). Thus $J^n = (J^n)J$ and so, since J is left T-nilpotent, $J^n = 0$ by Lemma B.29. Now by Lemma 5.65 all the factors $R/J, J/J^2, \ldots, J^{n-1}/J^n$ are right finitely generated. But each J^k/J^{k+1} is semisimple (because R/J is semisimple) and so is artinian, and it follows that R is right artinian. □

We can now prove the finitely generated case of our main theorem.

Proposition 6.51. *Let R be a left perfect, right simple injective ring. Then $soc_2({}_R R) = soc_2(R_R)$, which we write as $soc_2(R)$. If $soc_2(R)$ is finitely generated as a left R-module, then R is quasi-Frobenius.*

Proof. Since $S_r \subseteq^{ess} R_R$ (see Lemma B.31 and Theorem B.32), we have $S_r = S_l$ by Theorem 6.16. Hence $\mathtt{l}(J) = S_r = S_l = \mathtt{r}(J)$ because R is semilocal.

Claim. $\mathtt{l}(J^2) = \mathtt{r}(J^2) = soc_2({}_RR) = soc_2(R_R)$.

Proof. First $soc_2(R_R) \subseteq \mathtt{l}(J^2)$ because $soc_2(R_R)/S_r$ is right R-semisimple. However, if $aJ^2 = 0$ then $aJ \subseteq \mathtt{l}(J) = S_r$, so $[(aR + S_r)/S_r]J = 0$. Hence $(aR + S_r)/S_r$ is a semisimple right R-module (because R/J is semisimple) and it follows that $aR \subseteq soc_2(R_R)$. This proves that $soc_2(R_R) = \mathtt{l}(J^2)$. Similarly, $soc_2({}_RR) = \mathtt{r}(J^2)$. Finally, $\mathtt{l}(J^2) = \mathtt{r}(J^2)$ follows easily from $\mathtt{l}(J) = \mathtt{r}(J)$. This proves the Claim.

By hypothesis let $soc_2(R) = \sum_{i=1}^n Ra_i$, $a_i \in R$, and define $\varphi : R \to \oplus_{i=1}^n a_i R$ by $\varphi(r) = (a_1 r, \ldots, a_n r)$. Then $ker\varphi = \mathtt{r}[soc_2(R)] = \mathtt{rl}(J^2) = J^2$, using Theorem 6.16. Thus $R/J^2 \hookrightarrow \oplus_{i=1}^n a_i R$ and it follows that J/J^2 is right finite dimensional (R is right finite dimensional, again by Theorem 6.16). Hence J/J^2 is right finitely generated (it is semisimple because R/J is semisimple), and so R is right artinian by Osofsky's lemma (Lemma 6.50). □

The essence of our main theorem is in the following result.

Proposition 6.52. *Suppose R is a semiperfect, right simple injective ring with $S_r \subseteq^{ess} R_R$. If $soc_2(R)$ is countably generated as a left R-module, then J/J^2 is finitely generated as a right R-module.*

Proof. As $soc(eR) \neq 0$ for all local $e^2 = e \in R$, it follows from Theorem 6.16 that R is right and left Kasch, and that $S_r = S_l$ is finitely generated and essential in both ${}_RR$ and R_R. Write $S = S_r = S_l$. As R/J is semisimple, $S = \mathtt{l}(J) = \mathtt{r}(J)$. Then the proof of the Claim in Proposition 6.51 goes through to show that $\mathtt{l}(J^2) = \mathtt{r}(J^2) = soc_2({}_RR) = soc_2(R_R)$.

Claim. $hom_R(J/J^2, R_R) \cong \mathtt{l}(J^2)/\mathtt{l}(J)$.

Proof. If $b \in \mathtt{l}(J^2)$ then $\lambda_b : J/J^2 \to R$ is well defined by $\lambda_b(a + J^2) = ba$. Hence the map $b \longmapsto \lambda_b$ is an R-homomorphism $\mathtt{l}(J^2) \to hom_R(J/J^2, R_R)$ with kernel $\mathtt{l}(J)$. So it remains to show that this map is onto. If $\lambda : J/J^2 \to R$ is R-linear, define $\gamma : J \to R$ by $\gamma(a) = \lambda(a + J^2)$. Then $im(\gamma) = im(\lambda)$ is right semisimple because this is true of J/J^2. Furthermore, $im(\gamma)$ is finitely generated because S is right finite dimensional. But then $\gamma = c\cdot$ for $c \in R$ by Lemma 6.1. Thus $c \in \mathtt{l}(J^2)$ so, for $a \in J$, $\lambda(a + J^2) = \gamma(a) = ca = \lambda_c(a + J^2)$. Hence $\lambda = \lambda_c$ as required. This proves the Claim.

We now show that J/J^2 is finitely generated as a right R-module. Suppose not. Then, since R has finitely many isomorphism classes of simple right modules, and since J/J^2 is semisimple, there exists a simple right R-module K such that $K^{(\mathbb{N})}$ is a direct summand of J/J^2 (where $K^{(\mathbb{N})}$ denotes the direct sum of a countably infinite number of copies of K). Now write ${}_RT = hom_R(K, R)$. Then T is a simple left R-module by Theorem 2.31 because R is right mininjective and right Kasch. Thus

$$T^{\mathbb{N}} = hom_R(K, R)^{\mathbb{N}} \cong hom_R(K^{(\mathbb{N})}, R)$$

where $T^{\mathbb{N}}$ is the direct product of countably many copies of T. But if we write $J/J^2 = K^{(\mathbb{N})} \oplus Q$, the Claim gives

$$\frac{soc_2({}_RR)}{S} = \frac{1(J)}{1(J^2)} \cong hom\left(\frac{J}{J^2}, R\right) = hom(K^{(\mathbb{N})} \oplus Q, R)$$
$$\cong hom(K^{(\mathbb{N})}, R) \oplus hom(Q, R).$$

Thus $T^{\mathbb{N}}$ is isomorphic to a summand of $soc_2({}_RR)/S_r$. But $T^{\mathbb{N}}$ has Goldie dimension $|T|^{|\mathbb{N}|} > |\mathbb{N}|$, according to a well-known theorem of Erdös and Kaplansky [23, p. 276]. This is contradiction since $soc_2({}_RR)$ is countably generated. □

Observe that the proof of Proposition 6.52 actually yields the following: If R is a semiperfect, right simple injective ring in which $S_r \subseteq^{ess} R_R$ and $soc_2(R)$ is generated on the left by χ elements, where χ is any ordinal number, then $(J/J^2)_R$ is generated by fewer than χ elements. For if this is not the case we can use the same argument to show that $soc_2(R)/S$ contains a direct sum of $2^\chi > \chi$ simple modules, which is a contradiction. In particular, if $soc_2(R)$ is generated on the left by ω elements (where ω is the first infinite ordinal), then $(J/J^2)_R$ is finitely generated.

Theorem 6.53. *Suppose R is a left perfect, right simple injective ring. Then R is quasi-Frobenius if and only if $soc_2(R)$ is countably generated as a left R-module.*

Proof. Proposition 6.52 and its proof show that $soc_2(R_R) = soc_2({}_RR)$, and that J/J^2 is finitely generated as a right R-module. Hence R is right artinian by Osofsky's lemma (Lemma 6.50). Then R is right self-injective by Theorem 6.47. Thus R is quasi-Frobenius. □

Remarks. Similar arguments give the following results:

(1) If R is a semiperfect, right simple injective ring with $S_r \subseteq^{ess} R_R$, and if J/J^2 is countably generated as a left R-module, then $soc_2(R)$ is finitely generated as a right R-module.
(2) If R is a left perfect, right self-injective ring, and if J/J^2 is countably generated as a left R-module, then $soc_2(R)$ is finitely generated as a right R-module, and so it is quasi-Frobenius by Theorem 5.66.

Conjecture. *Every left perfect, right simple injective ring is right self-injective.*

The proof of Theorem 6.47 shows that the conjecture is true if the Baba–Oshiro lemma remains true when "semiprimary" is replaced by "left perfect." Note that the Baba–Oshiro lemma is *not* true when "semiprimary" is replaced by "semiperfect" because, if it were, then Theorem 6.47 would assert that every semiperfect, right simple injective ring is right self-injective. This would imply that the localization $R = \mathbb{Z}_{(p)} = \{\frac{m}{n} \mid p \nmid n\}$ of $\mathbb{Z}$ at the prime p is self-injective (R is a local domain and so is semiperfect and simple injective). But the map $Rp^2 \to R$ given by $rp^2 \mapsto rp$ does not extend to $R \to R$ because R is a domain.

Notes on Chapter 6

The concept of a right simple injective ring is due to Harada [91] (see also [92]).

Example 6.6 was first given in 1986 by Clark [35] to show that if every ideal of a commutative ring has a commutative endomorphism ring, the ring need not have a self-injective maximal ring of quotients. Proposition 6.7, Proposition 6.8, and Theorem 6.9 were established in [174].

Conditions (4), (5), and (6) in Lemma 6.13 are equivalent without the simple injective hypothesis. Rings satisfying these conditions are called *quasi-dual* rings and were studied by Page and Zhou [189].

The term "dual ring" goes back to Kaplansky [118] who, in 1948, was interested in a duality between the closed right ideals and the closed left ideals in a topological ring. Dual rings arose in Nakayama's work on finite dimensional algebras and led in the artinian case to the notion of a quasi-Frobenius ring. These ideas were extended by Azumaya, Osofsky, and Utumi in their work on PF rings, which have come to be called rings with *perfect duality*.

A thorough investigation of dual rings was carried out by Hajarnavis and Norton in 1985 [89], where most of the results on the subject were established. In particular Proposition 6.17, Theorem 6.29, and (1), (2), (4), (6), and (7) of Theorem 6.19 are due to Hajarnavis and Norton. For results on $AB5^*$ see

Lemonnier [135], Xue [227], Herbera and Shamsuddin [95], and Ánh, Herbera, and Menini [5].

Theorem 6.31, Corollary 6.33, Example 6.34, Lemma 6.37, and Theorem 6.38 are all established in [27]. Proposition 6.36 was first proved by Chen and Ding [33].

In Theorem 6.39, the fact that every left perfect, left and right self-injective ring is quasi-Frobenius is due to Osofsky [182].

In 1993, Baba and Oshiro [13, Proposition 2] proved Lemma 6.46, which extends a fundamental result of Fuller [70] on duality in artinian rings. The notion of a simple M-injective module is due to Harada [94].

In connection with Example 6.49, we do not know if a left perfect, right simple injective ring is right self-injective.

As background to Remark (2) following Theorem 6.53, note that Clark and Huynh [36] show that a right and left perfect, right self-injective ring R is quasi-Frobenius if and only if $soc_2(R)$ is finitely generated as a right R-module.

7
FGF Rings

A theorem of Faith and Walker asserts that a ring R is quasi-Frobenius if and only if every injective right R-module is projective and hence that every right module over a quasi-Frobenius ring embeds in a free module. There is an open problem here. If we call a ring R a right *FGF ring* if every finitely generated right R-module can be embedded in a free right R-module, it is not known if the following assertion is true:

The FGF-Conjecture. *Every right FGF ring is quasi-Frobenius*

Here are four important results on the conjecture:

(1) Every left Kasch, right FGF ring is quasi-Frobenius.
(2) Every right self-injective, right FGF ring is quasi-Frobenius.
(3) Every right perfect, right FGF ring is quasi-Frobenius.
(4) Every right CS, right FGF ring is quasi-Frobenius.

We prove all these assertions; in fact we capture all of (1), (2), and (3) in Theorem 7.19: If $M_n(R)$ is a right C2 ring for each $n \geq 1$ and every 2-generated right R-module embeds in a free module then R is quasi-Frobenius. This theorem also implies that the FGF-conjecture is true for right FP-injective rings, and it reformulates the conjecture by showing that it suffices to prove that every right FGF ring is a right C2 ring. Furthermore, the theorem shows that the conjecture is true for semiregular rings with $Z_r = J$. We call these rings right weakly continuous, and investigate their basic properties.

We then turn to the fundamental work of Gómez Pardo and Guil Asensio. A ring R is called a right *CF ring* if every cyclic (that is principal) right R-module can be embedded in a free module.[1] The open problem here is as follows:

[1] Historically, principal modules are referred to as *cyclic* modules in this context.

The CF-Conjecture. *Every right CF ring is right artinian.*

Referring to (4) from the preceding list, Gómez Pardo and Guil Asensio prove that every right CS, right CF ring is right artinian. Note that the Björk example is a left and right artinian, left CF ring that is neither a right CF ring nor a left FGF ring.

The chapter concludes with a self-contained discussion of the Faith–Walker theorems. The proofs require several results about injective modules over noetherian rings that are of interest in their own right.

7.1. FGF Rings and CF Rings

We begin by showing that the FGF-conjecture is true if we insist that every right *and* every left finitely generated module embeds in a free module. In fact we have the following theorem:

Theorem 7.1. *A ring is quasi-Frobenius if and only if every right and every left cyclic module can be embedded in a free module.*

Proof. If the condition holds and T is any right ideal of R, let $\sigma : R/T \to R^{(n)}$ be an embedding. If $\sigma(1+T) = (a_1, \ldots, a_n)$ then $T = \mathtt{r}\{a_1, \ldots, a_n\}$, so $T = \mathtt{rl}(T)$. Similarly, $L = \mathtt{lr}(L)$ for every left ideal L of R, so R is a dual ring. Hence Theorem 6.19 shows that R is left and right finitely cogenerated. Thus every cyclic right (or left) R-module is finitely cogenerated (because it is embedded in a finite direct sum of copies of R). Hence R is right and left artinian by Vámos' lemma (Lemma 1.52). But R is right and left P-injective (being a dual ring), and so is right and left mininjective. Thus R is quasi-Frobenius by Ikeda's theorem (Theorem 2.30). □

The following result records some properties of the right CF and FGF rings for reference.

Lemma 7.2. *Let R denote a ring.*

(1) R is a right CF ring if and only if every right ideal T has the form $T = \mathtt{r}(F)$ for some finite set $F \subseteq R$. In this case R is left P-injective and right Kasch.
(2) If R is a right FGF ring, then R is left FP-injective.
(3) The condition "right FGF ring" is a Morita invariant.

Proof. (1). If T is a right ideal and $\sigma : R/T \to R^n$ is an embedding, then $T = \mathtt{r}\{a_1, \ldots, a_n\}$, where $\sigma(1+T) = (a_1, \ldots, a_n)$. Conversely, if $T = \mathtt{r}\{a_1, \ldots, a_n\}$ then $r \mapsto (a_1 r, \ldots, a_n r)$ is an R-linear mapping $R \to R^n$

with kernel T. This proves the first sentence, and the rest follows by taking T to be principal and maximal, respectively.

(2) and (3). Observe that a ring is a right FGF ring if and only if every finitely generated right module can be embedded in a projective module. This statement is categorical by Proposition A.4 and Corollary A.8, and so (3) follows from the Morita equivalence theorem (Theorem A.20). Hence, if R is a right FGF ring then $M_n(R)$ is also a right FGF ring, and so it is left P-injective by (1). Hence R is left FP-injective by Theorem 5.41, proving (2). □

Clearly, every right FGF ring is a right CF ring. The converse is not true:

Example 7.3. The Björk example is a left CF ring that is not a left FGF ring and not a right CF ring.

Proof. Using the notation of Example 2.5, we have that the cyclic left R-modules are 0, R, and $R/J \cong J$. Since each is embedded in ${}_RR$, R is a left CF ring. However, R is not left mininjective (see Example 2.5), so it is not a right FGF ring by Lemma 7.2. Since R is not quasi-Frobenius, it is not a right CF ring by Theorem 7.1. □

Lemma 7.2 is useful in proving the following characterizations of the right FGF rings that clarifies the relationship between right FGF and right CF rings. Recall that R_n and R^n denote the column and row matrices, respectively, over the ring R.

Theorem 7.4. *The following conditions are equivalent for a ring R:*

(1) R is a right FGF ring.
(2) For all $n \geq 1$, every submodule $X_R \subseteq R_n$ has the form $X = \mathbf{r}_{R_n}(A)$ for some $m \times n$ matrix A over R.
(3) For all $n \geq 1$, every submodule $X_R \subseteq R_n$ has the form $X = \mathbf{r}_{R_n}\{\bar{r}_1, \ldots, \bar{r}_m\}$ for $\bar{r}_j \in R^n$.
(4) $M_n(R)$ is a right CF ring for every $n \geq 1$.

Proof. (1)⇒(2). We have $R_n/X \hookrightarrow R_m$ for some $m \geq 1$ by (1), so let $\gamma : R_n \to R_m$ have $ker(\gamma) = X$. If $\{\underline{e}_1, \ldots, \underline{e}_n\}$ is the standard basis of R_n, write $\gamma(\underline{e}_i) = \underline{a}_i$ for each i and let A denote the $m \times n$ matrix with these $\underline{a}_i$ as its columns. Then (2) follows from the fact that $\gamma(\underline{r}) = A\underline{r}$ for every $\underline{r}$ in R_n.

(2)⇒(3). Given X, choose A as in (2), and take $\bar{r}_i$ to be row i of A.

(3)⇒(4). If T is a right ideal of $M_n(R)$, we must (by Lemma 7.2) show that $T = \mathbf{r}(\mathcal{F})$ for some finite subset $\mathcal{F}$ of $M_n(R)$. Now T has the form $T =$

$[X\ X\ \cdots\ X]$, where $X_R \subseteq R_n$ consists of all (all first) columns of matrices in T. By (3) let $X = \mathbf{r}_{R_n}\{\bar{r}_1, \ldots, \bar{r}_m\}$ for $\bar{r}_j \in R^n$. For each $j = 1, \ldots, m$, let B_j denote the $n \times n$ matrix with $\bar{r}_j$ as row 1 and all other rows zero, and let $\mathcal{F} = \{B_1, \ldots, B_m\}$. Then $X = \mathbf{r}_{R_n}(\mathcal{F})$, and so $T = \mathbf{r}_{M_n(R)}(\mathcal{F})$, as required.

(4)⇒(1). If $X_R \subseteq R_n$ it suffices to find an R-morphism $\gamma : R_n \to R_m$ with $ker(\gamma) = X$. Write $S = M_n(R)$. By (4) and Lemma 7.2, the right ideal $T = [X\ X \cdots X]$ of S has the form $T = \mathbf{r}_S(\mathcal{F})$, where $\mathcal{F}$ is a finite subset of S. Let $\{\bar{r}_1, \bar{r}_2, \ldots, \bar{r}_m\}$ denote the rows of the matrices in $\mathcal{F}$. Then $T = \mathbf{r}_S\{\bar{r}_1, \bar{r}_2, \ldots, \bar{r}_m\}$, so $X = \mathbf{r}_{R_n}\{\bar{r}_1, \bar{r}_2, \ldots, \bar{r}_m\}$, and it follows that the map $\gamma : R_n \to R_m$ given by $\gamma(\underline{r}) = (\bar{r}_1\underline{r}, \bar{r}_2\underline{r}, \ldots, \bar{r}_m\underline{r})^T$ has $ker(\gamma) = X$, as required. □

It is interesting to note the parallel between (1)⇔(4) in Theorem 7.4 and the corresponding characterization of the right FP-injective rings R as those for which $M_n(R)$ is right P-injective for each n.

7.2. C2 Rings

The right FGF rings are closely related to the right C2 rings, and this section is devoted to a study of these rings. Recall that a ring R is called a right *C2 ring* if R_R has the C2-condition (if aR is isomorphic to a summand of R_R, $a \in R$, then aR is itself a summand). Every regular ring is a left and right C2 ring, and every right continuous ring is a right C2 ring. The only C2 domains are the division rings. In fact, we have the following example. Recall that a ring is called *I-finite* if it contains no infinite set of orthogonal idempotents.

Example 7.5. Let R be an I-finite ring. If R is a right C2 ring then every monomorphism $R_R \to R_R$ is epic. The converse is true if 0 and 1 are the only idempotents in R.

Proof. Suppose that $\mathbf{r}(a) = 0$, $a \in R$, and consider $R \supseteq aR \supseteq a^2R \supseteq a^3R \supseteq \cdots$. Since $a^kR \cong R$, we have $a^kR = e_kR$ for some $e_k^2 = e_k$. Hence $a^kR = a^{k+1}R$ for some k by the I-finite hypothesis, whence $R = aR$ because $\mathbf{r}(a) = 0$. This proves that monomorphisms $R_R \to R_R$ are epic. Conversely, given that monomorphisms are epic, assume that 0 and 1 are the only idempotents in R and let $aR \cong P$, where P is a summand of R. Then either $P = 0$ (so $aR = 0$ is a summand) or $P = R$. In the second case, if $\sigma : R \to aR$ is an isomorphism let $\sigma(1) = ab$. Then $\mathbf{r}(b) = 0$, so b is a unit by hypothesis, whence a is a unit and again $aR = R$ is a summand. Thus R is a right C2 ring. □

If F is a field, the ring $R = \begin{bmatrix} F & F \\ 0 & F \end{bmatrix}$ is a right and left artinian ring, so monomorphisms are epic (on either side). However, R is not a right C2 ring because $J = \begin{bmatrix} 0 & F \\ 0 & 0 \end{bmatrix} \cong \begin{bmatrix} 0 & 0 \\ 0 & F \end{bmatrix} = \begin{bmatrix} 0 & 0 \\ 0 & 1 \end{bmatrix} R$, but J is not a summand of R. Similarly, R is not a left C2 ring.

Example 7.6. Every left Kasch ring is a right C2 ring, but the converse is false.

Proof. The first part is Proposition 1.46. Any regular, right self-injective ring is right C2, but it is neither left nor right Kasch if it is not artinian. A specific example is an infinite product of division rings. □

Before giving more examples of C2 rings we derive some basic characterizations of these rings that will be used later.

Lemma 7.7. *The following conditions are equivalent for a ring R:*

(1) R is a right C2 ring.
(2) Every R-isomorphism $aR \to eR$, $a \in R$, $e^2 = e \in R$, extends to R.
(3) If $\mathrm{r}(a) = \mathrm{r}(e)$, $a \in R$, $e^2 = e \in R$, then $e \in Ra$.
(4) If $\mathrm{r}(a) = \mathrm{r}(e)$, $a \in R$, $e^2 = e \in R$, then $Re = Ra$.
(5) If $Ra \subseteq Re \subseteq \mathrm{lr}(a)$, $a \in R$, $e^2 = e \in R$, then $Re = Ra$.
(6) If aR is projective, $a \in R$, then aR is a direct summand of R_R.

Proof. (6)⇒(1)⇒(2)⇒(3)⇒(4)⇒(5) are routine computations. Assume that (5) holds. If aR is projective then $\mathrm{r}(a)$ is a direct summand of R, say $\mathrm{r}(a) = \mathrm{r}(e)$ for $e^2 = e$. Thus $a = ae$, so $Ra \subseteq Re$. But $e \in \mathrm{lr}(a)$ [because $\mathrm{r}(a) \subseteq \mathrm{r}(e)$], so we have $Ra \subseteq Re \subseteq \mathrm{lr}(a)$. Thus $Ra = Re$ by (5), so Ra is a direct summand of R, whence aR is a summand, proving (6). □

Condition (3) in Lemma 7.7 gives

Corollary 7.8. *A direct product of rings is a right C2 ring if and only if each factor is a C2 ring.*

Corollary 7.9. *The following conditions are equivalent for a local ring R:*

(1) R is a right C2 ring.
(2) Every monomorphism $R_R \to R_R$ is epic.
(3) $J = \{a \in R \mid \mathrm{r}(a) \neq 0\}$.

In particular, any local ring with nil radical is a right and left C2 ring.

Proof. Example 7.5 shows that (1)⇔(2) for any ring in which 0 and 1 are the only idempotents. Given (2), it is clear that $J \subseteq \{a \in R \mid \mathtt{r}(a) \neq 0\}$; this is equality in a local ring. Hence (2)⇒(3). Finally, if (3) holds, suppose $\mathtt{r}(a) = \mathtt{r}(e)$, $a \in R$, $e^2 = e \in R$. By Lemma 7.7 we must show that $e \in Ra$. This is clear if $e = 0$. If $e = 1$ then $\mathtt{r}(a) = 0$, so $a \notin J$ by (3). Hence $Ra = R$ because R is local, and so $e \in Ra$ as required. Thus (3)⇒(1). Finally, the last statement follows from (3) because R is local. □

Using Lemma 5.1, condition (3) of Lemma 7.7 gives immediately

Corollary 7.10. *Every right P-injective ring is a right C2 ring.*

The converse of Corollary 7.10 is false. Indeed, if V is a two-dimensional vector space over a field F, consider the trivial extension $R = T(F, V) = F \oplus V$. This is a commutative, local, artinian ring with $J^2 = 0$, so R is a C2 ring by Corollary 7.9. However, R is not P-injective. For if $V = vF \oplus wF$, let $\theta : V \to V$ be a linear transformation with $\theta(v) = w$. Then $(0, x) \mapsto [0, \theta(x)]$ is an R-linear map $R \to R$ that does not extend to $R \to R$ because $w \notin vF$.

Example 7.11. There exists a left C2 ring that is not a right C2 ring.

Proof. Faith and Menal give an example (Example 8.16 below) of a right noetherian ring R in which J is nilpotent and every right ideal is an annihilator, but which is not right artinian. Thus R is left P-injective and hence left C2. Suppose that R is a right C2 ring. Since R is I-finite, every monomorphism $R_R \to R_R$ is epic (see Example 7.5), so R is semilocal by the Camps–Dicks theorem (Theorem C.2) because it is right finite dimensional. Since J is nilpotent, R is right artinian by the Hopkins–Levitzki theorem. But this is a contradiction and hence R is not a right C2 ring. □

Proposition 7.12. *If R is a right C2 ring, so is eRe for any $e^2 = e \in R$ such that $ReR = R$.*

Proof. Write $S = eRe$ and suppose that $\mathtt{r}_S(a) = \mathtt{r}_S(f)$, $a \in S$, $f^2 = f \in S$. By Lemma 7.7 we must show that $f \in Sa$. It suffices to show that $f \in Ra$, so, by Lemma 7.7, we show that $\mathtt{r}_R(a) = \mathtt{r}_R(f)$. If $r \in \mathtt{r}_R(a)$ then, for all $x \in R$, $a(erxe) = arxe = 0$ and so $erxe \in \mathtt{r}_S(a) = \mathtt{r}_S(f)$. Thus $frxe = 0$ for all $x \in R$, so $fr = 0$ because $ReR = R$. Thus $\mathtt{r}_R(a) \subseteq \mathtt{r}_R(f)$; the other inclusion is proved in the same way. □

Proposition 7.12 is half of the proof that a "right C2 ring" is a Morita invariant. We are going to characterize when this is true (in Theorem 7.16), but we must first develop some results about endomorphism rings. Recall that a module is called a C2 module if it satisfies the C2-condition.

Theorem 7.13. *The following conditions are equivalent for a module M_R with $E = end(M_R)$:*

(1) M_R is a C2 module.
(2) If $\sigma : N \to P$ is an R-isomorphism, where $N \subseteq M$ and P is a direct summand of M, then σ extends to some $\beta \in E$.
(3) If $\alpha : P \to M$ is R-monic, where P is a direct summand of M, there exists $\beta \in E$ with $\beta \circ \alpha = \iota$, where $\iota : P \to M$ is the inclusion.
(4) If $\alpha : P \to M$ is R-monic, where P is a direct summand of M, and if $\pi^2 = \pi \in E$ satisfies $\pi(M) = P$, there exists $\beta \in E$ with $\pi \circ \beta \circ \alpha = 1_P$.

Proof. (1)⇒(2). If σ is as in (2), let $M = N \oplus N'$ by (1). Then $(n+n') \mapsto \sigma(n)$ extends σ.

(2)⇒(3). If α is as in (3) then $\sigma : \alpha(P) \to P$ is an R-isomorphism if we define $\sigma[\alpha(p)] = p$ for all $p \in P$. By (2) let $\beta \in E$ extend σ. Then $\beta \circ \alpha = \iota$.

(3)⇒(4). If α is as in (4), let $\beta \circ \alpha = \iota$ by (3), where $\beta \in E$. Then $\pi \circ \beta \circ \alpha = 1_P$.

(4)⇒(1). Suppose a submodule $N \subseteq M$ is isomorphic to P, where P is a direct summand of M, say $\alpha : P \to N$ is an R-isomorphism. We must show that N is a direct summand of M. If $\pi^2 = \pi \in E$ satisfies $\pi(M) = P$, (4) provides $\beta \in E$ such that $\pi \circ \beta \circ \alpha = 1_P$. Define $\theta = \alpha \circ \pi \circ \beta \in E$. Then $\theta^2 = \theta$ and $\theta(M) \subseteq N$, so we are done if we can show that $N \subseteq \theta(M)$. But $\theta \circ \alpha = \alpha$ and so $N = \alpha(P) = \theta[\alpha(P)] \subseteq \theta(M)$, as required. □

It is easy to verify that direct summands of a C2 module are again C2 modules. But the direct sum of C2 modules need not be a C2 module: If $R = \begin{bmatrix} F & F \\ 0 & F \end{bmatrix}$, $A = \begin{bmatrix} F & F \\ 0 & 0 \end{bmatrix}$, and $B = \begin{bmatrix} 0 & 0 \\ 0 & F \end{bmatrix}$, where F is a field, then R_R is not a C2 module, but $R = A \oplus B$ and both A_R and B_R are C2 modules. (B_R is simple, and A_R has exactly one proper submodule $J \ncong A$.)

Theorem 7.14. *Let M_R be a module and write $E = end(M_R)$. Then the following hold:*

(1) If E is a right C2 ring then M_R is a C2 module.
(2) The converse in (1) holds if $ker(\alpha)$ is generated by M whenever $\alpha \in E$ is such that $\mathbf{r}_E(\alpha)$ is a direct summand of E_E.

Proof. (1). Let $\alpha : P \to M$ be R-monic, where P is a direct summand of M, let $\pi^2 = \pi \in E$ satisfy $\pi(M) = P$, and write $ker(\pi) = Q$. Hence $M = P \oplus Q$ and we extend α to $\bar{\alpha} \in E$ by defining $\bar{\alpha}(p + q) = \alpha(p)$. Since α is monic, $ker(\bar{\alpha}) = Q = ker(\pi)$. It follows that

$$\mathrm{r}_E(\bar{\alpha}) = \{\lambda \in E \mid \lambda(M) \subseteq Q\} = \mathrm{r}_E(\pi).$$

Since E is a right C2 ring, Lemma 7.7 gives $\pi \in E\bar{\alpha}$, say $\pi = \beta \circ \bar{\alpha}$ with $\beta \in E$. Then $\pi \circ \beta \circ \alpha = 1_P$, so M_R is a C2 module by Theorem 7.13.

(2) Let $\mathrm{r}_E(\alpha) = \mathrm{r}_E(\pi)$, where α and $\pi^2 = \pi$ are in E. By Lemma 7.7, we must show that $\pi \in E\alpha$.

Claim. $ker(\alpha) = ker(\pi)$.

Proof. $1 - \pi \in \mathrm{r}_E(\pi) = \mathrm{r}_E(\alpha)$, so $\alpha = \alpha \circ \pi$, whence $ker(\pi) \subseteq ker(\alpha)$. However, we have $ker(\alpha) = \Sigma\{\theta(M) \mid \theta \in E, \theta(M) \subseteq ker(\alpha)\}$ by hypothesis. Since $\theta(M) \subseteq ker(\alpha)$ implies $\theta \in \mathrm{r}_E(\alpha) = \mathrm{r}_E(\pi)$, it follows that $\theta(M) \subseteq ker(\pi)$. Hence $ker(\alpha) \subseteq ker(\pi)$, proving the Claim.

Now write $\pi(M) = P$ and $ker(\pi) = Q$. Then $P \cap ker(\alpha) = 0$ by the Claim, so $\alpha_{|P}$ is monic. Since M has the C2 condition, Theorem 7.13 provides $\beta \in E$ such that $\beta \circ (\alpha_{|P}) = \iota$, where $\iota : P \to M$ is the inclusion. We claim that $\beta \circ \alpha = \pi$, which proves (2). If $q \in Q$ then $(\beta \circ \alpha)(q) = 0 = \pi(q)$ by the Claim; if $p \in P$ then $(\beta \circ \alpha)(p) = p = \pi(p)$. As $M = P \oplus Q$ this shows that $\pi = \beta\alpha \in E\alpha$, as required. □

Since a free module generates all of its submodules, we obtain

Theorem 7.15. *If M_R is free then M is a C2 module if and only if $end(M_R)$ is a right C2 ring. In particular R^n is a right C2 module if and only if $M_n(R)$ is a right C2 ring.*

By Proposition 7.12, being a right C2 ring is a Morita invariant if we can show that $M_2(R)$ is a right C2 ring whenever R is a right C2 ring. Hence Theorem 7.15 gives

Theorem 7.16. *The following conditions are equivalent:*

(1) "Right C2 ring" is a Morita invariant.
(2) If R is a right C2 ring then $(R \oplus R)_R$ is a C2 module.

We call a ring R a *strongly right C2 ring* if $M_n(R)$ is a right C2 ring for every $n \geq 1$. This is a Morita invariant class of rings by Proposition 7.12, and it contains every Morita invariant class of right C2 rings.

Example 7.17. Every right FP-injective ring is a strongly right C2 ring.

Proof. A ring R is right FP-injective if and only if $M_n(R)$ is right P-injective for every $n \geq 1$ (Theorem 5.41). Hence we are done by Corollary 7.10. □

Example 7.18. Every semiregular ring with $J = Z_r$ is a strongly right C2 ring.

Proof. By Theorem 7.15 it is enough to show that R^n has the right C2-condition. Let $A \cong B$, where A and B are submodules of R^n and B is a direct summand. Then A is finitely generated (and projective) so, since R is semiregular, there is a decomposition $R^n = P \oplus Q$, where $P \subseteq A$ and $A \cap Q$ is small in R^n (see Theorem B.46). Thus $A = P \oplus (A \cap Q)$, where $A \cap Q \subseteq rad(R^n) = J^n = Z_r^n$ by hypothesis. This means that $A \cap Q$ is both projective and singular, and so $A \cap Q = 0$ by Lemma 4.24. Thus A is a summand of R^n, as required. □

Note that every right continuous ring is semiregular with $J = Z_r$ by Utumi's theorem (Theorem 1.26). Consequently, we call these latter rings right weakly continuous, and we return to them in Section 7.4.

Theorem 7.19. *Suppose that R is a strongly right C2 ring such that every 2-generated right R-module embeds in a free module. Then R is quasi-Frobenius.*

Proof. Let $a \in E(R_R)$, where $E(M)$ denotes the injective hull of a module M. By hypothesis, let $\sigma : R + aR \to (R^n)_R$ be monic. Since R is a right C2 ring and $\sigma(R) \cong R$, it follows that $\sigma(R)$ is a summand of R^n and hence of $\sigma(R + aR)$. But $\sigma(R) \subseteq^{ess} \sigma(R + aR)$ because $R \subseteq^{ess} R + aR$. This implies that $a \in R$, so $R = E(R_R)$ is right self-injective. Hence R_R is a cogenerator by Proposition 1.44 (since R is right Kasch by hypothesis), so R is a right PF ring by Osofsky's theorem (Theorem 1.57; see Corollary 7.33 in the next section). In particular, R_R is finitely cogenerated by Theorem 1.56, and so every cyclic module has finitely generated essential socle (it embeds in R^n for some n). Thus R is right artinian by Vámos' lemma (Lemma 1.52) and so is quasi-Frobenius by Theorem 1.50. □

Theorem 7.19 gives a uniform proof that the FGF-conjecture is true for right self-injective rings, left Kasch rings, and left perfect rings (as listed at the beginning of this chapter), as well as in other cases. In fact we need only assume that every 2-generated right module can be embedded in a free module.

Corollary 7.20. *Suppose a ring R has the property that every 2-generated right R-module embeds in a free module. Then R is quasi-Frobenius if it has any of the following properties:*

(1) R is semiregular with $J = Z_r$.
(2) R is left Kasch.
(3) R is semiperfect with $soc(Re) \neq 0$ for every local idempotent $e \in R$ (for example if R is right perfect).
(4) R is right FP-injective.

Proof. In each case we show that R is a strongly right C2 ring and apply Theorem 7.19.

(1) R is a strongly right C2 ring by Example 7.18.

(2) Since "left Kasch" is Morita invariant, this follows because left Kasch rings are right C2 (Example 7.6).

(3) We have $\mathtt{rl}(T) = T$ for every right ideal T of R because R/T embeds in a free module. In particular R is left mininjective and so is left minfull. Hence R is left Kasch by Theorem 3.12, so (3) follows from (2).

(4) This follows from Example 7.17. □

Since being a right FGF ring is a Morita invariant, Theorem 7.19 gives the following reduction in what is required to prove the FGF-conjecture.

Theorem 7.21. *The following statements are equivalent:*

(1) Every right FGF ring is a right C2 ring.
(2) Every right FGF ring is quasi-Frobenius.

Proof. Given (1), let R be a right FGF ring. Then $M_n(R)$ is also a right FGF ring by Theorem 7.4, and so it is a right C2 ring by (1). Hence R is a strongly right C2 ring and so is quasi-Frobenius by Theorem 7.19. □

7.3. The Gómez Pardo–Guil Asensio Theorem

Given a right R-module M we will denote by $\Omega(M)$ [respectively $C(M)$] a set of representatives of the isomorphism classes of the simple quotient modules (respectively simple submodules) of M. In particular, when $M = R_R$, then $\Omega(R)$ is a set of representatives of the isomorphism classes of simple right R-modules.

We will say that two modules M and N are *essentially equivalent* when they contain isomorphic essential submodules. Observe that if M and N are essentially equivalent, say with $M_0 \subseteq^{ess} M$, $N_0 \subseteq^{ess} N$, and $\sigma : M_0 \to N_0$, then σ induces an isomorphism $soc(M) \cong soc(N)$, and we have $C(M) = C(N)$. Moreover, it is clear that the property of being finitely cogenerated is preserved under essential equivalence.

We are going to characterize finitely cogenerated modules in terms of essential equivalence, and the proof proceeds via a series of lemmas. One of the key points of the proof will be to pass to the endomorphism ring of the quasi-injective hull of a finitely generated quasi-projective CS module essentially equivalent to M.

Lemma 7.22. *Let E be a right R-module, and let $S = end(E_R)$. If L is a direct summand of ${}_SE$ then $hom_R(E, L) \otimes_S E \cong L$.*

Proof. The group $hom_R(E, L)$ is a right S-module via $\lambda\alpha = \lambda \circ \alpha$ for $\lambda \in hom_R(E, L)$ and $\alpha \in S$. Hence the map $hom_R(E, L) \times E \to L$ given by $(\lambda, e) \mapsto \lambda(e)$ (where $e \in E$) is balanced and so induces a homomorphism $\sigma : hom_R(E, L) \otimes_S E \to L$, where $\sigma(\lambda \otimes e) \mapsto \lambda(e)$ for all λ and e. However, if $\pi : E \to L$ is the projection, the map $\tau : L \to hom_R(E, L) \otimes_S E$, where $\tau(x) = \pi \otimes x$ for $x \in L$, is $\mathbb{Z}$-linear and it is routine to verify that both composites $\sigma \circ \tau$ and $\tau \circ \sigma$ are identity maps. Hence σ is a $\mathbb{Z}$-isomorphism; it is clearly R-linear. □

Lemma 7.23. *Let E_R be an injective right R-module and let $S = end(E_R)$. If A is a direct summand of S_S then $hom_R(E, A \otimes_S E) \cong A$.*

Proof. Note first that there is $\mathbb{Z}$-morphism $\theta : A \otimes_S E \to E$ with $\theta(\alpha \otimes e) = \alpha(e)$. Write $A = \pi S$, $\pi^2 = \pi \in S$, and define $\sigma : A \to hom_R(E, A \otimes_S E)$ by

$$\sigma(\alpha) = \gamma_\alpha : E \to A \otimes_S E, \text{ where } \gamma_\alpha(e) = \pi \otimes \alpha(e) \text{ for } e \in E.$$

Then σ is $\mathbb{Z}$-linear. Note that $\theta[\gamma_\alpha(e)] = \pi\alpha(e) = \alpha(e)$ for all $e \in E$ because $\alpha \in A = \pi S$. It follows that σ is monic. However, if $\gamma : E \to A \otimes_S E$ then $\gamma(e) = \Sigma(\pi\alpha_i \otimes e_i) = \pi \otimes \bar{e}$ for $\bar{e} \in E$. Moreover, if also $\gamma(e) = \pi \otimes e_1$, $e_1 \in E$, then $\pi\bar{e} = \theta[\gamma(e)] = \pi e_1$. Hence $\alpha : E \to E$ is well defined by $\alpha(e) = \bar{e}$, where $\gamma(e) = \pi \otimes \bar{e}$, and then $\alpha \in S$ and $\gamma_\alpha(e) = \pi \otimes \alpha(e) = \pi \otimes \bar{e} = \gamma(e)$ for all $e \in E$. It follows that $\gamma = \gamma_\alpha = \sigma(\alpha)$, proving that σ is epic. □

Lemma 7.24. *Let E_R be quasi-injective with $S = end(E_R)$. Then there is a bijection between the set of isomorphism classes of indecomposable direct summands of E_R and the set of isomorphism classes of minimal right ideals of*

$S/J(S)$. In particular there is an injection from the set of isomorphism classes of simple submodules of E_R into the set of isomorphism classes of minimal right ideals of $S/J(S)$.

Proof. Let L_R be a direct summand of E_R. Then by Lemma 7.22 we have a canonical isomorphism $hom_R(E, L) \otimes_S E \cong L$. Similarly, if N_S is a direct summand of S_S, then $hom_R(E, N \otimes_S E) \cong N$ by Lemma 7.23. Thus, the assignments $L \mapsto hom_R(E, L)$ and $N \mapsto N \otimes_S E$ define inverse bijections between the sets of isomorphism classes of direct summands of E_R and S_S. It is also clear that these bijections preserve the property of being indecomposable. In fact, if $e \in S$ is any idempotent such that $L = eE \cong eS \otimes_S E$, then the corresponding direct summand of S_S is precisely $hom_R(E, eE) \cong eS$. Since $end(eE_R) \cong eSe \cong end(eS_S)$, we have that if eE_R is indecomposable then $end(eE)$ is local because E is quasi-injective. Hence the corresponding direct summand eS of S_S is an indecomposable projective module with local endomorphism ring. Writing $J = J(S)$, we know by Lemma B.2 that eSe is local if and only if eS/eJ is a simple right S-module. But eS/eJ is isomorphic, as a right S/J-module, to the minimal right ideal $\bar{e}(S/J)$ of S/J, where $\bar{e} = e + J$, and so we may assign to eS this minimal right ideal.

Conversely, since S is semiregular by Theorem 1.25, each minimal right ideal of S/J is of the form $\bar{e}(S/J)$, where $e \in S$ is an idempotent and $\bar{e} = e + J$. Then we can assign to $\bar{e}(S/J)$ the right ideal eS of S that is, clearly, a projective cover in $modS$ of $\bar{e}(S/J)$, so that eSe is local. Hence, we get a bijection between isomorphism classes of direct summands of S_S with local endomorphism rings and isomorphism classes of minimal right ideals of S/J, which completes the proof. $\square$

Let Q be a ring and $\{C_k \mid k \in K\}$ a family of pairwise nonisomorphic simple right Q-modules. This family is said to be *idempotent-orthogonal* (respectively *idempotent-semiorthogonal*) when there exists a family $\{e_k \mid k \in K\}$ of idempotents of Q satisfying $C_k e_k \neq 0$ for each $k \in K$ and $C_j e_k = 0$ if $j \neq k$ (respectively, $C_j e_k = 0$ or $C_k e_j = 0$). The proof of Lemma 7.26 requires the following set-theoretic result (see [127]).

Lemma 7.25 (Tarski's Lemma). *Let I be an infinite set. Then I can be decomposed as the union of a class K of subsets of I such that the following conditions are satisfied:*

(1) $|K| > |I|$.
(2) $|X| = |Y|$ *for each* X *and* Y *from* K.
(3) $|X \cap Y| < |X| = |Y|$ *if* $X \neq Y$; X, Y *from* K.

Lemma 7.26 (Osofsky). *Let Q be a regular right self-injective ring, and let $\{T_i \mid i \in I\}$ be a set of pairwise nonisomorphic minimal right ideals of Q. If I is an infinite set, there exists an idempotent-orthogonal family $\{C_k \mid k \in K\}$ of simple right Q-modules such that $|I| < |K|$.*

Proof. For each subset $J \subseteq I$, define $\Sigma(J) = \Sigma\{T \subseteq Q_Q | T \cong T_i$ for some $i \in J\}$. Since Q is regular and hence right nonsingular, $\Sigma(J)$ has a unique closure $E[\Sigma(J)]$ in Q_Q by Lemma 1.28 and, because Q is right self-injective, $E[\Sigma(J)]$ is actually the unique injective hull of $\Sigma(J)$ inside Q_Q and hence is a direct summand of Q_Q. Thus there exists an idempotent $e_J \in Q$ such that $E[\Sigma(J)] = e_J Q$. We first show that e_J is central. It is enough to prove that, for every $x \in Q$, $e_J x = e_J x e_J$ and $x e_J = e_J x e_J$. Assume, then, that $e_J x(1 - e_J) \neq 0$. Since $\Sigma(J)$ is essential in $e_J Q$, we see that there exists an element $q \in Q$ and a simple right ideal T of Q such that $T \cong T_i$ for some $i \in J$ and $0 \neq e_J x(1 - e_J)q \in T$. Thus the homomorphism $(1 - e_J)qQ \to T$ given by left multiplication by $e_J x$ is nonzero and, since T_Q is a projective module it is, actually, a split epimorphism. Therefore, $(1 - e_J)Q$ contains a simple submodule isomorphic to T_i, for some $i \in J$, contradicting the fact that all these submodules are contained in $e_J Q$. Thus we have proved that $e_J Q(1 - e_J) = 0$.

Assume now that $(1 - e_J)xe_J \neq 0$ for some $x \in Q$. Then, left multiplication by $(1 - e_J)x$ gives a nonzero homomorphism $e_J Q \to (1 - e_J)Q$. Its image $(1 - e_J)xe_J Q$ is a principal right ideal of Q contained in $(1 - e_J)Q$ and, since Q is regular, it is a projective module. Hence $e_J Q$ contains a submodule isomorphic to $(1 - e_J)xe_J Q$. But the latter module has zero intersection with $e_J Q$ and thus it contains no submodule isomorphic to T_i, for any $i \in J$, which contradicts the fact that $\Sigma(J)$ is essential in $e_J Q$. Thus we see that all the e_J are indeed central idempotents.

For subsets X and Y of I we have $\Sigma(X) = \Sigma(X - Y) \oplus \Sigma(X \cap Y)$, so that $E[\Sigma(X)] = E[\Sigma(X - Y)] \oplus E[\Sigma(X \cap Y)]$, and hence $e_X Q = e_{X \setminus Y} Q \oplus e_{X \cap Y} Q$. Thus we see that $e_{X \cap Y} \in e_X Q$ and, similarly, $e_{X \cap Y} \in e_Y Q$. Consequently, $e_{X \cap Y} \in e_X Q \cap e_Y Q \subseteq e_X e_Y Q = e_Y e_X Q$. However, if T is a simple right Q-module contained in $e_X e_Y Q$, then $T_i \cong T \cong T_j$ for some $i \in X$ and $j \in Y$, and so $i = j \in X \cap Y$. Therefore, $T \subseteq \Sigma(X \cap Y)$. Because $e_X e_Y Q$ has essential socle, we have that $e_X e_Y Q \subseteq E[\Sigma(X \cap Y)] = e_{X \cap Y} Q$, and so $e_{X \cap Y} Q = e_X e_Y Q$. Since $e_X e_Y$ and $e_{X \cap Y}$ are both central idempotents, we see that $e_X e_Y = e_{X \cap Y}$. Observe also that, similarly, $e_X Q = e_Y Q$ implies $e_X = e_Y$ so that the idempotent e_X is uniquely determined by the set X. Furthermore, if $X \subseteq Y \subseteq I$, then $e_X Q \subseteq e_Y Q$ and $e_X e_Y = e_Y e_X = e_X$, whereas if $X \cap Y = \emptyset$, then $e_X e_Y = 0$.

Next, we construct the idempotent-orthogonal family $\{C_k \mid k \in K\}$. By Lemma 7.25, the infinite set I can be decomposed as the union of a class K of subsets of I such that the following conditions are satisfied: (1) $|K| > |I|$; (2) $|X| = |Y|$ for each $X, Y \in K$; and (3) $|X \cap Y| < |X| = |Y|$ if $X \neq Y$ $(X, Y \in K)$.

Now let $N = \Sigma\{e_Z Q | Z \subseteq I \text{ and } |Z| < |X| \text{ for each } X \in K\}$ and define, for each $X \in K$, $N_X = (1 - e_X)Q + N$. Assume that $e_X \in N_X$. Then we may write $e_X = (1 - e_X)x_0 + n$, with $x_0 \in Q$, $n \in N$. Thus there exist sets $Z_1, \ldots, Z_r \subseteq I$ such that $|Z_j| < |X|$ for $j = 1, \ldots, r$, and $n = \sum_{j=1}^{r} e_{Z_j} x_j$, with $x_j \in Q$, so that $e_X Q \subseteq (1 - e_X)Q + \sum_{j=1}^{r} e_{Z_j} Q$. Since $|X|$ is infinite, we have that $|\cup_{j=1}^{r} Z_j| < |X|$ and hence there exists an element $i \in X \backslash (\cup_{j=1}^{r} Z_j)$. Setting $e_i = e_{\{i\}}$ we see that, as $i \in X$, $e_i \in e_X Q$ and so there exist elements $q, q_j \in Q$ such that $e_i = (1 - e_X)q + \sum_{j=1}^{r} e_{Z_j} q_j$. Because $i \notin Z_j$ we see that $e_i e_{Z_j} = 0$ for each $j = 1, \ldots, r$ and hence $e_i = e_i(1 - e_X)q = 0$, which is a contradiction and shows that $e_X \notin N_X$. However, if $X, Y \in K$ and $X \neq Y$, then $e_Y = (1 - e_X)e_Y + e_X e_Y = (1 - e_X)e_Y + e_{X\cap Y}$. Since $|X \cap Y| < |X|$, for each $X \in K$, we see that $e_{X\cap Y} \in N$ and so $e_Y \in N_X$.

Now, let M_X be a maximal right ideal of Q containing N_X. Since $e_Y \in N_X \subseteq M_X$, we see that the simple right Q-module Q/M_X is annihilated by e_Y, for each $Y \in K$, $Y \neq X$. It is clear, however, that $e_X \notin M_X$ for, otherwise, we would have $1 = e_X + (1 - e_X) \in M_X$. Then we can define a set of simple right Q-modules indexed by K, by setting $C_k = Q/M_k$ for each $k \in K$, and we have that $C_k e_k \neq 0$ and $C_j e_k = 0$ for each $k, j \in K$, $j \neq k$. Furthermore, since the e_k are central idempotents for all $k \in K$, we have that $C_j \cong C_k$ implies $j = k$. □

Lemma 7.26 will be applied to the case in which $Q = S/J$, where S is the endomorphism ring of the quasi-injective hull $E_R = E(P_R)$ of a finitely generated quasi-projective module P_R essentially equivalent to M_R. To obtain useful information about E, and hence about M, it is necessary to establish some connection between the simple right S/J-modules and the simple quotients of P_R. This is the object of the next lemma.

If P and M are R-modules, P is called *M-projective* if for R-morphisms $\lambda : M \to N$ and $\phi : P \to N$ with λ epic, there exists $\hat{\phi} : P \to M$ such that $\phi = \lambda \circ \hat{\phi}$. The module P is called *quasi-projective* if it is P-projective. As in Lemma 1.12, if $K \subseteq M$ are modules and P is M-projective then P is both K-projective and M/K-projective.

If M is a module, the *quasi-injective hull* $\hat{M}$ of M is defined to be

$$\hat{M} = \Sigma\{\lambda(M) \mid \lambda : end[E(M)]\}.$$

Hence $M \subseteq^{ess} \hat{M}$, and $M = \hat{M}$ if and only if M is quasi-injective by Theorem 1.15.

Lemma 7.27. *Let P_R be a finitely generated, quasi-projective CS module, let E_R be its quasi-injective hull, and let $S = end(E_R)$. If $\{C_k \mid k \in K\}$ is an idempotent-semiorthogonal family of simple right $S/J(S)$-modules, then we have $|K| \leq |\Omega(P)|$.*

Proof. We show first that each direct summand of E_R contains an essential submodule that is isomorphic to a direct summand of P_R and hence a finitely generated E-projective module. Indeed, let X be a direct summand of E and $Y = X \cap P$. Since P is a CS module, there exists a direct summand Z of P such that Y is essential in Z. Moreover, since X is E-injective (being a direct summand of E), the inclusion of Y into X has an extension to Z that is a monomorphism because $Y \subseteq^{ess} Z$. Thus we see that X contains an essential submodule isomorphic to Z that is finitely generated and E-projective because it is a direct summand of P.

Consider an idempotent-semiorthogonal family $\{C_k\}_{k \in K}$ of simple right $S/J(S)$-modules. Since idempotents of $S/J(S)$ lift modulo $J(S)$ by Theorem 1.25, there exist idempotents $\{s_k\}_{k \in K}$ of S such that $C_k s_k \neq 0$ for any $k \in K$ and either $C_j s_k = 0$ or $C_k s_j = 0$ for $k \neq j$. Let, for each $k \in K$, $c_k \in C_k$ be an element such that $c_k s_k \neq 0$ and call $p_k : S_S \to C_k$ the homomorphism defined by $p_k(1) = c_k s_k$. If $s_{k*} = hom_R(E, s_k)$ is the endomorphism of S_S given by left multiplication with s_k, we have that $(p_k \circ s_{k*})(1) = c_k s_k^2 = c_k s_k = p_k(1)$, and so $p_k \circ s_{k*} = p_k$, from which it follows that

$$(p_k \otimes_S E) \circ s_k = (p_k \otimes_S E) \circ (s_{k*} \otimes_S E) = (p_k \circ s_{k*}) \otimes_S E = p_k \otimes_S E.$$

If we set $E_k = s_k(E)$ and $E_k^{'} = (1 - s_k)(E)$, then as we have observed at the beginning of the proof, E_k and $E_k^{'}$ have finitely generated E-projective essential submodules P_k and $P_k^{'}$, respectively, which are isomorphic to direct summands of P. Thus $T = P_k \oplus P_k^{'}$ is essential in $E = E_k \oplus E_k^{'}$ and, moreover, $(p_k \otimes_S E)(P_k^{'}) \subseteq [(p_k \otimes_S E) \circ s_k \circ (1 - s_k)](E) = 0$. Suppose that $(p_k \otimes_S E)(P_k) = 0$. Then $(p_k \otimes_S E)(P_k \oplus P_k^{'}) = 0$ and if $N_k = ker\ p_k$, then $ker\ (p_k \otimes_S E) = N_k E$, so that $T \subseteq N_k E$. Since T is finitely generated, there exist elements $h_1, \ldots, h_n \in N_k$ such that $T \subseteq \Sigma_{i=1}^n im\ h_i$. Denote by $\pi_i : E^n \to E$ the canonical projections and set $h = \Sigma_{i=1}^n (h_i \circ \pi_i) : E^n \to E$. Then $im\ h = \Sigma_{i=1}^n im\ h_i$ and hence $T \subseteq im\ h$. Thus if we set $X = im\ h$, with canonical projection $\beta : E^n \to X$ and canonical injection $\alpha : X \to E$, and let $u : T \to E$ and $v : T \to X$ be the canonical inclusions, we have that $u = \alpha \circ v$. Since T is E-projective, there exists a morphism $g : P \to E^n$ such that $\beta \circ g = v$. Using the quasi-injectivity of E^n we also obtain an extension $t : E \to E^n$ of g, so that $t \circ u = g$. Thus we have

$$u = \alpha \circ v = \alpha \circ \beta \circ g = h \circ g = h \circ t \circ u,$$

so that $(1 - h \circ t) \circ u = 0$ and hence $ker\ (1 - h \circ t)$ is essential in E_R. This entails that $1 - h \circ t \in J(S)$ and so $h \circ t$ is an isomorphism. But since $h \circ t = \Sigma_{i=1}^n (h_i \circ \pi_i) \circ t = \Sigma_{i=1}^n h_i \circ (\pi_i \circ t) \in N_k$, we see that $N_k = S$, a contradiction that shows that $(p_k \otimes_S E)(P_k) \neq 0$.

Let $h_k : P_k \to E_k, i_k : E_k \to E$, and $t_k = i_k \circ h_k : P_k \to E$ be the inclusions, and set $L_k := im\ [(p_k \otimes_S E) \circ t_k]$, with canonical projection $q_k : P_k \to L_k$ and inclusion $w_k : L_k \to C_k \otimes_S E$. As we have just seen, L_k is a (finitely generated) nonzero module, and hence we can choose for each $k \in K$ a simple quotient U_k of L_k with canonical projection $\pi_k : L_k \to U_k$. Denoting by $[X]$ the isomorphism class of a right R-module X, we see that $[U_k] \in \Omega(P)$ and so we can define a map from K to $\Omega(P)$ by assigning $k \mapsto [U_k]$.

To complete the proof of the lemma, we show that the map just defined is injective. Suppose that $[U_j] = [U_k]$ for $j, k \in K$. Since $\{C_k \mid k \in K\}$ is an idempotent-semiorthogonal family of simple right $S/J(S)$-modules, we can assume that, say, $C_k s_j = 0$. Let $\varphi : U_j \to U_k$ be an isomorphism. If $\alpha_k : U_k \to E(U_k)$ denotes the inclusion, for each $k \in K$ [where $E(U_k)$ is the injective hull of U_k], we obtain by injectivity an R-homomorphism $\phi : E(U_j) \to E(U_k)$ satisfying $\phi \circ \alpha_j = \alpha_k \circ \varphi$. Also, $\alpha_k \circ \pi_k$ has an extension π'_k to $C_k \otimes_S E$, so that $\alpha_k \circ \pi_k = \pi'_k \circ w_k$. However, because P_j is P_k-projective, we obtain a homomorphism $\psi : P_j \to P_k$ such that $\pi_k \circ q_k \circ \psi = \varphi \circ \pi_j \circ q_j$. Now the quasi-injectivity of E gives an endomorphism $\tau : E \to E$, that is, an element $\tau \in S$, such that $\tau \circ t_j = t_k \circ \psi$. Observe then that $\phi \circ \alpha_j = \alpha_k \circ \varphi$ is a monomorphism, and hence the morphism $\phi \circ \alpha_j \circ \pi_j \circ q_j : P_j \to E(U_k)$ is nonzero (with image isomorphic to U_j). Thus we see that

$$\begin{aligned} 0 \neq \phi \circ \alpha_j \circ \pi_j \circ q_j &= \alpha_k \circ \varphi \circ \pi_j \circ q_j = \alpha_k \circ \pi_k \circ q_k \circ \psi \\ &= \pi'_k \circ w_k \circ q_k \circ \psi = \pi'_k \circ (p_k \otimes_S E) \circ t_k \circ \psi \\ &= \pi'_k \circ (p_k \otimes_S E) \circ \tau \circ t_j = \pi'_k \circ (p_k \otimes_S E) \circ \tau \circ i_j \circ h_j. \end{aligned}$$

Assume now that $j \neq k$ and consider the homomorphism $p_k \circ \tau_* \circ i_{j*} : s_j S \to C_k$, where $\tau_* = hom_R(E, \tau)$ and $i_{j*} = hom_R(E, i_j)$. If we set $x := (p_k \circ \tau_*)(1) \in C_k$, we have that $(p_k \circ \tau_* \circ i_{j*})(s_j) = (p_k \circ \tau_*)(s_j) = x s_j \in C_k s_j = 0$. Tensoring with ${}_S E$ we then see that $(p_k \otimes_S E) \circ (\tau_* \otimes_S E) \circ (i_{j*} \otimes_S E) = 0$ and, since $\tau_* \otimes_S E \cong \tau$ and $i_{j*} \otimes_S E \cong i_j$, that $(p_k \otimes_S E) \circ \tau \circ i_j = 0$, which gives a contradiction and shows that we must have $j = k$. □

Lemma 7.28. *Let R be a ring, and let P_R be a finitely generated quasi-projective, CS module, such that $|\Omega(P)| \leq |C(P)|$. Then $|\Omega(P)| = |C(P)|$, and P_R has finitely generated essential socle.*

Proof. Let E_R be the quasi-injective hull of P_R, $S = end(E_R)$, and $J = J(S)$. Let $\mathcal{S}$ be the set of isomorphism classes of minimal right ideals of S/J. Then $C(P) = C(E)$ and so $|C(P)| \leq |\mathcal{S}|$ by Lemma 7.24 and, if $\mathcal{S}$ is infinite, we have by Lemma 7.26 an idempotent-orthogonal family $\{C_k \mid k \in K\}$ of simple right S/J-modules such that $|\mathcal{S}| < |K|$. From Lemma 7.27 it follows that $|K| \leq |\Omega(P)|$ and so we have a chain of inequalities

$$|\Omega(P)| \leq |C(P)| \leq |\mathcal{S}| < |K| \leq |\Omega(P)|$$

that shows that $\mathcal{S}$, and hence $\Omega(P)$, must be finite with, say, $|\Omega(P)| = n \leq |\mathcal{S}| = r$.

Let $C_1, \ldots, C_r$ be representatives of the elements of $\mathcal{S}$. For each $i = 1, \ldots, r$ there exist idempotent elements $e_1, e_2, \ldots, e_r \in S$ such that, if $\bar{e}_i = e_i + J$, then $C_i = \bar{e}_i(S/J)$. Since $X\bar{e}_i \cong hom_{S/J}[\bar{e}_i(S/J), X]$ for $X \in mod(S/J)$, we see that $\bar{e}_i(S/J)\bar{e}_j = 0$ for $i, j \leq r, i \neq j$, and $\bar{e}_i(S/J)\bar{e}_i \neq 0$ for all $i = 1, \ldots, r$. Thus $\{C_1, \ldots, C_r\}$ is an idempotent-orthogonal family of simple right S/J-modules with respect to $\{e_1, \ldots, e_r\}$ and so $r \leq n$ by Lemma 7.27. Then it follows that $r = n$ and hence $|\Omega(P)| = n = |C(P)|$.

Next we show that S/J is a semisimple ring. Suppose that this is not the case. Then there exists a simple right S-module $C = C_{r+1}$ that is not isomorphic to any of the $C_1, \ldots, C_r$. Moreover, $C\bar{e}_i \cong hom_{S/J}[\bar{e}_i(S/J), C] = 0$ for all $i = 1, \ldots, r$ and so the family $\{C_i \mid i = 1, \ldots, r+1\}$ of simple right S/J-modules is idempotent-semiorthogonal with respect to the idempotents $\{\bar{e}_1, \ldots, \bar{e}_r, 1\}$. Then, again by Lemma 7.27, we obtain that $n + 1 = r + 1 \leq n$; this is a contradiction that shows that C cannot exist, and hence that S/J is a semisimple ring. Therefore, S is semiperfect and hence it has no infinite families of orthogonal idempotents, from which it follows that E_R is finite dimensional. Since E_R is quasi-injective we see that E_R is actually a finite direct sum of indecomposable submodules. The number of isomorphism classes of indecomposable direct summands of E_R equals r by Lemma 7.24 and so it is also equal to $|C(P)| = |C(E)|$. Since each quasi-injective indecomposable module is uniform, we see that each indecomposable direct summand of E_R has an essential simple submodule and so is finitely cogenerated, Hence E_R and hence also P_R are finitely cogenerated. □

Theorem 7.29. *The following are equivalent for a right module M_R:*

(1) M is finitely cogenerated.
(2) M is essentially equivalent to a finitely generated quasi-projective CS module P that cogenerates every simple quotient of P.
(3) M is essentially equivalent to a finitely generated quasi-projective CS module P such that $|\Omega(P)| \leq |C(P)|$.

Proof. The fact that (1) implies (2) is clear for if M is finitely cogenerated, then $soc(M)$ is a finitely generated essential submodule that is quasi-projective, CS, and cogenerates its simple quotients. The implication (2)⇒(3) is also clear and so we only need to prove that (3) implies (1). Suppose that M is essentially equivalent to a finitely generated quasi-projective CS module P such that $|\Omega(P)| \leq |C(P)|$. Then P is finitely cogenerated by Lemma 7.28, and hence so is M. □

Lemma 7.30. *If M is a finitely generated module and $E(M)$ is projective, then $E(M)$ is finitely generated.*

Proof. Let $E(M) \oplus X = F$, where F is free. Since M is finitely generated, we have $M \subseteq F_0$, where $F = F_0 \oplus F_1$ and F_0 is a finitely generated submodule of F. Let $\pi : F \to F_0$ be the projection with $ker(\pi) = F_1$. Then $ker(\pi) \cap M = 0$, so $ker(\pi) \cap E(M) = 0$. Thus $\pi_{|E(M)} : E(M) \to F_0$ is monic, so we are done because the injective module $\pi[E(M)]$ is a direct summand of F_0. □

Corollary 7.31. *Let R be a ring such that $E(R_R)$ is a projective cogenerator of $modR$. Then R is a right PF ring.*

Proof. By Lemma 7.30, $E(R_R)$ is finitely generated. Since R_R and $E(R_R)$ are essentially equivalent, and $E(R_R)$ is finitely generated and projective and is an injective cogenerator, it follows from Theorem 7.29 that R_R has finitely generated essential socle. Hence R is right finite dimensional and right Kasch [by Proposition 1.44 because $E(R_R)$ is a cogenerator], and so R has only finitely many isomorphism classes of simple right modules; let $\{K_1, \ldots, K_n\}$ be an irredundant set. If we write $E_i = E(K_i)$, then $E_1, \ldots, E_n$ are pairwise nonisomorphic, indecomposable injective modules that are projective. Hence $end(E_i)$ is local for each i, and so Lemma 1.54 shows that $rad(E_i)$ is maximal and small in E_i. Hence $T_i = E_i/rad(E_i)$ is simple and E_i is a projective cover of T_i. Moreover, if $T_i \cong T_j$ then $E_i \cong E_j$ by Corollary B.17, and hence $i = j$. Thus

$\{T_1, \ldots, T_n\}$ is a set of distinct representatives of the simple right R-modules and it follows that every simple right R-module has a projective cover. Thus R is semiperfect by Theorem B.21.

Let $\{e_1, \ldots, e_n\}$ be a basic set of local idempotents in R. Since each E_i is indecomposable and projective we have $E_i \cong e_{\sigma i}R$ for some $\sigma i \in \{1, \ldots, n\}$. Since the E_i are pairwise nonisomorphic, it follows that σ is a bijection and hence that each e_iR is injective. But then eR is injective for every local idempotent in R. Since R is semiperfect, it follows that R is right self-injective and so is a right PF ring by Theorem 1.56. □

The next corollary extends Osofsky's theorem (Theorem 1.57) that every right self-injective, right cogenerator ring has finitely generated essential right socle.

Corollary 7.32. *Let R be a right CS, right Kasch ring. Then R has finitely generated, essential right socle.*

Proof. The proof is immediate from (2) of Theorem 7.29. □

It is interesting to observe that, although the rings of the preceding corollary have finitely generated, essential right socle, we do not know whether they must be semiperfect. This is in stark contrast with what happens when the CS and Kasch conditions are on opposite sides for, although a right CS left Kasch ring is semiperfect by Lemma 4.1, it is not known whether a right self-injective, left Kasch ring has finite essential right socle. If this were the case, then these rings would be precisely the right PF rings.

Strengthening the hypotheses of Corollary 7.32, we obtain the following characterization of right PF rings, which gives a different extension of Osofsky's theorem (Theorem 1.57).

Corollary 7.33. *Let R be a ring. Then R is a right PF ring if and only if R_R is a cogenerator that is a right CS ring.*

Proof. Every right PF ring is right self-injective and is a right cogenerator by Theorem 1.56. Conversely, if R_R is a CS cogenerator then R has finitely generated, essential right socle by Corollary 7.32. Since R is right finite dimensional and right Kasch, let $\{K_1, \ldots, K_n\}$ be a set of representatives of the simple right R-modules. If we write $E_i = E(K_i)$, then $E_1, \ldots, E_n$ are pairwise nonisomorphic, indecomposable injective modules. Since R_R is a cogenerator, there

replace left PP by right PP. Indeed, such a ring is right weakly continuous. But $Z_r = 0$ because R is a right PP ring, so $J = 0$ and R is regular by Theorem 7.38. Hence $Z_l = 0$, so R is semisimple.

Right continuity is not a Morita invariant [because the matrix ring $M_2(R)$ is right continuous if and only if R is right self-injective by Theorem 1.35]. However, right weak continuity *is* a Morita invariant. The proof requires the following lemma, which is of independent interest.

Lemma 7.39. *Let R denote a ring and, as usual, denote $Z_r = Z(R_R)$.*

(1) If $S = eRe$, where $e^2 = e \in R$ satisfies $ReR = R$, then $Z(S_S) = eZ_re$.
(2) If $S = M_n(R)$ where $n \geq 1$, then $Z(S_S) = M_n(Z_r)$.
(3) Each of the following is a Morita invariant property of rings:
(i) $J \subseteq Z_r$.
(ii) $Z_r \subseteq J$.
(iii) $J = Z_r$.

Proof. (1). Let $z \in eZ_re$ and $0 \neq b \in S$. Then $bR \cap \mathbf{r}_R(z) \neq 0$, say $zbr = 0$, where $br \neq 0$. Then $0 \neq brR = brReR$, so $brse \neq 0$ for some $s \in R$. Since $be = b$ we have $0 \neq berse \in bS \cap \mathbf{r}_S(z)$, and hence $\mathbf{r}_S(z) \subseteq^{ess} S_S$. This proves that $eZ_re \subseteq Z(S_S)$. Conversely, suppose $z \in Z(S_S)$. Since $z = eze$ it suffices to show that $z \in Z_r$. So let $0 \neq b \in R$; we must show that $bR \cap \mathbf{r}_R(z) \neq 0$. If $eb = 0$ then $zb = 0$, so $bR \cap \mathbf{r}_R(z) = bR \neq 0$. If $eb \neq 0$ then $ebReR \neq 0$, so let $ebre \neq 0$, where $r \in R$. Then $ebreS \cap \mathbf{r}_S(z) \neq 0$ by hypothesis, say $z(ebres) = 0$, but $ebres \neq 0$, where $s \in S$. As $ze = z$, we have $0 \neq bres \in bR \cap \mathbf{r}_R(z)$.

(2). Let $\alpha = [a_{ij}] \in Z(S_S)$; we must show that each $a_{ij} \in Z_r$. Since $\varepsilon_{pq}\alpha\varepsilon_{kl} \in Z(S_S)$ for any matrix units ε_{pq} and ε_{kl}, it suffices to show that $\alpha = \left[\begin{smallmatrix} a & 0 \\ 0 & 0 \end{smallmatrix}\right] \in Z(S_S)$ implies that $a \in Z_r$. Thus, if $0 \neq b \in R$, we must show that $\mathbf{r}_R(a) \cap bR \neq 0$ that is, $br \neq 0$ but $abr = 0$ for some $r \in R$. Write $\beta = \left[\begin{smallmatrix} b & 0 \\ 0 & 0 \end{smallmatrix}\right]$. Then $\mathbf{r}_S(\alpha) \cap \beta S \neq 0$, so $\beta\rho \neq 0$ and $\alpha\beta\rho = 0$ for some $\rho \in S$. If $\rho = \left[\begin{smallmatrix} r & X \\ Y & C \end{smallmatrix}\right]$ this implies that either $br \neq 0$ or $bX \neq 0$, while $abr = 0 = abX$. If $br \neq 0$ we are done; otherwise $bx \neq 0$ for some $x \in X$, and again we are done because $abx = 0$.

Conversely, if $\alpha = [a_{ij}] \in M_2(Z_r)$, then $T = \cap_{i,j}\mathbf{r}_R(a_{ij})$ is essential in R_R. Write $\overline{T} = M_n(T)$, so that $\overline{T} \subseteq \mathbf{r}_R(a)$; we show that $\overline{T} \subseteq^{ess} S_S$. If $0 \neq \omega \in S$ let X be a nonzero column of ω, say column k. If R_n denotes the set of columns from R, then $T_n \subseteq^{ess} R_n$ by Lemma 1.1, so let $0 \neq Xr \subseteq T_n$, $r \in R$. If $\rho = r\varepsilon_{kk}$, where ε_{kk} is the matrix unit, then $0 \neq \omega\rho \in T \cap \omega S$, as required.

(3). This follows from (1) and (2) and the well-known facts that $J[M_n(R)] = M_n[J(R)]$ for any $n \geq 1$ and $J(eRe) = eJ(R)e$ for any $e^2 = e \in R$. □

Theorem 7.40. *Being right weakly continuous is a Morita invariant property of rings.*

Proof. Semiregularity is a Morita invariant by Corollary B.55, as is the condition that $J = Z_r$ by Lemma 7.39. □

With this we obtain a strengthening of a theorem of Utumi that R is right self-injective if and only if $R \oplus R$ is continuous (equivalently quasi-continuous) as a right R-module (Theorem 1.35).

Corollary 7.41. *The following are equivalent for a ring R:*

(1) R is right self-injective.
(2) R is right weakly continuous and $R \oplus R$ is CS as a right module.
(3) R is a right C2 ring and $R \oplus R$ is CS as a right module.

Proof. (1)⇒(3) is clear and (3)⇒(2) because summands of CS modules are CS.

(2)⇒(1). If R is right weakly continuous, so also is $M_2(R) \cong end(R \oplus R)$ by Theorem 7.40. In particular $end(R \oplus R)$ is a right C2 ring, and this implies that $R \oplus R$ has the right C2-condition by Theorem 7.15. Hence $R \oplus R$ is continuous, and this implies that R is right self-injective by Theorem 1.35. □

We now turn to weak continuity of endomorphism rings. Recall that a module M_R is said to *generate* a submodule K if $K = \Sigma\{\theta(M) \mid \theta : M_R \to K_R\}$, and that M is called *quasi-projective* if, whenever $\beta : M_R \to N_R$ is onto and $\theta : M_R \to N_R$, there exists $\lambda : M_R \to M_R$ such that $\theta = \beta \circ \lambda$.

Theorem 7.42. *If M_R is a continuous, quasi-projective module that generates each of its submodules, then $end(M_R)$ is weakly continuous.*

Proof. If we write $E = end(M_R)$, then Theorem 1.25 shows that E is semiregular and $J(E) = \{\alpha \in E \mid ker(\alpha) \subseteq^{ess} M\}$. Thus $Z(E_E) \subseteq J(E)$ by Corollary B.41. Conversely, let $\alpha \in J(E)$. If $0 \neq \beta \in E$, we must show that $\beta E \cap \mathrm{r}_E(\alpha) \neq 0$. We have $ker(\alpha) \cap \beta(M) \neq 0$; for convenience write $K = ker(\alpha) \cap \beta(M)$. Since M generates K, we have $K = \Sigma\{\theta(M) \mid \theta : M_R \to K_R\}$. Hence $\theta(M) \neq 0$ for some $\theta : M \to K$. Since M is quasi-projective, there exists $\lambda \in E$ such that $\theta = \beta\lambda$. Hence $\beta\lambda \neq 0$ because $\theta(M) \neq 0$ and

$\beta\lambda(M) = \theta(M) \subseteq K \subseteq ker(\alpha)$. It follows that $0 \neq \beta\lambda \in \beta E \cap \mathbf{r}_E(\alpha)$, as required.

$$\begin{array}{ccc} & M & \\ \overset{\lambda}{\swarrow} & \downarrow \theta & \\ M \overset{\beta}{\to} & \beta(M) \to 0 & \end{array}$$

□

Since a free module is quasi-projective and generates each of its submodules, we obtain

Corollary 7.43. *If M_R is free and continuous then $end(M)$ is weakly continuous.*

Note that, if $R = \left[\begin{smallmatrix} F & F \\ 0 & F \end{smallmatrix}\right]$, where F is a field, then R is a left and right artinian, semiregular ring that is not right weakly continuous as we have seen. Note further that R shows that a projective module need not generate its submodules. Indeed, consider $e = \left[\begin{smallmatrix} 1 & 0 \\ 0 & 0 \end{smallmatrix}\right]$ in R. Then $eR = \left[\begin{smallmatrix} F & F \\ 0 & 0 \end{smallmatrix}\right]$ is projective with submodule $J = \left[\begin{smallmatrix} 0 & F \\ 0 & 0 \end{smallmatrix}\right]$, but the only R-morphism $\lambda : eR \to J$ is $\lambda = 0$.

We conclude this section with a brief discussion of right ACS rings; that is rings R for which, $a \in R$, then $\mathbf{r}(a) \subseteq^{ess} eR$ for some $e^2 = e \in R$. This class of rings includes all domains, all right uniform rings, and all right CS rings; and every regular ring is a right and left ACS ring. If 0 and 1 are the only idempotents in R, then R is a right ACS ring if and only if every element $a \notin Z_r$ satisfies $\mathbf{r}(a) = 0$. In particular, the localization $\mathbb{Z}_{(p)}$ of $\mathbb{Z}$ at the prime p is a commutative, local (hence semiregular) ACS ring in which $Z_r \neq J$.

A direct product $R = \Pi_{i \in I} R_i$ of rings is a right ACS ring if and only if each R_i is a right ACS ring.

A ring R is called a right PP ring if every principal right ideal is projective, equivalently if $\mathbf{r}(a) \subseteq^{\oplus} R_R$ for every $a \in R$. Hence Proposition 7.36 shows that the right PP rings are precisely the right nonsingular, right ACS rings. A result of Small [205] shows that an I-finite, right PP ring R is a Baer ring; that is, every left (equivalently right) annihilator is generated by an idempotent. (In particular R is left PP and has ACC and DCC on right and left annihilators.) Small's theorem is the nonsingular case of the next result.

Proposition 7.44. *Let R be a right ACS ring. Then the following hold:*

(1) Every left annihilator $L \nsubseteq Z_r$ contains a nonzero idempotent.

(2) If R is I-finite, every left annihilator L has the form $L = Re \oplus S$, where $e^2 = e$ and $S \subseteq Z_r$ is a left ideal.

Proof. (1). If $L = \mathtt{l}(X)$, choose $a \in L$, $a \notin Z_r$. By hypothesis, $\mathtt{r}(a) \subseteq^{ess} eR$, where $e^2 = e$, and $e \neq 1$ because $a \notin Z_r$. Hence $X \subseteq \mathtt{r}(a) \subseteq eR$, so $0 \neq (1 - e) \in \mathtt{l}(X) = L$.

(2). If $L \subseteq Z_r$ take $e = 0$ and $S = L$. Otherwise use (1) and the I-finite hypothesis to choose e maximal in $\{e \mid 0 \neq e^2 = e \in L\}$, where $e \leq f$ means $e \in fRf$. Then $L = Re \oplus [L \cap R(1 - e)]$, so it suffices to show that $L \cap R(1-e) \subseteq Z_r$. If not let $0 \neq f^2 = f \in L \cap R(1-e)$ by (1). Then $fe = 0$, so $g = e + f - ef$ satisfies $g^2 = g \in L$ and $g \geq e$. Thus $g = e$ by the choice of e, so $f = ef$ and

$$f = f^2 = f(ef) = 0,$$

which is a contradiction. □

The proof of Proposition 7.36 goes through as written to prove the following module theoretic version.

Lemma 7.45. *If M_R is a module, the following conditions are equivalent for $m \in M$:*

(1) $\mathtt{r}(m) \subseteq^{ess} eR$ for some $e^2 = e \in R$.
(2) $mR = P \oplus S$, where P_R is projective and S_R is singular.

If R is a ring, we say that a right R-module M_R is an *ACS module* if the conditions in Lemma 7.45 are satisfied for every element $m \in M$. Hence a ring R is a right ACS ring if and only if R_R is an ACS module. The next result gives a similar characterization of the right CS rings.

Proposition 7.46. *A ring R is a right CS ring if and only if every principal right R-module (respectively every right R-module) is an ACS module.*

Proof. If R is a right CS ring, let $m \in M_R$. Then $\mathtt{r}(m)$ is a right ideal of R, so $\mathtt{r}(m) \subseteq^{ess} eR$ for some $e^2 = e \in R$ by the CS-condition. Conversely, let T be a right ideal of R and write $M = R/T = mR$, where $m = 1 + T$. By hypothesis $M = P \oplus S$, where P_R is projective and S_R is singular. Hence Lemma 7.45 gives $\mathtt{r}(m) \subseteq^{ess} eR$ for some $e^2 = e \in R$, and we are done because $\mathtt{r}(m) = T$. □

for each $i = 1, 2, \ldots, n$. Set $K = J_1 \cup \cdots \cup J_n$ and observe that $|\oplus_{k\in K} E_k| \leq n\mathfrak{c}^2 = \mathfrak{c}$. For each $b \in B$ let $\iota_b : E \to E^B$ be the natural injection. Since $\{(\oplus_{k\in K} E_k) \cap \iota_b(E) \mid b \in B\}$ is an independent set of submodules of $\oplus_{k\in K} E_k$, and since $\oplus_{k\in K} E_k$ has at most $2^{\mathfrak{c}} < |B|$ subsets, there exists $b \in B$ such that $(\oplus_{k\in K} E_k) \cap \iota_b(E) = 0$. Thus the projection of E^B onto $\oplus_{k\in J-K} E_k$ is monic on $\iota_b(E)$. In particular, $\oplus_{k\in J-K} E_k = Q \oplus V$ for some $Q \cong E_R$. By Lemma 7.50 there is a subset $J_{n+1} \subseteq J - K$ such that $|J_{n+1}| \leq \mathfrak{c}$ and $Q \subseteq \oplus_{j\in J_{n+1}} E_j$. Now a standard induction argument establishes the existence of $J_1, J_2, \ldots$ in the Claim, and we are done with $J_0 = J - \cup_{n=1}^{\infty} J_n$. □

In preparation for the next Faith–Walker theorem, we need the following theorem. A module M is called countably generated if it has a countable spanning set.

Theorem 7.52. *Let $\mathfrak{c}$ be an infinite cardinal number. If a module M is a direct sum of $\mathfrak{c}$-generated submodules, so also is every direct summand of M.*

Proof (Anderson and Fuller [1]). Let $M = \oplus_{i\in I} M_i$, where each M_i is $\mathfrak{c}$-generated. Suppose that $M = P \oplus Q$, and let $\{P_j \mid j \in J\}$ and $\{Q_k \mid k \in K\}$ denote the $\mathfrak{c}$-generated submodules of P and Q, respectively. Let $\mathcal{P}$ denote the set of ordered triples (I', J', K') such that

(i) $I' \subseteq I,\ J' \subseteq J,\ K' \subseteq K$ and

(ii) $\oplus_{i\in I'} M_i = (\oplus_{j\in J'} P_j) \oplus (\oplus_{k\in K'} Q_k)$.

Define a partial ordering $\leq$ on $\mathcal{P}$ by

$$(I', J', K') \leq (I'', J'', K'') \quad \text{if} \quad I' \subseteq I'',\ J' \subseteq J'' \quad \text{and} \quad K' \subseteq K''.$$

Then $(\mathcal{P}, \leq)$ is inductive, so let (I', J', K') be a maximal element in $\mathcal{P}$; we show that $I' = I$ (and hence that $\mathcal{P} = \oplus_{j\in J'} \mathcal{P}_j$).

Let π and $\tau = 1_M - \pi$ be idempotents in $end(M_R)$ satisfying $\pi(M) = P$ and $\tau(M) = Q$. Assume that $I' \neq I$ and let $i \in I - I'$. By Lemma 7.50 each $\mathfrak{c}$-generated submodule of M is contained in a sum of at most $\mathfrak{c}$ of the M_i. In particular if $D \subseteq I$ is of cardinality at most $\mathfrak{c}$, then both $\pi(\oplus_{d\in D} M_d)$ and $\tau(\oplus_{d\in D} M_d)$ are $\mathfrak{c}$-generated submodules of M, and so their sum (also $\mathfrak{c}$-generated) is contained in a sum of at most $\mathfrak{c}$ of the M_i. So by an induction argument it follows that there exists an increasing sequence $D_1 \subseteq D_2 \subseteq \cdots$ of subsets of I, each of cardinality at most $\mathfrak{c}$, such that

$$\begin{aligned}
M_i &\subseteq \pi(M_i) + \tau(M_i) \subseteq \oplus_{d\in D_1} M_d,\\
\oplus_{d\in D_1} M_d &\subseteq \pi(\oplus_{d\in D_1} M_d) \oplus \tau(\oplus_{d\in D_1} M_d) \subseteq \oplus_{d\in D_2} M_d,\\
&\vdots\\
\oplus_{d\in D_n} M_d &\subseteq \pi(\oplus_{d\in D_n} M_d) \oplus \tau(\oplus_{d\in D_n} M_d) \subseteq \oplus_{d\in D_{n+1}} M_d,\\
&\vdots
\end{aligned}$$

Define $D = \cup_{n=1}^{\infty} D_n$. Since the M_d are independent and $i \notin I'$, it is clear that $D \nsubseteq I'$. Note also that $\oplus_{d\in D} M_d$ is $\mathfrak{c}$-generated (as $\mathfrak{c}^2 = \mathfrak{c}$) and that

$$\pi(\oplus_{d\in D} M_d) \subseteq \oplus_{d\in D} M_d \quad \text{and} \quad \tau(\oplus_{d\in D} M_d) \subseteq \oplus_{d\in D} M_d.$$

Now define

$$M' = \oplus_{i\in I'} M_i, \quad P' = \oplus_{j\in J'} P_j, \quad \text{and} \quad Q' = \oplus_{k\in K'} Q_k.$$

Then $M' = P' \oplus Q'$ because (I', J', K') is in $\mathcal{P}$. Also set

$$M'' = \oplus_{t\in I'\cup D} M_t, \quad P'' = \pi(M''), \quad \text{and} \quad Q'' = \tau(M'').$$

Then

$$\begin{aligned}
P'' &= \pi(P' + Q' + \oplus_{d\in D} M_d) \subseteq P' + \oplus_{d\in D} M_d \subseteq M'',\\
Q'' &= \tau(P' + Q' + \oplus_{d\in D} M_d) \subseteq Q' + \oplus_{d\in D} M_d \subseteq M''.
\end{aligned}$$

So, since $M'' \subseteq \pi(M'') \oplus \tau(M'') = P'' \oplus Q''$ we have $M'' = P'' \oplus Q''$. Since P' and Q' are direct summands of M contained in P'' and Q'', respectively, we have $P'' = P' \oplus P_1'$ and $Q'' = Q' \oplus Q_1'$. It follows that

$$M'' = P'' \oplus Q'' = M' \oplus (P_1' \oplus Q_1')$$

because $M' = P' \oplus Q'$. But then $P_1' \oplus Q_1' \cong M''/M' \cong \oplus_{d\in D-I'} M_d$ is nonzero and $\mathfrak{c}$-generated. This contradicts the maximality of (I', J', K') in $\mathcal{P}$ and so completes the proof of the theorem. □

Since every free module is a direct sum of countably generated (indeed principal) modules, and every projective module is a direct summand of a free module, it follows that every projective module is a direct sum of countably generated modules, a fact first proved by Kaplansky [119].

Combining Theorems 7.51 and 7.52, we obtain

Lemma 7.53. *A ring R is right noetherian if and only if a right R-module H exists such that every right R-module embeds in a direct sum of copies of H.*

Proof. Since every module embeds in an injective module, the necessity of the condition follows from Theorem 7.51 and the fact that every $\mathfrak{c}$-generated module is isomorphic to a submodule of the direct sum of the set of images of $R^{(C)}$, where $|C| = \mathfrak{c}$. Conversely, if H satisfies the condition then every injective right R-module is isomorphic to a direct summand of a direct sum of copies of H. Hence we are done by Theorems 7.51 and 7.52. □

Proposition 7.54. *Let R be a quasi-Frobenius ring. A right (or left) R-module is injective if and only if it is projective.*

Proof. We prove the proposition for right modules; the other case is analogous. If M_R is projective it is a direct summand of a free module $F = R^{(I)}$, and F is injective by Theorem 7.48 because R is right noetherian and right self-injective.

Conversely, if M is injective we may assume that M is indecomposable by Lemma 7.49. Let $\{e_1, \ldots, e_n\}$ be basic orthogonal idempotents in R, so that $\{e_iR/e_iJ \mid 1 \leq i \leq n\}$ is a complete set of simple right R-modules. Since R is right artinian, M contains a simple submodule K, so $M = E(K)$ because M is indecomposable. Hence $M \cong E(e_iR/e_iJ)$ for some i. Since the $E(e_iR/e_iJ)$ are pairwise nonisomorphic, it follows that

$$\{E(e_iR/e_iJ) \mid 1 \leq i \leq n\}$$

is a complete set of indecomposable injective right R-modules. But the right ideals $e_1R, \ldots, e_nR$ are also pairwise nonisomorphic, indecomposable, and injective. Hence $E(e_iR/e_iJ) \cong e_{\sigma i}R$ for $1 \leq i \leq n$, where σ is a permutation of $\{1, 2, \ldots, n\}$. It follows that each $E(e_iR/e_iJ)$, and hence M, is projective. □

The condition that all projective modules are injective actually characterizes the quasi-Frobenius rings, as does the condition that injectives are projective.

Theorem 7.55. *A ring R is quasi-Frobenius if and only if every projective right (or left) R-module is injective.*

Proof. If the condition holds then R is right self-injective. In fact $R^{(\mathbb{N})}$ is right injective, so R has ACC on right annihilators by Lemma 7.47. Hence R is quasi-Frobenius by Theorem 1.50. The converse follows by Proposition 7.54. □

We come finally to the theorem of Faith and Walker (referred to at the beginning of this section) that prompts the FGF-conjecture.

Theorem 7.56 (Faith–Walker Theorem). *The following conditions are equivalent for a ring R:*

(1) R is quasi-Frobenius.
(2) Every injective right (left) module is projective.
(3) Every right (left) module embeds in a free module.

Proof. (1)⇒(2). This is by Proposition 7.54.

(2)⇒(3). Every module M embeds in $E(M)$, which is projective by (2), and so embeds in a free module.

(3)⇒(1). We prove it for right modules; the other case is analogous. Since every module embeds in a direct sum of copies of R, R is right noetherian (by Lemma 7.53) and right Kasch (since simple modules embed in R). It follows that R has only finitely many isomorphism classes of simple right modules; let $\{K_1, \dots, K_n\}$ be an irredundant set. If we write $E_i = E(K_i)$, then $E_1, \dots, E_n$ are pairwise nonisomorphic, indecomposable, injective modules that are projective by (3). Hence $end(E_i)$ is local for each i, and so Lemma 1.54 shows that $rad(E_i)$ is maximal and small in E_i. Hence $T_i = E_i/rad(E_i)$ is simple and E_i is a projective cover of T_i. Moreover, if $T_i \cong T_j$ then $E_i \cong E_j$ by Corollary B.17, and hence $i = j$. Thus $\{T_1, \dots, T_n\}$ is a complete set of representatives of the simple right R-modules, and it follows that every simple right R-module has a projective cover. Thus R is semiperfect by Theorem B.21.

Let $\{e_1, \dots, e_n\}$ be a basic set of local idempotents in R. Since each E_i is indecomposable and projective we have $E_i \cong e_{\sigma i}R$ for some $\sigma i \in \{1, \dots, n\}$. Since the E_i are pairwise nonisomorphic, it follows that σ is a bijection and hence that each $e_i R$ is injective. But then eR is injective for every local idempotent in R. Since R is semiperfect, it follows that R is right self-injective and so (being right noetherian) is quasi-Frobenius by Theorem 1.50. □

Notes on Chapter 7

The FGF problem originated in the work of Levy [137] and received its name and current form from Faith [58]. In [142] Menal considered a modified version

Proof. First, Z_r is nilpotent by the ACC on right annihilators (Lemma 3.29), and so $Z_r \subseteq J$. Hence R is semilocal with $J = Z_r$ by Lemma 8.1, and so R is semiprimary. □

Note that if R is as in Lemma 8.2 and if $S_r = S_l$ is finite dimensional as a left R-module (for example if R is commutative), then R is left artinian by Lemma 3.30.

Theorem 8.3. *Every right noetherian ring with $S_r \subseteq S_l$ and $S_r \subseteq^{ess} R_R$ is right artinian.*

Proof. Since a right noetherian, semiprimary ring is right artinian by the Hopkins–Levitzki theorem, the result is an immediate consequence of Lemma 8.2. □

Note that the Faith–Menal counterexample (Example 8.16) shows that the hypothesis $S_r \subseteq S_l$ cannot be removed from Theorem 8.3; the ring $\mathbb{Z}$ of integers shows that the essential socle hypothesis cannot be removed.

Theorem 3.31 implies that a right artinian, right and left mininjective ring is quasi-Frobenius. The next theorem extends this by replacing right artinian by right noetherian.

Theorem 8.4. *The following are equivalent for a ring R:*

(1) R is right and left mininjective, right noetherian, and $S_r \subseteq^{ess} R_R$.
(2) R is right and left mininjective, right finitely cogenerated, with ACC on right annihilators.
(3) R is quasi-Frobenius.

Proof. (1)⇒(2) and (3)⇒(1) are clear. Given (2), R is semiprimary by Lemma 8.2 because right mininjective rings have $S_r \subseteq S_l$. In particular, R is semilocal, so (3) follows from Theorem 3.31. □

We can now combine Theorems 3.31 and 8.4 into one result.

Theorem 8.5. *Let R be a right and left mininjective ring with ACC on right annihilators, in which $S_r \subseteq^{ess} R_R$. Then the following are equivalent:*

(1) R is quasi-Frobenius.
(2) R is semilocal.
(3) S_r is right finitely generated.

In particular, Theorem 8.5 shows that a right and left mininjective, right Goldie ring with essential right socle is quasi-Frobenius.

8.2. Right Johns Rings

Recall that a ring R is called a right *CF ring* if every cyclic right R-module embeds in a free module. If T is a right ideal of R, this certainly implies that R/T embeds in R^I for some set I, equivalently that $\mathtt{rl}(T) = T$. Hence every right noetherian, right CF ring is right Johns. In the next theorem we clarify the relationship between right Johns rings and right CF rings. We need two lemmas. Since every right Johns ring is left P-injective, the first lemma investigates the right noetherian, left P-injective rings.

Lemma 8.6. *Let R be a right noetherian, left P-injective ring. Then the following hold:*

(1) J is nilpotent.
(2) $\mathtt{l}(J) \subseteq^{ess} {}_RR$.
(3) $\mathtt{l}(J) \subseteq^{ess} R_R$.

Proof. We prove (2), then (1), and finally (3).

(2). If $0 \neq x \in R$ we must show that $\mathtt{l}(J) \cap Rx \neq 0$. Choose $y \in R$ such that $yx \neq 0$ and $\mathtt{r}(yx)$ is maximal in $\{\mathtt{r}(ax) | a \in R, ax \neq 0\}$.

Claim. *yxR is a simple right ideal of R.*

Proof. If $0 \neq yxtR \subset yxR$ then (as R is left P-injective) $\mathtt{rl}(yxtR) = yxtR \neq yxR = \mathtt{rl}(yxR)$. Hence $\mathtt{l}(yxR) \subset \mathtt{l}(yxtR)$, so there exists $b \in R$ such that $byxt = 0$ but $byx \neq 0$. But then $t \in \mathtt{r}(byx) - \mathtt{r}(yx)$, contradicting the choice of y. This proves the Claim.

Thus $yxJ = 0$, so $0 \neq yx \in Rx \cap \mathtt{l}(J)$, proving (2).

(1). There exists $k \geq 1$ such that $\mathtt{l}(J^k) = \mathtt{l}(J^{k+1}) = \cdots$. If J is not nilpotent, choose $\mathtt{r}(x)$ maximal in $\{\mathtt{r}(y) \mid yJ^k \neq 0\}$. Then $xJ^{2k} \neq 0$ because $\mathtt{l}(J^{2k}) = \mathtt{l}(J^k)$, so there exists $b \in J^k$ with $xbJ^k \neq 0$. Since $\mathtt{l}(J) \subseteq \mathtt{l}(J^k)$ we have $\mathtt{l}(J^k) \subseteq^{ess} {}_RR$ by (2). Thus $Rxb \cap \mathtt{l}(J^k) \neq 0$, say $0 \neq cxb \in \mathtt{l}(J^k)$. Hence $\mathtt{r}(x) \subset \mathtt{r}(cx)$ because $xb \neq 0$, which contradicts the maximality of $\mathtt{r}(x)$. This proves (1).

(3). If $0 \neq d \in R$, we must show that $dR \cap \mathtt{l}(J) \neq 0$. This is clear if $dJ = 0$. Otherwise, since J is nilpotent by (1), there exists $m \geq 1$ such that $dJ^m \neq 0$ but $dJ^{m+1} = 0$. Then $0 \neq dJ^m \subseteq dR \cap \mathtt{l}(J)$, as required. □

Recall again that a ring is called a right Johns ring if it is right noetherian and every right ideal is an annihilator. The next lemma contains some interesting properties of these rings that we will need. A ring R is called a right *V ring* if every simple right R-module is injective.

Lemma 8.7. *Let R be a right Johns ring. Then the following hold:*

(1) J is nilpotent.
(2) $\mathtt{r}(J) = \mathtt{l}(J) = S_r$.
(3) $\mathtt{r}(S_r) = \mathtt{l}(S_r) = J$.
(4) $S_r \subseteq^{ess} R_R$ and $S_r \subseteq^{ess} {}_RR$.
(5) $J = Z_r = Z_l$.
(6) R/J is a right V ring.

Proof. We frequently use the fact that $\mathtt{rl}(T) = T$ for every right ideal T of R. Note that this implies that R is left P-injective.

(1). This follows from Lemma 8.6.

(2). By Lemma 8.6 we have $\mathtt{l}(J) \subseteq^{ess} R_R$. It follows that $\mathtt{ll}(J) \subseteq Z_r$. But Z_r is nilpotent (as R is right noetherian), so $\mathtt{ll}(J) \subseteq J$. If we write $\mathtt{l}^1(J) = \mathtt{l}(J)$ and define $\mathtt{l}^{k+1}(J) = \mathtt{l}[\mathtt{l}^k(J)]$ for each k, it follows that $\mathtt{l}(J) \subseteq \mathtt{l}^3(J) \subseteq \mathtt{l}^5(J) \subseteq \cdots$, whence $\mathtt{l}^k(J) = \mathtt{l}^{k+2}(J)$ for some $k \geq 2$. But then $\mathtt{rl}^k(J) = \mathtt{rl}^{k+2}(J)$, so $\mathtt{l}^{k-1}(J) = \mathtt{l}^{k+1}(J)$. Continuing in this way, we get $J = \mathtt{l}^2(J)$, and finally $\mathtt{r}(J) = \mathtt{rl}[\mathtt{l}(J)] = \mathtt{l}(J)$.

Clearly $S_r \subseteq \mathtt{l}(J)$. If $T \subseteq^{ess} R_R$ then $\mathtt{l}(T) \subseteq Z_r$, and $Z_r \subseteq J$ because Z_r is nilpotent (since R is right noetherian). Hence $\mathtt{l}(T) \subseteq J$, so $\mathtt{r}(J) \subseteq \mathtt{rl}(T) = T$. It follows that $\mathtt{r}(J) \subseteq \cap\{T \mid T \subseteq^{ess} R_R\} = S_r$, proving (2).

(3). Since $S_r = \mathtt{l}(J)$ by (2), we have $\mathtt{r}(S_r) = \mathtt{rl}(J) = J$. Moreover, $J = \mathtt{ll}(J)$ by the proof of (2), so (2) gives $J \subseteq \mathtt{l}(S_r) = \mathtt{ll}(J) = J$. Hence $J = \mathtt{l}(S_r)$.

(4). This follows from (2) and Lemma 8.6.

(5). $J = Z_l$ because R is left P-injective (Theorem 5.14). Since $Z_r S_r = 0$ always holds, we have $Z_r \subseteq \mathtt{l}(S_r) = J$ by (3). However, $J \subseteq Z_r$ because $\mathtt{r}(J) \subseteq^{ess} R_R$ by (2) and (4).

(6). S_r is a right R/J-module via $x(r + J) = xr$, $x \in S_r$, $r \in R$. Suppose that $K_{R/J}$ is simple; we must show that $K_{R/J}$ is injective. Now K_R is simple and so is isomorphic to a summand of S_r (since R is right Kasch, being right Johns). Hence (6) follows from the

Claim. S_r is injective as a right R/J-module.

Proof. Let $E = E(S_r)$ be the injective hull of S_r as a right R/J-module, and view $E = E_R$ via $xr = x(r + J)$ for $x \in E$, $r \in R$. We must show that $S_r = E$.

Suppose on the contrary that there exists $x \in E - S_r$, and write $T = \mathtt{r}_R(x)$. Then $T \supseteq J$, so $\mathtt{l}_R(T) \subseteq \mathtt{l}_R(J) = S_r$ by (2). It follows that $\mathtt{l}_R(T) = \mathtt{l}_{S_r}(T)$. But R is right Johns, so $T = \mathtt{r}_R[\mathtt{l}_R(T)] = \mathtt{r}_R[\mathtt{l}_{S_r}(T)]$, whence $T = \mathtt{r}_R\{a_i \mid i \in I\}$ for some set I, where each $a_i \in S_r$. Hence the map $\sigma : R/T \to S_r^I$ given by $\sigma(r + T) = \langle a_i r \rangle$ is a (well-defined) embedding. Thus the right (R/J)-module $M = x(R/J) = xR \cong R/T$ embeds into a direct product of copies of S_r. But R is right noetherian, so S_r is finitely generated and [by (4)] essential in R_R. Since $soc(M) \subseteq soc(E) = S_r$, it follows that M is finitely cogenerated. Hence, by Lemma 1.51, M can be embedded in a finite direct sum of copies of S_r. This means that M_R is semisimple, so $M = soc(M) \subseteq soc(E) = S_r$, which is a contradiction. Thus $E = S_r$ and the Claim [and hence (6)] is proved. □

Lemma 8.8. *Let M_R be a finite dimensional module.*

(1) If M satisfies the C2-condition then monomorphisms in $end(M)$ are isomorphisms.
(2) In this case, $end(M)$ is semilocal.

Proof. If $\sigma : M \to M$ is monic, then $\sigma(M)$ is a direct summand of M by the C2-condition, say $M = \sigma(M) \oplus K$. If $K \neq 0$ then

$$dim(M) \geq dim[\sigma(M)] + dim(K) > dim[\sigma(M)] = dim(M),$$

which is a contradiction. Hence $K = 0$, and so σ is an isomorphism. This proves (1), and then (2) follows from a result of Camps and Dicks (Corollary C.3) because M is finite dimensional. □

Theorem 8.9. *The following are equivalent for a ring R:*

(1) R is a right Johns, left Kasch ring.
(2) R is a right Johns, right C2 ring.
(3) R is a semilocal, right Johns ring.
(4) R is right artinian and every right ideal is an annihilator.
(5) R is both a right CS ring and a right CF ring.
(6) R is a semiperfect, right CF ring with $S_l \subseteq^{ess} {}_R R$.
(7) R is a semilocal, right CF ring with $S_l \subseteq^{ess} R_R$.
(8) R is a left Kasch, right CF ring.

Proof. (1)⇒(2). This follows from Proposition 1.46.

(2)⇒(3). Since R is right noetherian, it follows from Lemma 8.8 that R is semilocal.

(3)⇒(4). R is semiprimary because J is nilpotent by Lemma 8.7, so it is right artinian by the Hopkins–Levitzki theorem.

(4)⇒(5). R is left minfull (since it is semiprimary and left mininjective), so we have $S_r = S_l$ (by Theorem 3.12), and hence $S_l \subseteq^{ess} R_R$. If T is a right ideal of R, then $T = \mathtt{rl}(T)$ by hypothesis, and $\mathtt{rl}(T) \subseteq^{ess} eR$ for some $e^2 = e \in R$ by Lemma 4.2. Hence R is a right CS ring. If C is a cyclic right R-module, then C is torsionless [because $T = \mathtt{rl}(T)$ for each right ideal T] and C is finitely cogenerated by Vámos' lemma (Lemma 1.52) because R is right artinian. Hence C embeds in a free module by Lemma 1.51. Thus R is a right CF ring.

(5)⇒(6). R is right finitely cogenerated by Corollary 7.32 because it is right Kasch (being a right CF ring). Then the right CF-condition shows that every cyclic right R-module is finitely cogenerated. This implies that R is right artinian by Vámos' lemma, and so $S_l \subseteq^{ess} {}_RR$ (as R is semiprimary).

(6)⇒(7). We must show that $S_l \subseteq^{ess} R_R$. Since right CF rings are left P-injective, R is a left GPF ring, and so $S_r = S_l \subseteq^{ess} R_R$ by Theorem 5.31.

(7)⇒(1). We have $\mathtt{rl}(T) = T$ for each right ideal T because R is a right CF ring. Hence R is left P-injective, and so $S_l \subseteq S_r$ by Theorem 2.21. Thus $S_l = S_r$ because $S_l \subseteq^{ess} R_R$. Since R is semilocal, S_l is finitely generated as a right R-module by Theorem 5.52, whence R_R is finitely cogenerated. Thus every cyclic right R-module is finitely cogenerated (it embeds in R^n for some $n \geq 1$ because R is a right CF ring), so R is right artinian by Vámos' lemma. Hence R is right Johns. Finally, since R is semiperfect and $S_l \subseteq^{ess} R_R$, Lemma 1.48 shows that R is left Kasch.

(1)⇒(8). This is because (1)⇒(5).

(8)⇒(1). We have $T = \mathtt{rl}(T)$ for each right ideal T because R is a right CF ring, so it remains to show that R is right noetherian.

Claim. R is right quasi-continuous.

Proof. Let T_1 and T_2 be right ideals of R such that $T_1 \cap T_2 = 0$. Then $0 = \mathtt{rl}(T_1) \cap \mathtt{rl}(T_2) = \mathtt{r}[\mathtt{l}(T_1) + \mathtt{l}(T_2)]$. Since R is left Kasch, $\mathtt{l}(T_1) + \mathtt{l}(T_2) = R$, and the Claim follows by Theorem 6.31.

In particular, R is a right CS ring, so R is semiperfect with $S_l \subseteq^{ess} R_R$ by Theorem 4.10. But $S_l \subseteq S_r$ by Theorem 2.21 because R is left P-injective (being right CF), so we have $S_l = S_r$. In particular, $S_r \subseteq^{ess} R_R$. Since R left Kasch it is a right C2 ring (Proposition 1.46), so R is right continuous. By Lemma 4.11, R is right finitely cogenerated. Thus every cyclic right R-module is finitely cogenerated (because R is right CF), so R is right artinian by Vámos' lemma. In particular, R is right noetherian, as required. □

If we replace right Johns by strongly right Johns in Theorem 8.9, we obtain the quasi-Frobenius rings. We need a preliminary lemma that is of interest in itself.

Lemma 8.10. *Let R be a ring and let $e^2 = e \in R$ satisfy $ReR = R$. If every right ideal of R is an annihilator, the same is true of eRe.*

Proof. Write $S = eRe$ and let T be a right ideal of S. We must show that $\mathtt{r}_S\mathtt{l}_S(T) \subseteq T$. Let $a \in \mathtt{r}_S\mathtt{l}_S(T)$ so that $\mathtt{l}_S(T)a = 0$. If we write $\bar{T} = TR$, it suffices to show that $a \in \bar{T}$. Since $\bar{T} = \mathtt{r}_R\mathtt{l}_R(\bar{T})$ by hypothesis, we must show that $\mathtt{l}_R(\bar{T})a = 0$. If $x \in \mathtt{l}_R(\bar{T})$ then, for all $r \in R$, $0 = rx\bar{T} = rxTR$, so $erxe \in \mathtt{l}_S(T)$. Hence $erxea = 0$ for all r and so, since $ReR = R$, $0 = xea = xa$, as required. □

Theorem 8.11. *The following are equivalent for a ring R:*

(1) R is quasi-Frobenius.
(2) R is a strongly right Johns, left Kasch ring.
(3) R is a strongly right Johns, right C2 ring.
(4) R is strongly right Johns and $S_r \subseteq S_l$.
(5) $M_2(R)$ is right Johns and $S_r \subseteq S_l$.
(6) R is a right Johns, right mininjective ring.
(7) R is a semilocal, right mininjective right CF ring.
(8) R is a right CS ring and every 2-generated right R-module embeds in a free module.
(9) R is a right FP-injective, right CF ring.
(10) R is a right CF ring and $\mathtt{lr}(F) = F$ for all finitely generated left ideals F of R.

Proof. (1)⇔(8). If (8) is satisfied, then R is right artinian by (5)⇒(4) of Theorem 8.9. Hence R is quasi-Frobenius by (3) of Corollary 7.20.

(1)⇒(2). This is obvious.

(2)⇒(3). This is by Proposition 1.46.

(3)⇒(4). R is right artinian by Theorem 8.9. Since R is left P-injective, it is a left GPF ring and so $S_r = S_l$ by Theorem 5.31.

(4)⇒(5). This is obvious.

(5)⇒(6). Since $M_2(R)$ is right Johns, Lemma 8.10 shows that R is also right Johns. Moreover, $M_2(R)$ is left P-injective, so R is left 2-injective by Proposition 5.36. To show that R is right mininjective, let kR be a simple right ideal of R; by Lemma 2.1 we must show that $\mathtt{lr}(k) = Rk$. We have $S_r = \mathtt{l}(J)$ by Lemma 8.7, so $\mathtt{lr}(S_r) = S_r$. Hence $Rk \subseteq \mathtt{lr}(k) \subseteq \mathtt{lr}(S_r) = S_r \subseteq S_l$, so it suffices to show that $Rk \subseteq^{ess} \mathtt{lr}(k)$. To this end, suppose that $Rt \cap Rk = 0$, where $t \in \mathtt{lr}(k)$. As R is left 2-injective, Lemma 1.36 gives $R = \mathtt{r}[Rt \cap Rk] = \mathtt{r}(t) + \mathtt{r}(k)$. But $\mathtt{r}(k) \subseteq \mathtt{r}(t)$ because $t \in \mathtt{lr}(k)$, so $R = \mathtt{r}(t)$. This means $t = 0$, as required.

(6)⇒(7). We have $S_r \subseteq S_l$ by Theorem 2.21, and $S_r \subseteq^{ess} R_R$ by Lemma 8.7. Hence R is right artinian by Theorem 8.3 and so R is a right CF ring by (3)⇒(5) of Theorem 8.9.

(7)⇒(1). R is left P-injective and right Kasch (it is a right CF ring), so it suffices to show that R is right artinian. (It is then quasi-Frobenius by Theorem 8.5.) We have $S_r = S_l$ by Theorem 2.21, and R is left Kasch by Lemma 5.49. By (7)⇒(4) in Theorem 8.9, it remains to show that $S_r = S_l \subseteq^{ess} R_R$. Let $0 \neq a \in R$, and suppose that $\mathtt{l}(a) \subseteq L \subseteq^{max} {}_RR$. Then $\mathtt{r}(L) \subseteq \mathtt{rl}(a) = aR$. Since R is left Kasch and left mininjective, it follows from Theorem 2.31 that $\mathtt{r}(L)$ is a simple right ideal. Thus $aR \cap S_r \neq 0$, as required.

(1)⇒(9). This is clear.

(9)⇒(10). By Corollary 5.43.

(10)⇒(1). Because (6)⇒(1), we prove (6). The hypothesis that $\mathtt{lr}(F) = F$ for all finitely generated left ideals F of R shows that R is right mininjective, so it remains to show that R is right Johns. By the CF hypothesis and (5) of Theorem 8.9, it is enough to show that R is a right CS ring; we show in fact that R is right quasi-continuous. For this it suffices, by Theorem 6.31, to show that if $P_1 \cap P_2 = 0$, where P_1 and P_2 are right ideals of R, then $\mathtt{l}(P_1) + \mathtt{l}(P_2) = R$. Since R is a right CF ring each R/P_i embeds in a finite direct sum of copies of R, and hence $P_i = \mathtt{r}(F_i)$, where F_i is a finitely generated left ideal of R for $i = 1, 2$. Hence $\mathtt{r}(F_1 + F_2) = \mathtt{r}(F_1) \cap \mathtt{r}(F_2) = P_1 \cap P_2 = 0$ and so, since F_1, F_2, and $F_1 + F_2$ are all finitely generated,

$$R = \mathtt{l}(0) = \mathtt{lr}(F_1 + F_2) = F_1 + F_2 = \mathtt{lr}(F_1) + \mathtt{lr}(F_2) = \mathtt{l}(P_1) + \mathtt{l}(P_2),$$

as required. □

8.3. The Faith–Menal Counterexample

We begin with a characterization of right V rings, where a ring R is called a right *V ring* if every simple right R-module is injective. Recall that if M is a module, the radical of M is defined by $rad(M) = \cap\{N \mid N \subseteq^{max} M\}$, where we take $rad(M) = M$ if M has no maximal submodules.

Lemma 8.12. *The following are equivalent for a ring R:*

(1) R is a right V ring.
(2) $rad(M_R) = 0$ for every right R-module $M \neq 0$.
(3) Every right ideal $T \neq R$ of R is an intersection of maximal right ideals.

Proof. (1)⇒(2). If $0 \neq m \in M$, let $\mathbf{r}_R(m) \subseteq T \subseteq^{max} R_R$. Then $\varphi : mR \to R/T$ is well defined by $\varphi(mr) = r + T$. Since the simple module R/T is injective by (1), φ extends to $\hat{\varphi} : M \to R/T$. Then $m \notin ker(\hat{\varphi})$, and so $m \notin rad(M)$.

(2)⇒(3). If $T \neq R$ is a right ideal of R, then $rad(R/T) = 0$.

(3)⇒(1). If K_R is a simple module and $\alpha : T \to K$ is R-linear, where T is a right ideal of R, we must extend α to $R \to K$. We may assume that $\alpha \neq 0$. Hence, by (3), there exists $X \subseteq^{max} R_R$ such that $ker(\alpha) \subseteq X$ but $T \nsubseteq X$. Then $X + T = R$ and $X \cap T = ker(\alpha)$ because $ker(\alpha) \subseteq^{max} T$. Thus $\hat{\alpha} : R \to K$ is well defined by $\hat{\alpha}(x + t) = \alpha(t)$, where $x \in X$ and $t \in T$. Clearly, $\hat{\alpha}$ extends α. □

Conditions (2) and (3) in Lemma 8.12 give

Corollary 8.13. *Let R be a right V ring. Then the following hold:*

(1) $J(R) = 0$.
(2) If A is any ideal of R, then R/A is also a right V ring.

Lemma 8.14. *Let R be a right V ring that has only one simple right module up to isomorphism. Then R is a simple ring.*

Proof. If $A \neq R$ is an ideal and $M \subseteq^{max} R_R$, it suffices to show that $A \subseteq M$. Then $A \subseteq J(R) = 0$, as required. Write $A = \bigcap_{i \in I} N_i$, where each $N_i \subseteq^{max} R_R$. For each $i \in I$ let $\sigma_i : R/M \to R/N_i$ be an isomorphism of right R-modules. If $\sigma_i(1 + M) = b_i + N_i$ then $M = \{r \mid b_i r \in N_i\}$. Now given $a \in A$, we have $b_i a \in A \subseteq N_i$ for each i because A is a left ideal. It follows that $a \in M$ and hence that $A \subseteq M$, as required. □

If R is a ring and ${}_RW_R$ is a bimodule, recall that the *trivial extension* of W by R is the additive group $T(R, W) = R \oplus W$ endowed with the multiplication $(a, w)(a', w') = (aa', aw' + wa')$. This is a ring and $(0, W)$ is an ideal of $T(R, W)$ such that $T(R, W)/(0, W) \cong R$.

Theorem 8.15. *Let R be a right noetherian domain that has a bimodule ${}_RW_R$ such that W_R is simple. The following are equivalent:*

(1) $T(R, W)$ is right Johns.
(2) R is a right V ring, and W_R is the only simple right R-module up to isomorphism.

Proposition 9.4. *The following are equivalent for* $R = [D, V, P]$:

(1) R *is right mininjective.*
(2) $\mathtt{l}_V(V) = 0$ *and* $dim(_D P) = 1$.

Proof. (1)⇒(2). If $0 \neq p_0 \in P$ and $u \in \mathtt{l}_V(V)$, and if $\gamma : p_0 D \to (u + p_0)D$ is given by $\gamma(p_0 d) = (u + p_0)d$, then γ is R-linear by Lemma 9.1. By (1), $\gamma = c\cdot$ is left multiplication by $c \in R$, so $u + p_0 = \gamma(p_0) = cp_0 \in P$. Thus $u = 0$, whence $\mathtt{l}_V(V) = 0$. If $0 \neq p \in P$ then $pR = pD$ is simple, so $\mathtt{lr}(p) = Rp$ by (1). Hence Lemma 9.1 gives

$$Dp = Rp = \mathtt{lr}(p) = \mathtt{l}(J) = \mathtt{l}_V(V) \oplus P = P.$$

Thus $dim(_D P) = 1$.

(2)⇒(1). Let $\gamma : K_R \to R_R$ be R-linear, where K_R is a simple right ideal; we must show that $\gamma = c\cdot$ for $c \in R$. We may assume that $\gamma \neq 0$. We have $S_r = P$ by (2), so $K \subseteq P$. It follows from Lemma 9.1 that $dim(K_D) = 1$, so write $K = p_0 D$, where $p_0 \in P$. Since $\gamma(K)$ is simple we have $\gamma(K) \subseteq S_r = P = Dp_0$ by (2), say $\gamma(p_0) = d_0\, p_0$, where $d_0 \in D$. Then for all $d \in D$, $\gamma(p_0\, d) = \gamma(p_0)\, d = (d_0 p_0)\, d = d_0(p_0\, d)$. This shows that $\gamma = d_0\cdot$, as required. □

It is worth noting that, since we are assuming that $P \neq 0$, (4) and (7) of Lemma 9.1 give

$$S_r \text{ is a simple right ideal} \quad \text{if and only if} \quad \mathtt{l}_V(V) = 0 \text{ and } dim(P_D) = 1.$$

The condition that $dim(P_D) = 1$ holds if $R = [D, V, P]$ is right simple injective. The next lemma will be used later and strengthens the condition in Proposition 9.4.

Lemma 9.5. *Suppose the ring* $R = [D, V, P]$ *is right simple injective. Then*

$$\mathtt{l}_V(V) = 0 \quad \text{and} \quad dim(P_D) = 1 = dim(_D P).$$

Proof. Since R is right mininjective, $\mathtt{l}_V(V) = 0$ and $dim(_D P) = 1$ by Proposition 9.4. Suppose that $dim(P_D) \geq 2$ and let $\{p_1, p_2, \ldots\}$ be a D-basis of P_D. Define $\alpha : P_D \to P_D$ by $\alpha(p_1) = p_2$ and $\alpha(p_i) = 0$ for all $i \geq 2$. Then α is R-linear by Lemma 9.1 so, since $im(\alpha) = p_2 D$ is simple, $\alpha = a\cdot$ for some $a \in R$ by hypothesis. If $a = d + v + p$ then $\alpha(p_i) = ap_i = dp_i$ for each i, so $d = 0$ because $\alpha(p_2) = 0$. But then $p_2 = \alpha(p_1) = dp_1 = 0$, which is a contradiction. □

The condition in Lemma 9.5 does not characterize when $R = [D, V, P]$ is right simple injective; surprisingly, this is equivalent to simple injectivity. This is part of our main result, a characterization of when $R = [D, V, P]$ is right self-injective. The following "separation" axiom will be referred to several times.

Condition S. *If $V = xD \oplus M_D$, $x \neq 0$, there exists $v_0 \in V$ such that $v_0 x \neq 0$ and $v_0 M = 0$.*

Observe that Condition S is equivalent to asking that, if $x \in V - X$, where $X_D \subseteq V$ is any subspace, there exists $v_0 \in V$ such that $v_0 x \neq 0$ and $v_0 X = 0$.

Theorem 9.6. *Let $R = [D, V, P]$. The following are equivalent:*

(1) R is right self-injective.
(2) R is right simple injective.
(3) $\mathtt{l}_V(V) = 0$, $dim(P_D) = 1 = dim({}_D P)$, and Condition S holds.

Proof. (1)⇒(2). This is clear.

(2)⇒(3). By Lemma 9.5 it remains to prove Condition S. Fix $0 \neq q \in P$ and let $V_D = xD \oplus M$, where $x \neq 0$ and $M \subseteq V_D$. Define

$$\beta : V \oplus P = xD \oplus M \oplus P \to P \qquad \text{by} \qquad \beta(xd + m + p) = qd.$$

This is well defined because D is a division ring, and it is R-linear because

$$\begin{aligned}\beta[(xd + m + p)(d_1 + v_1 + p_1)] &= \beta[xdd_1 + md_1 + (xdp_1 + mv_1 + pd_1)] \\ &= q(dd_1) \\ &= qd(d_1 + v_1 + p_1) \\ &= [\beta(xd + m + p)](d_1 + v_1 + p_1).\end{aligned}$$

Since $\beta[V \oplus P] = qD$ is simple, it follows from (2) that $\beta = b\cdot$ is left multiplication by $b \in R$. Write $b = d_0 + v_0 + p_0$, so that $q = \beta(x) = bx = d_0 x + v_0 x$. Hence $v_0 x = q \neq 0$ and $d_0 x = 0$. This means that $d_0 = 0$, so $v_0 m = bm = \beta(m) = 0$ for all $m \in M$, proving Condition S.

(3)⇒(1). If $T \subseteq R$ is a right ideal, let $\alpha : T \to R_R$ be R-linear; we must show that $\alpha = a\cdot$ for some $a \in R$. This is clear if $T = R$ or $T = 0$, so assume $0 \subset T \subseteq J$. Since $S_r = \mathtt{l}_V(V) \oplus P = P$ is simple by (3), it follows from Lemma 9.2 that

$$T = X \oplus P$$

a ring homomorphism $\rho : D \to RFM_I(D)$, given for $d \in D$ by

$$\rho(d) = [\rho_{ij}(d)], \quad \text{where} \quad e_i d = \Sigma_{k \in I} \rho_{ik}(d) e_k.$$

Conversely, every bimodule structure ${}_D V_D$ arises in this way from such a representation ρ.

Given ρ we get a bimodule ${}_D V_D$, so, if $\{f_k \mid k \in K\}$ is a basis of V_D, we obtain the "adjoint" representation $\psi : D \to CFM_K(D)$ – the column finite matrices – given for $d \in D$ by

$$\psi(d) = [\psi_{ij}(d)], \quad \text{where} \quad d f_k = \Sigma_{l \in K} f_l \psi_{lk}(d).$$

If $A \in M_{I \times K}(D)$ is an arbitrary $I \times K$ matrix, we get a product $V \times V \to D$, written $(v, w) \mapsto v \cdot w$, given by

$$v \cdot w = \Sigma_{i,k} v_i a_{ik} w_k, \quad \text{where} \quad v = \Sigma_i v_i e_i \text{ and } w = \Sigma_k f_k w_k. \tag{1}$$

As before, this satisfies all the bimap axioms except possibly $(vd)w = v(dw)$. Since $e_i \cdot f_k = a_{ik}$ we have

$$(e_i d) \cdot f_k = e_i \cdot (d f_k) \quad \text{if and only if} \quad \Sigma_j \rho_{ij}(d) a_{jk} = \Sigma_m a_{im} \psi_{mk}(d).$$

It follows that (1) defines a bimap on ${}_D V_D$ if and only if

$$\rho(d) A = A \psi(d) \qquad \text{for all } d \in D. \tag{2}$$

Theorem 9.13. *Given a bimodule ${}_D V_D$, let $\{e_i \mid i \in I\}$ and $\{f_k \mid k \in K\}$ be bases of ${}_D V$ and V_D, respectively, and assume that an $I \times K$ matrix A satisfies $\rho(d) A = A \psi(d)$ for all $d \in D$ as in the preceding. Then the following are equivalent:*

(1) $R = [D, V, D]$ is a counterexample to the Faith conjecture.
(2) The rows of A are a basis of the direct product D^K.

Proof. In view of Theorem 9.7, it suffices to prove the following:

(1) $1_V(V) = 0$ if and only if the rows of A are independent.
(2) Condition S is satisfied if and only if the rows of A span ${}_D(D^K)$.

Given $v = \Sigma_i v_i e_i$ in V write $\bar{v} = \langle v_i \rangle \in D^{(I)}$. Observe that $v \cdot f_k = \Sigma_i v_i (e_i \cdot f_k) = \Sigma_i v_i a_{ij}$, so

$$\langle v \cdot f_k \rangle = \bar{v} A. \tag{3}$$

Hence if $v \in V$ then $v \cdot V = 0$ if and only if $v \cdot f_k = 0$ for all $k \in K$, if and only if $\bar{v}A = 0$. Now (a) follows because the rows of A are independent if and only if $\bar{v}A = 0$ implies $\bar{v} = 0$.

If Condition S holds and $0 \neq \bar{b} = \langle b_k \rangle \in D^K$ is given, let $P \in CFM_K(D)$ be an invertible matrix with $\bar{b}$ as row 0. Define

$$\langle f_k' \rangle = \langle f_k \rangle P^{-1}$$

so that $\{f_k' \mid k \in K\}$ is a basis of V_D. By Condition S let $v_0 \in V$ satisfy

$$v_0 \cdot f_k' = \begin{cases} 1 & \text{if } k = 0, \\ 0 & \text{if } k \neq 0. \end{cases} \tag{4}$$

Then observe that

$$v_0 \cdot f_k = v_0 \cdot (\Sigma_l f_l' p_{lk}) = \Sigma_l (v_0 \cdot f_l') p_{lk} = p_{0k} = b_k.$$

Hence (3) shows that $\bar{b} = \bar{v}_0 A$ is a linear combination of the rows of A.

Finally, assume that the rows of A span ${}_D(D^K)$. If $\{f_k' \mid k \in K\}$ is any basis of V_D it suffices to find $v_0 \in V$ such that (4) holds. If $\bar{e}_0$ is row 0 of the $K \times K$ identity matrix, this asks for $v_0 \in V$ such that $\bar{e}_0 = \langle v_0 \cdot f_k' \rangle$. But there exists an invertible matrix $P \in CFM_K(D)$ such that $\langle f_k \rangle = \langle f_k' \rangle P$. By hypothesis row 0 of P is a linear combination of the rows of A; that is, $\bar{e}_0 P = \bar{v}_0 A$ for some $v_0 \in V$. But then (3) gives

$$\bar{e}_0 P = \bar{v}_0 A = \langle v_0 \cdot f_k \rangle = \langle v_0 \cdot f_k' \rangle P$$

using the fact that $\langle f_k \rangle = \langle f_k' \rangle P$. Since P is invertible, $\bar{e}_0 = \langle v_0 \cdot f_k \rangle$ as required. This completes the proof of (b). □

One difficulty with applying Theorem 9.13 is that, for a bimodule ${}_D V_D$, we cannot define the map ρ in terms of A and ψ. In a concrete example we have to first find ρ and ψ and then ask for the matrix A. However, A need not exist in general, even in the finite dimensional case. For example, let $D = F$ be a commutative field with endomorphism $\sigma : F \to F$, and consider $V = F^n$, where the right structure V_F is as usual and the left structure is defined by $f \cdot v = \sigma(f)v$. Then an invertible A exists such that (2) is satisfied if and only if $\sigma^2 = 1_F$. This example illustrates that the structure of A depends heavily on the particular bimodule structure, and not only on the dimensions.

9.4. Other Properties of $R = [D, V, P]$

Many other properties of the ring $R = [D, V, P]$ can be characterized as in Theorem 9.6 in terms of vector space properties of V and P. Several of these are collected in this section.

Recall that a ring R is called right Kasch if every simple right R-module embeds in R_R. The ring $R = [D, V, P]$ is local and so has only one simple module. Since $P \neq 0$ we have $S_r \neq 0$ (and $S_l \neq 0$), whence

Proposition 9.14. *$R = [D, V, P]$ is right and left Kasch.*

The next result follows from Lemma 9.1.

Proposition 9.15. *$R = [D, V, P]$ has finite right uniform dimension if and only if $dim(P_D) < \infty$ and $dim[l_V(V)_D] < \infty$.*

Recall that a ring R is called a left minannihilator ring if $\mathtt{lr}(K) = K$ for all simple left ideals K. These rings are closely related to the right mininjective rings (see Proposition 2.33) and the following result (with Proposition 9.4) shows that if $R = [D, V, P]$ is left minannihilator then it is right mininjective.

Proposition 9.16. *The following are equivalent for $R = [D, V, P]$:*

(1) R is a left minannihilator ring.
(2) $\mathtt{l}_V(V) = 0 = \mathtt{r}_V(V)$ and $dim({}_DP) = 1$.
(3) $S_r = S_l$ is simple as a left R-module.

Proof. (1)⇒(2). If $0 \neq p \in P$ then $\mathtt{r}(p) \supseteq \mathtt{r}(P) = J$, so $\mathtt{r}(p) = J$ because R is local. As $Dp = Rp$ is simple, (1) gives

$$Dp = \mathtt{lr}(p) = \mathtt{l}(J) = S_r = \mathtt{l}_V(V) \oplus P.$$

As $P \neq 0$, this gives $\mathtt{l}_V(V) = 0$ and $dim({}_DP) = 1$. Finally, if $w \in \mathtt{r}_V(V)$ and $0 \neq p \in P$ then $w + p$ and p are in S_r, so $\mathtt{r}(w + p) = J = \mathtt{r}(p)$. As before, (1) gives $D(w + p) = \mathtt{lr}(w + p) = \mathtt{lr}(p) = Dp$. Since $V \oplus P$ is direct, this implies $w = 0$, whence $\mathtt{r}_V(V) = 0$.

(2)⇒(3). We have $S_l = \mathtt{r}(J)$ because R is semilocal, so $S_l = \mathtt{r}_V(V) \oplus P$. Hence (2) and Lemma 9.1 show that $S_l = P = S_r$. This is left simple because $dim({}_DP) = 1$.

(3)⇒(1). Write $S = S_l = S_r$. This is the only simple left ideal by (3), so $S = P$ and (1) follows from $\mathtt{lr}(S) = \mathtt{l}(J) = S_r = S$. □

Turning to right continuity, we have the following proposition:

Proposition 9.17. *Let $R = [D, V, P]$.*

(1) R always satisfies the left and right C2-conditions.
(2) The following are equivalent:
   (a) R is right continuous.
   (b) R_R is uniform.
   (c) S_r is simple.
   (d) $\mathtt{l}_V(V) = 0$ and $dim(P_D) = 1$.
   (e) $P \subseteq T$ for all right ideals $T \neq 0$.
   (f) Every right ideal $T \neq 0$, R has the form $T = X \oplus P$, where $X_D \subseteq V_D$.

Proof. Let $T \cong eR$, $e^2 = e$. As R is local, either $e = 0$ (so $T = 0$ is a summand) or $e = 1$. In the last case, $T = aR$, $a \in R$, where $r(a) = 0$. Thus $a \notin J$, so $T = R$ is a summand. This proves half of (1); the rest follows by symmetry.

(a)⇒(b). If $T \neq 0$ is a right ideal then $T \subseteq^{ess} R_R$ by the C1-condition because R is local.

(b)⇒(c). This is clear since $S_r \neq 0$ by our standing assumption that $P \neq 0$.

(c)⇒(d). This follows from (4) and (7) of Lemma 9.1 because $P \neq 0$.

(d)⇒(e). Suppose $T \neq 0$ and $P \nsubseteq T$. Then $T \cap P = 0$ because $dim_D(P_D) = 1$. We may assume that $T \subseteq J$ because R is local. Let $t = v + p \in T$. If $v_1 \in V$ we have $t\, v_1 = v\, v_1 \in T \cap P = 0$, so $v \in \mathtt{l}_V(V) = 0$. Thus $T \subseteq P$, which is a contradiction.

(e)⇒(f). This is clear by Lemma 9.1.

(f)⇒(a). If $T \neq 0$ is a right ideal, then $0 \neq P \subseteq T$ by (f). It follows that R_R is uniform, so $T \subseteq^{ess} R_R$. Hence R satisfies the C1-condition, so (a) follows from (1). □

We now turn to a discussion of annihilators. Observe first that, if $X_D = \mathtt{r}_V(Y)$, where Y is a subset of V, we may assume that $Y = {}_DY$ is actually a submodule because $X = \mathtt{r}_V[\mathtt{l}_V\mathtt{r}_V(Y)]$. Similarly, if ${}_DX = \mathtt{l}_V(Y)$ we may assume that $Y = Y_D$ because $X = \mathtt{l}_V[\mathtt{r}_V\mathtt{l}_V(Y)]$.

Lemma 9.18. *Let $R = [D, V, P]$.*

(1) If $T = X_D \oplus P$, where $X \subseteq V$, then $\mathtt{l}(T) = \mathtt{l}_V(X) \oplus P$.
(2) If $L = {}_DY \oplus P$, where $Y \subseteq V$, then $\mathtt{r}(L) = \mathtt{r}_V(Y) \oplus P$.

Proof. We prove (1); (2) is similar. We have $\mathtt{l}(T) \subseteq J$ as $T \neq 0$. If $v + p \in \mathtt{l}(T)$ then $vx = (v + p)x = 0$ for all $x \in X$; that is, $v \in \mathtt{l}_V(X)$. Thus

$\mathtt{l}(T) \subseteq \mathtt{l}_V(X) \oplus P$. Conversely, if $v \in \mathtt{l}_V(X)$ then $(v + p)(x + p_1) = vx = 0$ for all $x + p_1$ in T, so $\mathtt{l}_V(X) \oplus P \subseteq \mathtt{l}(T)$. □

Lemma 9.19. *Let $R = [D, V, P]$ and suppose $T \neq 0, R$ and $L \neq 0, R$ are right and left ideals of R respectively.*

(1) T is a right annihilator in R if and only if $T = \mathtt{r}_V(Y) \oplus P$ for some ${}_DY \subseteq V$.
(2) L is a left annihilator in R if and only if $L = \mathtt{l}_V(X) \oplus P$ for some $X_D \subseteq V$.

Proof. Again we prove only (1) as (2) is analogous. If $T = \mathtt{r}_V(Y) \oplus P$ then $T = \mathtt{r}(Y \oplus P)$ by Lemma 9.18. Conversely, if T is a right annihilator then $T = \mathtt{rl}(T)$. Now $T \neq R$ means $T \subseteq J$, so $P \subseteq \mathtt{l}(T)$. Hence $\mathtt{l}(T) = Y \oplus P$ for some ${}_DY \subseteq V$ by Lemma 9.1, so $T = \mathtt{rl}(T) = \mathtt{r}(Y \oplus P) = \mathtt{r}_V(Y) \oplus P$ by Lemma 9.18. □

We say that V has *ACC on left annihilators* if it has ACC on subspaces of the form $\mathtt{l}_V(X)$, where $X \subseteq V$, with similar terminology for the DCC and for right annihilators.

Proposition 9.20. *Let $R = [D, V, P]$. Then R has ACC (DCC) on right (left) annihilators if and only if the same is true for V.*

Proof. We do the proof for the ACC on right annihilators; the other three cases are analogous. By Lemma 9.19, every ascending chain of right annihilators in R has the form $\mathtt{r}_V(Y_1) \oplus P \subseteq \mathtt{r}_V(Y_2) \oplus P \subseteq \cdots$. This gives $\mathtt{r}_V(Y_1) \subseteq \mathtt{r}_V(Y_2) \subseteq \cdots$, so, if V has the ACC, $\mathtt{r}_V(Y_n) = \mathtt{r}_V(Y_{n+1}) = \cdots$ for some n. Hence the chain in R terminates. Conversely, if $\mathtt{r}_V(Y_1) \subseteq \mathtt{r}_V(Y_2) \subseteq \cdots$ in V, then $\mathtt{r}(Y_1 \oplus P) \subseteq \mathtt{r}(Y_2 \oplus P) \subseteq \cdots$ by Lemma 9.18. If $\mathtt{r}(Y_n \oplus P) = \mathtt{r}(Y_{n+1} \oplus P) = \cdots$ for some n, it follows by Lemma 9.18 that $\mathtt{r}_V(Y_n) = \mathtt{r}_V(Y_{n+1}) = \cdots$. □

We can locate the right singular ideal Z_r in $R = [D, V, P]$.

Proposition 9.21. *Let $R = [D, V, P]$ and assume that $dim(P_D) = 1$. Then:*

(1) $Z_r = \mathtt{l}_V\mathtt{l}_V(V) \oplus P = \mathtt{l}(S_r) \subseteq^{ess} R_R$.
(2) $S_r \subseteq Z_r$.
(3) $Z_r = J$ if and only if $\mathtt{l}_V(V) \subseteq \mathtt{r}_V(V)$ if and only if $S_r \subseteq S_l$.

Proof. Write $U = \mathtt{l}_V(V)$, so that $S_r = U \oplus P$ by Lemma 9.1.

(1). Always $Z_r \subseteq \mathtt{l}(S_r) = \mathtt{l}_V(U) \oplus P$. We claim that $\mathtt{l}_V(U) \oplus P \subseteq Z_r$. Let $y = v + p \in \mathtt{l}_V(U) \oplus P$. Since $v \in \mathtt{l}_V(U)$ we have $U \subseteq \mathtt{r}_V(v)$, so $S_r = U \oplus P \subseteq \mathtt{r}_V(v) \oplus P \subseteq \mathtt{r}(y)$. Thus $y \in Z_r$ because $S_r \subseteq^{ess} R_R$. This proves the equalities in (1). Finally, $U \subseteq \mathtt{l}_V(U)$ because $U^2 = 0$. Hence $S_r \subseteq \mathtt{l}_V(U) \oplus P = \mathtt{l}(S_r)$, and (1) follows.

(2). Since $U^2 = 0$ we have $(S_r)^2 = 0$, so $S_r \subseteq \mathtt{l}(S_r)$ and (2) follows from (1).

(3). Since $Z_r = \mathtt{l}_V(U) \oplus P$ and $J = V \oplus P$, we have $Z_r = J$ if and only if $\mathtt{l}_V(U) = V$, if and only if $VU = 0$ if, and only if $U \subseteq \mathtt{r}_V(V)$. The second equivalence holds because $S_r = U \oplus P$ and $S_l = \mathtt{r}_V(V) \oplus S$ (by the right–left analogue of Lemma 9.1). □

Finally, we characterize when the ring $R = [D, V, P]$ is right principally injective (P-injective). Such a ring is both right mininjective and left minannihilator, a fact reflected in the following result.

Proposition 9.22. *If $R = [D, V, P]$, then R is right P-injective if and only if it satisfies the following three conditions:*

(1) $dim({}_D P) = 1$.
(2) $\mathtt{l}_V(V) = 0 = \mathtt{r}_V(V)$.
(3) $\mathtt{l}_V\, \mathtt{r}_V(v) = Dv$ for all $v \in V$.

Proof. We begin with the following result.

Claim. *Assume that $dim({}_D P) = 1$ and $\mathtt{r}_V(V) = 0$. If $0 \neq v + p \in V \oplus P$ then $R(v + p) = Dv \oplus P$.*

Proof. If $v = 0$ the proof is clear because $Rp = Dp = P$. If $v \neq 0$ then $Vv = P$ by the hypotheses. Hence $R(v + p) = \{dv + (dp + v_1 v) \mid d \in D$ and $v_1 \in V\} = Dv \oplus P$, proving the Claim.

Assume first that R is right P-injective. Then Proposition 9.4 implies (a) and $\mathtt{l}_V(V) = 0$. To show that $\mathtt{r}_V(V) = 0$, suppose that $0 \neq w \in \mathtt{r}_V(V)$. Then $Vw = 0$, so $Rw = Dw$, and we have $\mathtt{lr}(w) = Rw = Dw \subseteq V$ by P-injectivity. But if $p \in P$ then $\mathtt{r}(w) \subseteq J = \mathtt{r}(p)$, so $p \in \mathtt{lr}(w)$. This implies $P \subseteq V$, a contradiction. Hence $\mathtt{r}_V(V) = 0$, proving (b). Finally, to show that $Dv = \mathtt{l}_V\, \mathtt{r}_V(v)$, we may assume that $v \neq 0$. Then the Claim and Lemma 9.18 give $\mathtt{r}(v) = \mathtt{r}(Rv) = \mathtt{r}[Dv \oplus P] = \mathtt{r}_V(v) \oplus P$. Hence $\mathtt{l}_V\, \mathtt{r}_V(v) \oplus P = \mathtt{lr}(v) = Rv = Dv \oplus P$, and (c) follows.

Conversely, assume (a), (b), and (c). If $a \in R$ we must show that $\mathtt{lr}(a) = Ra$. This is clear if $a = 0$ or if $a \notin J$ (because R is local). If $0 \neq a \in J$, say

$a = v + p$, where $v \in V$ and $p \in P$, then $Ra = Dv \oplus P$ by the Claim. Hence Lemma 9.18 gives $\mathtt{r}(a) = \mathtt{r}_V(v) \oplus P$, and then $\mathtt{lr}(a) = \mathtt{l}_V\,\mathtt{r}_V(v) \oplus P = Dv \oplus P = Ra$ by (c). □

Example 9.23. As in Examples 9.10–9.11 and 9.12, let $D = D^{(I)}$, where $I = \{1, 2, 3, \ldots\}$. If A is the $I \times I$ identity matrix, the bimap is $vw = v_1 w_1 + v_2 w_2 + \cdots$, where $v = \langle v_i \rangle$ and $w = \langle w_i \rangle$. Then $\mathtt{l}_V(V) = 0 = \mathtt{r}_V(V)$ is clear and it is a routine matter to verify that $\mathtt{l}_V \mathtt{r}_V(v) = Dv$ and $\mathtt{r}_V \mathtt{l}_V(v) = vD$ for all $v \in V$. Hence $R = [D, V, D]$ is a right and left P-injective ring that is neither right nor left artinian.

Notes on Chapter 9

The construction in this chapter is motivated by the 1966 paper of Osofsky [182], and the results appeared in [7]. More work on semiprimary, right self-injective rings in which the Jacobson radical cubes to zero was carried out by Koike [126].

Finally, our choice of σ guarantees $F(\sigma_X) = \tau_{FX}$ for each X_R. To see that $G(\tau_M) = \sigma_{GM}$ for each M_S, it suffices (since F is faithful) to show that $F[G(\tau_M)] = F(\sigma_{GM})$. But $F(\sigma_{GM}) = \tau_{FGM}$ by the definition of σ, and $\tau_{FGM} = FG(\tau_M)$ comes from the first diagram (replacing λ by τ_M, M by FGM, and N by M, and noting that τ_M is an isomorphism). □

A.2. Morita Invariants

Let p be a property of modules that is preserved by isomorphisms. Then p is called a *Morita invariant* if, for every additive equivalence $F : modR \to modS$, FX has p whenever X has p. Note that if FX has p then X has p because $GFX \cong X$ for any equivalence inverse G of F. Thus p is a Morita invariant means that X has p if and only if FX has p. Here is a proof that "injective" and "projective" are Morita invariants.

Proposition A.4. *Let $F : modR \to modS$ be an additive equivalence. If X_R and Y_R are modules, then X is Y-injective (Y-projective) if and only if FX is FY-injective (FY-projective). In particular, "injective" and "projective" are Morita invariants.*

Proof. We prove the injective part; the projective result is analogous. Let G be an equivalence inverse of F with natural isomorphisms $\sigma : GF \to 1_{modR}$ and $\tau : FG \to 1_{modS}$. Given λ and μ in $modS$ with λ monic as in the first diagram, we must find an S-morphism $\gamma : FY \to FX$ such that $\mu = \gamma \circ \lambda$. Apply G to

$$\begin{array}{ccc} M_S & \overset{\lambda}{\to} & FY \\ \downarrow \mu & \swarrow \gamma & \\ FX & & \end{array}$$

obtain the second diagram, where $G(\lambda)$ is monic by Lemma A.2. By hypothesis there exists $\alpha : Y \to X$ such that $\alpha \circ \sigma_Y \circ G(\lambda) = \sigma_X \circ G(\mu)$. We claim that

$$\begin{array}{ccccc} GM & \overset{G(\lambda)}{\to} & GFY & \overset{\sigma_Y}{\to} & Y \\ \downarrow G(\mu) & & & & \\ GFX & & & & \\ \downarrow \sigma_X & & & & \\ X & & & & \end{array}$$

$\gamma = F(\alpha)$ satisfies $\mu = F(\alpha) \circ \lambda$. As G is faithful, it suffices to show that $G(\mu) = GF(\alpha) \circ G(\lambda)$. However, $GF(\alpha) = \sigma_X^{-1} \circ \alpha \circ \sigma_Y$ because σ is natural, so this requirement reads $G(\mu) = \sigma_X^{-1} \circ \alpha \circ \sigma_Y \circ G(\lambda)$. But this is the defining property of α. Finally, the last sentence follows because every S-module has the form FX for some R-module X (Theorem A.3). □

We now give two technical results that describe the close relationship between the modules X_R and $(FX)_S$, where $F : modR \to modS$ is an additive equivalence. The first result is concerned with exact sequences. We say that the sequence $Z \overset{\alpha}{\to} X \overset{\beta}{\to} Y$ of R-morphisms is *exact* at X if $im(\alpha) = ker(\beta)$, and any sequence of modules is called exact if it is exact at every "interior" module. Thus $X \overset{\beta}{\to} Y$ is monic if and only if $0 \to X \overset{\beta}{\to} Y$ is exact, and β is epic if and only if $X \overset{\beta}{\to} Y \to 0$ is exact. The following result contains the key observation.

Proposition A.5. *Let $F : modR \to modS$ be an additive equivalence. Then a sequence $0 \to Z \overset{\alpha}{\to} X \overset{\beta}{\to} Y$ in $modR$ is exact if and only if the sequence $0 \to FZ \overset{F(\alpha)}{\to} FX \overset{F(\beta)}{\to} FY$ is exact in $modS$.*

Proof. Let $G : modS \to modR$ be a compatible equivalence inverse of F with natural isomorphisms $\sigma : GF \to 1_{modR}$ and $\tau : FG \to 1_{modS}$. It is only necessary to prove the forward implication because either row in the following commutative diagram is exact if and only if the other row is exact. Hence assume

$$\begin{array}{ccccccc} 0 \to & GFZ & \overset{GF(\alpha)}{\to} & GFX & \overset{GF(\beta)}{\to} & GFY \\ & \sigma_Z \downarrow & & \sigma_X \downarrow & & \sigma_Y \downarrow \\ 0 \to & Z & \overset{\alpha}{\to} & X & \overset{\beta}{\to} & Y \end{array}$$

that $0 \to Z \overset{\alpha}{\to} X \overset{\beta}{\to} Y$ is exact; we must show that the sequence $0 \to FZ \overset{F(\alpha)}{\to} FX \overset{F(\beta)}{\to} FY$ is also exact. First, $F(\alpha)$ is monic by Lemma A.2. Next, $im[F(\alpha)] \subseteq ker[F(\beta)]$ because $F(\beta) \circ F(\alpha) = F(\beta \circ \alpha) = F(0) = 0$. So if we write $K = ker[F(\beta)]$, it remains to show that $K \subseteq im[F(\alpha)]$. Let $i_K : K \to FX$ be the inclusion.

Claim. $\beta \circ \sigma_X \circ G(i_K) = 0$.

Proof. As F is faithful, it suffices to verify that $F(\beta) \circ F(\sigma_X) \circ FG(i_K) = 0$.

$$\begin{array}{ccc} FGK & \overset{FG(i_K)}{\to} & FGFX \\ \downarrow \tau_K & & \downarrow \tau_{FX} \\ K & \overset{i_K}{\to} & FX \end{array}$$

Now $F(\sigma_X) \circ FG(i_K) = \tau_{FX} \circ FG(i_K) = i_K \circ \tau_K$ by compatibility and the diagram. It follows that $F(\beta) \circ F(\sigma_X) \circ FG(i_K) = F(\beta) \circ i_K \circ \tau_K = 0 \circ \tau_K = 0$ because $K = ker\{F(\beta)\}$, proving the Claim.

For convenience, write $\delta = \sigma_X \circ G(i_K)$. Then $im(\delta) \subseteq ker(\beta)$ by the Claim, so $im(\delta) \subseteq im(\alpha)$ by hypothesis. Since α is monic, $\gamma : G(K) \to Z$ is well defined as follows: If $w \in G(K)$ then $\delta(w) = \alpha(z)$ for a unique element $z \in Z$,

$$\begin{array}{ccc} G(K) & \overset{\delta}{\to} & X \\ \downarrow\gamma & \nearrow\alpha & \\ Z & & \end{array}$$

so take $\gamma(w) = z$. Then $\delta = \alpha \circ \gamma$, so $F(\delta) = F(\alpha) \circ F(\gamma)$ and it follows that $im[F(\delta)] \subseteq im[F(\alpha)]$. Hence it remains to show that $im[F(\delta)] = K$. But $F(\delta) = F(\sigma_X) \circ FG(i_K) = \tau_{FX} \circ FG(i_K) = i_K \circ \tau_K$ by the first diagram, so $im[F(\delta)] = im\{i_K \circ \tau_K\} = im\{i_K\} = K$, as required. □

An exact sequence of the form $0 \to Z \overset{\alpha}{\to} X \overset{\beta}{\to} Y \to 0$ is called a *short exact* sequence, and it is said to *split* if any of the following equivalent conditions are satisfied:

(1) There exists $\beta' : Y \to X$ such that $\beta \circ \beta' = 1_Y$.
(2) $im(\alpha) = ker(\beta)$ is a direct summand of X.
(3) There exists $\alpha' : X \to Z$ such that $\alpha' \circ \alpha = 1_Z$.

These concepts are very useful in classifying rings, so the following result is of interest.

Theorem A.6. *Let $F : modR \to modS$ be an additive equivalence. Then a sequence $0 \to Z \overset{\alpha}{\to} X \overset{\beta}{\to} Y \to 0$ in $modR$ is exact (split) in $modR$ if and only if $0 \to FZ \overset{F(\alpha)}{\to} FX \overset{F(\beta)}{\to} FY \to 0$ has the same property in $modS$.*

Proof. Let $\sigma : GF \to 1_{modR}$ and $\tau : FG \to 1_{modS}$ be natural equivalences. Assume first that $0 \to Z \overset{\alpha}{\to} X \overset{\beta}{\to} Y \to 0$ is exact. Then $0 \to FZ \overset{F(\alpha)}{\to} FX \overset{F(\beta)}{\to} FY$ is exact at FX by Proposition A.5, and $F(\beta)$ is epic by Lemma A.2. Hence $0 \to FZ \overset{F(\alpha)}{\to} FX \overset{F(\beta)}{\to} FY \to 0$ is exact. Conversely, if this sequence is exact, then $0 \to Z \overset{\alpha}{\to} X \overset{\beta}{\to} Y$ is exact by Proposition A.5, and $\beta = \sigma_Y \circ GF(\beta) \circ \sigma_X^{-1}$ is epic, again by Lemma A.2.

If $0 \to Z \overset{\alpha}{\to} X \overset{\beta}{\to} Y \to 0$ is split then the conditions preceding this theorem show that $0 \to FZ \overset{F(\alpha)}{\to} FX \overset{F(\beta)}{\to} FY \to 0$ is also split. Conversely, if $\lambda : FY \to FX$ satisfies $F(\beta) \circ \lambda = 1_{FY}$, applying G gives $1_{GFY} =$

$GF(\beta) \circ G(\lambda) = (\sigma_Y^{-1} \circ \beta \circ \sigma_X) \circ G(\lambda)$, and it follows that $\sigma_Y = \beta \circ \sigma_X \circ G(\lambda)$. Hence $\beta \circ \beta' = 1_Y$ with $\beta' = \sigma_X \circ G(\lambda) \circ \sigma_Y^{-1}$. □

The second technical result concerns the lattice $lat_R(X)$ of submodules of X_R and the corresponding lattice for $(FX)_S$.

Proposition A.7. *Let $F : modR \to modS$ be an additive equivalence, and let X_R denote a module. Then $\Phi_X : lat_R(X) \to lat_S(FX)$ is a lattice isomorphism, where we define*

$$\Phi_X(K) = im\{F(i_K)\} \qquad \text{for all } K \subseteq X \text{ with inclusion } K \xrightarrow{i_K} X.$$

Proof. Let $G : modS \to modR$ be a compatible equivalence inverse of F with natural isomorphisms $\sigma : GF \to 1_{modR}$ and $\tau : FG \to 1_{modS}$. If $N \subseteq FX$, define $\Upsilon : lat_S(FX) \to lat_R(X)$ by

$$\Upsilon(N) = im[\sigma_X \circ G(j_N)] \qquad \text{for all } N \subseteq FX \text{ with inclusion } N \xrightarrow{j_N} FX.$$

We show that Υ is the inverse of Φ_X and that Φ_X and Υ preserve inclusions.

If $K \subseteq X$ write $N = im[F(i_K)] = \Phi_X(K) \subseteq FX$, and for convenience let $\phi : F(K) \to N$ denote the map $F(i_K)$ with codomain restricted to N. Then ϕ is an isomorphism [$F(i_K)$ is one-to-one] and $F(i_K) = j_N \circ \phi$. Hence $G(\phi)$ is also an isomorphism, so

$$\begin{aligned} \Upsilon\Phi(K) = \Upsilon(N) &= im[\sigma_X \circ G(j_N)] = im[\sigma_X \circ G(j_N) \circ G(\phi)] \\ &= im[\sigma_X \circ GF(i_K)] = im(i_K \circ \sigma_K) = im(i_K) = K, \end{aligned}$$

where we used the fact that σ is natural.

Similarly, starting with $N \subseteq FX$ write $K = \Upsilon(N) = im[\sigma_X \circ G(j_N)]$. Since $G(j_N)$ is monic, it follows that $\sigma_X \circ G(j_N) : GN \to K$ is an isomorphism, which we call $\delta : GN \to K$, and which makes the diagram commutative. Then

$$\begin{array}{ccc} GN & \xrightarrow{G(j_N)} & GFX \\ \downarrow \delta & & \downarrow \sigma_X \\ K & \xrightarrow{i_K} & X \end{array}$$

$F(\delta)$ is also an isomorphism and, since $F(\sigma_X) = \tau_{FX}$ by compatibility, we have

$$\begin{aligned} \Phi_X\Upsilon(N) = \Phi_X(K) &= im[F(i_K)] = im[F(i_K) \circ F(\delta)] \\ &= im[F(i_K \circ \delta)] = im[F(\sigma_X \circ G(j_N)] \\ &= im[\tau_{FX} \circ FG(j_N)] = im(j_N \circ \tau_N) = im(j_N) = N. \end{aligned}$$

Hence Φ is a bijection and $\Phi^{-1} = \Upsilon$.

To see that Φ_X preserves inclusions let $K_1 \subseteq K \subseteq X$ and let $\nu : K_1 \to K$ be the inclusion. Then $i_{K_1} = i_K \circ \nu$, so

$$\Phi_X(K_1) = im\{F(i_{K_1})\} = im\{F(i_K) \circ F(\nu)\} \subseteq im\{F(i_K)\} = \Phi_X(K),$$

as required. Similarly, if $N_1 \subseteq N \subseteq FX$ and $\lambda : N_1 \to N$ is the inclusion, then $j_{N_1} = j_N \circ \lambda$, so $G(j_{N_1}) = G(j_N) \circ G(\lambda)$. It follows in the same way that $\Upsilon(N_1) \subseteq \Upsilon(N)$, so Υ preserves inclusions. □

Corollary A.8. *Each of the following properties is a Morita invariant:*

simple,	*semisimple,*
Indecomposable,	*uniform,*
artinian,	*noetherian,*
composition length n,	*finite dimensional,*
finitely generated,	*finitely cogenerated.*

Proof. Only the last two properties are not clearly determined by $lat(X)$. But X is finitely generated if and only if $X = \Sigma_{i\in I} X_i$ implies that $X = \Sigma_{i\in J} X_i$ for a finite subset $J \subseteq I$; and X is finitely cogenerated if and only if it has a finitely generated essential socle. □

Propositions A.5 and A.7 combine to give more information. If Φ_X is the lattice isomorphism in Proposition A.7, the next result asserts that Φ_X preserves kernels and (to some extent) images.

Proposition A.9. *Let $F : modR \to modS$ be an additive equivalence. If $\alpha : X \to Y$ is an R-morphism then the following hold:*

(1) $ker[F(\alpha)] = \Phi_X[ker(\alpha)]$.
(2) If α is monic, $im[F(\alpha)] = \Phi_Y[im(\alpha)]$.

Proof. Write $K = ker(\alpha)$ and $U = im(\alpha)$. The sequence $0 \to K \overset{i_K}{\to} X \overset{\alpha}{\to} U \to 0$ is exact, so $0 \to FK \overset{F(i_K)}{\to} FX \overset{F(\alpha)}{\to} FU \to 0$ is also exact by Theorem A.6. Hence $ker[F(\alpha)] = im[F(i_K)] = \Phi_X(K)$, proving (1). Now consider $0 \to U \overset{i_U}{\to} Y \overset{\theta}{\to} Y/U \to 0$, where θ is the coset map. Applying F and Theorem A.6, we obtain $ker[F(\theta)] = im[F(i_U)] = \Phi_Y(U)$. However, we have $0 \to X \overset{\alpha}{\to} Y \overset{\theta}{\to} Y/U \to 0$, so $ker[F(\theta)] = im[F(\alpha)]$. This proves (2). □

Proposition A.10. *Let $F : modR \to modS$ be an additive equivalence. If $\alpha : X \to Y$ has essential image (small kernel) the same is true of $F(\alpha)$:*

$FX \to FY$. In particular, if $\alpha : X \to Y$ is an injective hull (projective cover), so also is $F(\alpha) : FX \to FY$.

Proof. Let $0 \to X \overset{\sigma}{\to} Y$ be an injective hull in $modR$ (that is, Y is injective and $im(\sigma) \subseteq^{ess} Y$). Then $0 \to FX \overset{F(\sigma)}{\to} FY$ is exact by Lemma A.2, FY is injective by Proposition A.4, and so it remains to show that $im[F(\sigma)] \subseteq^{ess} FY$. We have $im[F(\sigma)] = \Phi_Y[im(\sigma)]$ by Proposition A.9, where $\Phi : lat_R(Y) \to lat_S(FY)$ is as in Proposition A.7. If $\Phi_Y[im(\sigma)] + M = FY$, $M_S \subseteq FY$, then $M = \Phi_Y(Z)$ for some $Z \subseteq Y$, again by Lemma A.7. Since Φ_Y is a lattice isomorphism,

$$\Phi_Y[Y] = FY = \Phi_Y[im(\sigma)] + \Phi_Y(Z) = \Phi_Y[im(\sigma) + Z],$$

so $Y = im(\sigma) + Z$ because Φ_Y is one-to-one. Hence $Z = Y$ by hypothesis, so $M = \Phi_Y(Z) = FY$. This proves that $im[F(\sigma)] \subseteq^{ess} FY$. The proof for projective covers is analogous. □

There is clearly more that can be said about Morita invariants, but these results should provide the reader with the means to verify specific cases.

A.3. Tensor Products

If R is a ring, V_R and $_RW$ are modules, and A is a $\mathbb{Z}$-module, a function $\pi : V \times W \to A$ is called a *product* if it preserves addition in each variable and is *balanced* (or *middle associative*) in the sense that $\pi(vr, w) = \pi(v, rw)$ for all $r \in R$, $v \in V$, and $w \in W$.

Lemma A.11. *Given modules V_R and $_RW$ there is a special product $\tau : V \times W \to V \otimes_R W$ that is uniquely determined by the following universal property: Given any product $\lambda : V \times W \to A$ there exists a uniquely determined*

$$\begin{array}{ccc} V \times W & \overset{\tau}{\to} & V \otimes_R W \\ & \lambda \searrow & \downarrow \alpha \\ & & A \end{array}$$

$\mathbb{Z}$-morphism $\alpha : V \otimes_R W \to A$ such that $\lambda = \alpha \circ \tau$.

Proof. Let F be the free $\mathbb{Z}$-module on the set $V \times W$ with basis $\{f_{(v,w)} \mid (v, w) \in V \times W\}$, and let K be the submodule of F generated by all elements of the form $f_{(v+v',w)} - f_{(v,w)} - f_{(v',w)}$, $f_{(v,w+w')} - f_{(v,w)} - f_{(v,w')}$, and $f_{(vr,w)} - f_{(v,rw)}$.

shows that $f_n f_i = 0$ if $i < n$, and $f_i f_n = 0$ is clear. Hence $\{f_1, \ldots, f_{n-1}, f_n\}$ is orthogonal. Since $\bar{f}_i = \bar{e}_i - \bar{e}_i \bar{f}_n = \bar{r}_i - \bar{r}_i \bar{r}_n = \bar{r}_i$ for $i < n$, we can use the f_i in (2).

(3). If $\Sigma_{i=1}^n \bar{r}_i = \bar{1}$ in (2), then $1 - \Sigma_{i=1}^n e_i$ is an idempotent in $A \subseteq J$, so $1 = \Sigma_{i=1}^n e_i$. □

If R is any ring and e and f are idempotents in R, write $e \leq f$ if $eRe \subseteq fRf$, equivalently if $ef = e = fe$. This is a partial order on the set of idempotents in R, and the minimal nonzero elements (if they exist) are called *primitive* idempotents. Thus an idempotent $e \in R$ is primitive if and only if the ring eRe contains no idempotent except 0 and e; if and only if eR (respectively Re) is an indecomposable module. Here a module M is *indecomposable* if $M = K \oplus N$ implies $K = 0$ or $N = 0$. Clearly every local idempotent is primitive, but the converse is not true (consider $\mathbb{Z}$).

Lemma B.6. *The following conditions are equivalent for a ring R:*

(1) R contains no infinite orthogonal sets of idempotents.
(2) R has the ACC on direct summand right (left) ideals.
(3) R has the DCC on direct summand left (right) ideals.
(4) R has the DCC on idempotents.
(5) R has the ACC on idempotents.

Proof. (1)⇒(2). Suppose that $e_1 R \subseteq e_2 R \subseteq \cdots$, where each e_i is an idempotent. Construct idempotents $f_1, f_2, \ldots$ with $f_k \in e_k R$ inductively as follows: Put $f_1 = e_1$. Given $f_k \in e_k R \subseteq e_{k+1} R$, we have $f_k = e_{k+1} f_k$, so define $f_{k+1} = f_k + e_{k+1} - f_k e_{k+1}$. One verifies that $f_{k+1} \in e_{k+1} R$ is an idempotent and $f_k \leq f_{k+1}$. Hence $f_1 \leq f_2 \leq f_3 \leq \cdots$, from which it follows that $\{f_1, f_2 - f_1, f_3 - f_2, \ldots\}$ is an orthogonal family of idempotents. By (1) there exists $n \geq 1$ such that $f_{k+1} = f_k$ for all $k \geq n$. Thus $e_{k+1} = f_k e_{k+1} \in f_k R$ for each $k \geq n$, so $e_{k+1} R \subseteq f_k R \subseteq e_k R \subseteq e_{k+1} R$. It follows that $e_{k+1} R = f_k R$ for each $k \geq n$ and hence that $e_{n+1} R = e_{n+2} R = \cdots$, proving (2).

(2)⇒(3). If $e^2 = e$ and $f^2 = f$ then $Re \supseteq Rf$ if and only if $(1 - e)R \subseteq (1 - f)R$.

(3)⇒(4). If $e_1 \geq e_2 \geq \cdots$ then $Re_1 \supseteq Re_2 \supseteq \cdots$, so let $Re_k = Re_{k+1}$ for all $k \geq n$ by (4). Then $e_k = e_k e_{k+1}$ because $e_k \in Re_{k+1}$, whereas $e_k e_{k+1} = e_{k+1}$ because $e_{k+1} \leq e_k$. Hence $e_k = e_{k+1}$ for all $k \geq n$.

(4)⇒(5). If $e_1 \leq e_2 \leq \cdots$ are idempotents, we obtain a descending chain $(1 - e_1) \geq (1 - e_2) \geq \cdots$.

(4) R is I-finite and primitive idempotents in R are local.
(5) $1 = e_1 + \cdots + e_m$*, where the* e_i *are local, orthogonal idempotents.*

Proof. Through the proof we write $\bar{R} = R/J$ and $\bar{r} = r + J$ for each $r \in R$.

(1)⇒(2). Let $T \nsubseteq J$ be a right ideal and (since R/J is semipotent) let $\bar{0} \neq \bar{r}^2 = \bar{r} \in (T + J)/J$. We may assume that $r \in T$. Since $r^2 - r \in J$, there exists an idempotent f with $f - r \in J$. Then Lemma B.4 shows that there exists $e^2 = e \in T$ such that $e - r \in J$. Since $e \notin J$ this proves (2).

(2)⇒(3). Since R/J is I-finite, this follows from Lemma B.7.

(3)⇒(4). Let e be a primitive idempotent in R; we show that eJ is the unique maximal right ideal contained in eR, and invoke Proposition B.2. It suffices to show that $T \subseteq eJ$ for every right ideal $T \subset eR$. If not, let $t \in T - eJ$. Then $t \notin J$ so, by (3), let $0 \neq f^2 = f \in tR$. Thus $f \in T$, and $f = e$ because eR is indecomposable (e is primitive). But then $T = eR$, a contradiction.

(4)⇒(5). Since R is I-finite, we have $1 = e_1 + \cdots + e_m$, where the e_i are orthogonal and primitive. Now apply (4).

(5)⇒(1). Given the situation in (5), we have $\bar{R} = \bar{e}_1\bar{R} \oplus \cdots \oplus \bar{e}_m\bar{R}$, where each $\bar{e}_i\bar{R}$ is simple by Proposition B.2 because $\bar{e}_i\bar{R} \cong e_iR/e_iJ$. Hence $\bar{R}$ is semisimple. Now suppose $\bar{t}^2 = \bar{t}$, $t \in R$. Since $\bar{R}$ is I-finite, write $\bar{t} = \bar{f}_1 + \cdots + \bar{f}_k$ and $\bar{1} - \bar{t} = \bar{f}_{k+1} + \cdots + \bar{f}_n$, where the $\bar{f}_i$ are primitive orthogonal idempotents. But the $\bar{f}_i$ are local by the proof of (3)⇒(4), so $\bar{R} = \bar{f}_1\bar{R} \oplus \cdots \oplus \bar{f}_n\bar{R}$, where each $\bar{f}_i\bar{R}$ is simple. Hence the Jordan–Hölder theorem shows that $n = m$ and (after possible relabeling) $\bar{e}_i\bar{R} \cong \bar{f}_i\bar{R}$ for each i. By Lemma B.8, write $\bar{e}_i = \bar{a}_i\bar{b}_i$ and $\bar{f}_i = \bar{b}_i\bar{a}_i$, where $\bar{a}_i \in \bar{e}_i\bar{R}\bar{f}_i$ and $\bar{b}_i \in \bar{f}_i\bar{R}\bar{e}_i$ for each i. If $\bar{u} = \bar{a}_1 + \cdots + \bar{a}_n$ and $\bar{v} = \bar{b}_1 + \cdots + \bar{b}_n$, then $\bar{u}\bar{v} = \bar{1} = \bar{v}\bar{u}$ and $\bar{u}\bar{f}_i = \bar{a}_i = \bar{e}_i\bar{u}$ for each i. Thus $\bar{f}_i = \bar{u}^{-1}\bar{e}_i\bar{u}$ for each i. But u is necessarily a unit in R, and it follows that $f = u^{-1}(e_1 + \cdots + e_k)u$ is an idempotent in R with $\bar{f} = \bar{t}$. □

A ring R is called *semiprimary* if R/J is semisimple and J is nilpotent, so every semiprimary ring is semiperfect by Lemma B.3 (but not conversely – consider $\mathbb{Z}_{(p)}$). Of course every left or right artinian ring is semiprimary; however, the matrix ring $\begin{bmatrix} \mathbb{Q} & \mathbb{R} \\ 0 & \mathbb{Q} \end{bmatrix}$ is a semiprimary ring that is neither right nor left artinian.

Corollary B.10. *Let R be a semiperfect ring. Then the following hold:*

(1) eRe *is semiperfect for any* $e^2 = e \in R$*.*
(2) Any matrix ring $M_n(R)$ *is semiperfect.*
(3) Any homomorphic image R/A *of R is semiperfect.*

In particular, being semiperfect is a Morita invariant.

projective if it satisfies the following equivalent conditions:

(1) Every R-epimorphism $M \overset{\psi}{\to} P \to 0$ splits; that is, $ker(\psi) \subseteq^{\oplus} M$.
(2) If $M \overset{\alpha}{\to} N \to 0$ is R-epic then every R-homomorphism $P \overset{\beta}{\to} N$ factors in the form $\beta = \alpha \circ \gamma$ for some R-linear map $\gamma : P \to M$.

Every free module is projective, and a module P is projective if and only if it is isomorphic to a direct summand of a free module. The direct sum of a family of modules is projective if and only if each of them is projective.

Recall that an R-monomorphism $0 \to M \overset{\sigma}{\to} E$ is called an injective hull of the module M if E is injective and $im(\sigma) \subseteq^{ess} E$. Dually, a *projective cover* of M is an epimorphism $P \overset{\pi}{\to} M \to 0$, where P is projective and $ker(\pi) \subseteq^{sm} P$. As is customary in the literature, we will sometimes abuse the terminology and refer to the module P itself as a projective cover of M.

Although every module has an injective hull, projective covers seldom exist. The right perfect rings were first identified by Bass [16] as the rings for which every right module has a projective cover, and we return to this later. The semiperfect rings turn out to be those for which every finitely generated right (or left) module has a projective cover. To prove this, the following lemma is essential.

Lemma B.15 (Bass' Lemma). *The following conditions are equivalent for modules $K \subseteq P$ with P projective:*

(1) P/K has a projective cover.
(2) $P = Q \oplus P_0$, where $Q \subseteq K$ and $P_0 \cap K \subseteq^{sm} P_0$.

If (2) holds the restriction $P_0 \to P/K$ of the coset map is a projective cover.

Proof. (1)⇒(2). Let $P' \overset{\pi}{\to} P/K$ be a projective cover, and let $P \overset{\phi}{\to} P/K$ be the coset map. Since P is projective, there exists $P \overset{\alpha}{\to} P'$ such that $\pi \circ \alpha = \phi$.

$$\begin{array}{ccc} & P & \\ {\scriptstyle\alpha}\swarrow & \downarrow \phi & \\ P' & \overset{\pi}{\to} & P/K \end{array}$$

As ϕ is epic, it follows that $P' = \alpha(P) + ker(\pi)$, so $P' = \alpha(P)$ because $ker(\pi) \subseteq^{sm} P'$. But then $P \overset{\alpha}{\to} P'$ splits because P' is projective; that is, there exists $\beta : P' \to P$ such that $\alpha \circ \beta = 1_{P'}$. Hence $P = ker(\alpha) \oplus \beta(P')$. Moreover, $ker(\alpha) \subseteq ker(\phi) = K$ because $\pi \circ \alpha = \phi$, so it remains to show that $\beta(P') \cap K$ is small in $\beta(P')$. Since $\beta : P' \to \beta(P')$ is an isomorphism, we have $\beta[ker(\pi)] \subseteq^{sm} \beta(P')$. But $\phi \circ \beta = \pi \circ \alpha \circ \beta = \pi \circ 1_{P'} = \pi$, so $\beta[ker(\pi)] = \beta(P') \cap ker(\phi) = \beta(P') \cap K$, and (2) follows with $Q = ker(\alpha)$ and $P_0 = \beta(P')$.

(2)⇒(1). Given (2), the restriction of ϕ to $P_0 \to P/K$ is onto because $P = K + P_0$, and its kernel is $P_0 \cap ker(\phi) = P_0 \cap K$, which is small in P_0 by (2). This proves (1) and the last statement. □

It follows from the proof of Bass' lemma that, as for injective hulls, projective covers are unique in the following sense.

Corollary B.16. *If $P \overset{\phi}{\to} M$ and $P' \overset{\pi}{\to} M$ are both projective covers there exists an isomorphism $P \overset{\alpha}{\to} P'$ such that $\phi = \pi \circ \alpha$.*

Proof. The map α in the proof of Bass' lemma is epic because $ker(\pi)$ is small in P', and it is monic because $ker(\alpha)$ is a direct summand of P that is small in P [since $ker(\alpha) \subseteq K = ker(\phi) \subseteq^{sm} P$.] □

We need the following related result later.

Corollary B.17. *Let P and Q be projective modules.*

(1) If $P/N \cong Q/M$, where $N \subseteq^{sm} P$ and $M \subseteq^{sm} Q$, then $P \cong Q$.
(2) If P and Q have small radicals, then $P \cong Q$ if and only if $P/rad(P) \cong Q/rad(Q)$.

Proof. (1). Let $\sigma : Q/M \to P/N$ be an isomorphism, and let $\pi : P \to P/N$ and $\phi : Q \to Q/M$ be the coset maps. Then $\pi : P \to P/N$ and $\sigma \circ \phi : Q \to P/N$ are both projective covers, so Corollary B.16 applies.

(2). This comes from (1) and the fact that $\alpha[rad(M)] \subseteq rad(N)$ holds for any R-linear map $\alpha : M \to N$. □

The following alternate form of Bass' lemma will be needed. The proof requires the fact that if $X \subseteq N \subseteq M$ are modules and $X \subseteq^{sm} N$ then $X \subseteq^{sm} M$.

Corollary B.18. *If $K \subseteq P$ are modules with P projective, then P/K has a projective cover if and only if $K = Q + X$, where $Q \subseteq^{\oplus} P$ and $X \subseteq^{sm} P$.*

Proof. If P/K has a projective cover, take $X = K \cap P_0$ in Bass' lemma (it is small in P, being small in P_0). Conversely, if $K = Q + X$, where $Q \subseteq^{\oplus} P$ and $X \subseteq^{sm} P$, let $P = Q \oplus P_1$ and define $\phi : P_1 \to P/K$ by $\phi(p_1) = p_1 + K$.

Theorem B.27. *A finitely generated projective module P is semiperfect if and only if every maximal submodule has a supplement in P.*

Proof. Let A denote the sum of all supplemented submodules of P. If $A = P$ then P is supplemented by Lemma B.26 (because P is finitely generated). But if $A \neq P$ let $A \subseteq M \subseteq^{max} P$ (again since P is finitely generated), and let S be a supplement of M in P. We show that S is supplemented, (which is a contradiction because then $S \subseteq A \subseteq M$, so $P = M + S \subseteq M$). But $M \cap S \subseteq^{sm} S$ by Lemma B.22 and $M \cap S \subseteq^{max} S$ because $P = M + S$. It follows that $M \cap S$ is the only maximal submodule of S and that S is principal. Hence every submodule of S is contained in a maximal submodule, namely, $M \cap S$. It follows that S is supplemented, as required (in fact, S is a supplement of every proper submodule of S). □

Applying Theorems B.25 and B.27 with $P = R$ gives immediately

Theorem B.28. *The following conditions (and their left–right analogues) are equivalent for a ring R:*

(1) R is semiperfect.
(2) Every right ideal of R has a supplement in R.
(3) Every maximal right ideal of R has a supplement in R.

B.4. Perfect Rings

A one-sided ideal A of a ring R is called *right T-nilpotent* if for any sequence $a_1, a_2, \ldots$ from A we have

$$a_n a_{n-1} \cdots a_2 a_1 = 0 \quad \text{for some } n \geq 1,$$

and A is called *left T-nilpotent* if we insist instead that $a_1 a_2 \cdots a_{n-1} a_n = 0$ for some n. Clearly every nilpotent ideal is both right and left T-nilpotent, and every right or left T-nilpotent ideal is nil. The following notation is convenient: If ${}_RV$ is a left R-module, write $\mathrm{r}_V(A) = \{v \in V \mid Av = 0\}$.

Lemma B.29. *The following are equivalent for a right ideal A of R:*

(1) A is right T-nilpotent.
(2) If ${}_RV \neq 0$ is any left module then $\mathrm{r}_V(A) \subseteq^{ess} {}_RV$.
(3) If ${}_RV \neq 0$ is any left module then $\mathrm{r}_V(A) \neq 0$.

(4) *If* $M_R \neq 0$ *is any right module then* $MA \subseteq^{sm} M$.
(5) *If* $M_R \neq 0$ *is any right module then* $MA \neq M$.
(6) *If* F_R *is a countably generated free module then* $FA \subseteq^{sm} F$.

Proof. (2)⇒(3) and (4)⇒(5) are clear.

(1)⇒(2). Suppose $Rv \cap \mathrm{r}_V(A) = 0$, $v \in V$. If $v \neq 0$ then $v \notin \mathrm{r}_V(A)$, so $Av \neq 0$, say $a_1 v \neq 0$, $a_1 \in A$. But then $a_1 v \notin \mathrm{r}_V(A)$, so $a_2(a_1 v) \neq 0$ for some $a_2 \in A$. Continuing, we contradict (1). Hence $\mathrm{r}_V(A) \subseteq^{ess} {}_RV$.

(3)⇒(4). If MA is not small in M_R let $MA + X = M$, where $X \neq M$ is a submodule. If we write $N = M/X$ then $N \neq 0$ and $NA = N$. Let $B = \{b \in R \mid Nb = 0\}$, a two-sided ideal of R, and regard $V = R/B \neq 0$ as a left module. By (3), let $0 \neq v \in \mathrm{r}_V(A)$. Writing $v = r + B$, we have $r \notin B$ but $Ar \subseteq B$. Thus $Nr \neq 0$ while $NAr = 0$, which is a contradiction because $NA = N$.

(5)⇒(6). If $FA + X = F$, with X a submodule, then $(F/X)A = F/X$, so $X = F$ by (5).

(6)⇒(1). Let F_R have basis $\{f_1, f_2, \dots\}$. Given $a_1, a_2, \dots$ from A write $G = (f_1 - f_2 a_1)R + (f_2 - f_3 a_2)R + \cdots$. Then $G + FA = F$ because $f_i = (f_i - f_{i+1}a_i) + f_{i+1}a_i$ for each i, so $G = F$ by (6). In particular

$$f_1 = (f_1 - f_2 a_1)r_1 + (f_2 - f_3 a_2)r_2 + \cdots + (f_n - f_{n+1}a_n)r_n$$

for some n, where each $r_i \in R$. Hence $r_1 = 1$, $r_2 = a_1 r_1$, $r_3 = a_2 r_2, \dots$, $r_n = a_{n-1} r_{n-1}$, and $0 = a_n r_n$. Hence $a_n \cdots a_2 a_1 = 0$, proving (1). □

Corollary B.30. *If* A *and* B *are right T-nilpotent right ideals, so also is* $A + B$.

Proof. If ${}_RV \neq 0$ then $\mathrm{r}_V(A+B) = \mathrm{r}_V(A) \cap \mathrm{r}_V(B) \subseteq^{ess} {}_RV$. by (2) of Lemma B.29. Hence $A + B$ is right T-nilpotent by the same lemma. □

A module is called *semiartinian* if every nonzero factor module has nonzero socle (for example artinian and semisimple modules), and a ring R is called a right *semiartinian ring* if R_R is a semiartinian module. We need the following characterizations of these rings.

Lemma B.31. *The following are equivalent for a ring* R:

(1) *R is right semiartinian.*
(2) *Every nonzero right R-module has an essential socle.*
(3) *Every nonzero right R-module has a simple submodule.*
(4) *Every right R-module is semiartinian.*

(1)⇒(2). Given (1), Bass' lemma (Lemma B.15) shows that there exists $e^2 = e \in R$ such that $eR \subseteq aR$ and $aR \cap (1-e)R \subseteq J$. Since $e \in aR$, we have $aR \cap (1-e)R = (1-e)aR$, and (2) follows.

(2)⇒(3). If $e = ar$, $r \in R$, is as in (2) then $a - ara = (1-e)a \in J$ and $ara = (ara)r(ara)$.

(3)⇒(1). Let $a - b \in J$, where $b = brb$, and write $f = br$, so that $f^2 = f$ and $b = fb$. Then $f - ar = (b-a)r \in J$, so the element ar lifts to the idempotent f modulo J. Hence Lemma B.4 shows that there exists $e^2 = e \in aR$ such that $e - ar \in J$. Writing $\bar{x} = x + J$ for all $x \in R$ gives $\bar{e} = \bar{f}$, and so $\bar{a} - \bar{e}\bar{a} = \bar{b} - \bar{f}\bar{b} = \bar{0}$. Thus $(1-e)a \in J$, so $aR \cap (1-e)R \subseteq (1-e)aR \subseteq J$. Since $eR \subseteq aR$, (1) follows by Bass' lemma. □

An element a in a ring R is called *semiregular* if it satisfies the conditions in Lemma B.40, and R is called a *semiregular ring* if every element is semiregular. Semiregular rings are semipotent by (2) of Lemma B.40.

Corollary B.41. *If R is semiregular and $T \nsubseteq J$ is a one-sided ideal, there exists $e^2 = e \in T - J$.*

Corollary B.42. *Suppose that R is a semiregular ring. Then:*

(1) eRe is semiregular for every $e^2 = e \in R$.
(2) R/A is semiregular for any ideal A of R.

Proof. (1). Given $a \in eRe$ choose $f^2 = f \in aR$ with $(1-f)a \in J$. Then $ef = f$, so $g = fe$ is an idempotent in $a(eRe)$ and $(e-g)a = (1-f)a \in J \cap eRe = J(eRe)$.

(2). This is clear by (3) of Lemma B.40. □

The next theorem shows that these semiregular rings simultaneously generalize the semiperfect and regular rings. We need the following useful fact about regular rings.

Lemma B.43 (Von Neumann's Lemma). *If R is a regular ring every finitely generated right (left) ideal is a direct summand, and the intersection of two summands is a summand.*

Proof. If $T = a_1R + \cdots + a_nR$, $a_i \in R$, we use induction on n. If $n = 1$ and $a_1ba_1 = a_1$ then $T = e_1R$, where $e_1 = a_1b$ is an idempotent. In general write $a_2R + \cdots + a_nR = fR$, where $f^2 = f$, and then let $(1-f)a_1R = eR$, where

$e^2 = e$. It follows that $T = eR + fR$. But $fe = 0$, so $g = e + f - ef$ is an idempotent and $gR = eR + fR = T$.

Now let Re and Rf be summands, $e^2 = e$, $f^2 = f$. Then, by the first part of this proof, $(1 - e)R + (1 - f)R = gR$ for some $g^2 = g \in R$, so $Re \cap Rf = \mathbb{1}(gR) = R(1 - g)$ is a summand. Similarly, $eR \cap fR$ is a summand. □

Theorem B.44. *The following are equivalent for a ring R:*

(1) R is a semiregular ring.
(2) R/J is regular and idempotents can be lifted modulo J.
(3) R/T has a projective cover for every finitely generated right (left) ideal $T \subseteq R$.

Proof. As before write $\bar{x} = x + J$ for every $x \in R$.

(1)⇒(2). R/J is regular by (3) of Lemma B.40. Suppose that $a^2 - a \in J$ and, by Lemma B.40, choose $e^2 = e \in aR$ such that $(1 - e)a \in J$. If $f = e + ea(1 - e)$ then $f^2 = f$ and $\bar{f} = \bar{a}$ because $\bar{e} = \bar{a}\bar{e}$ and $\bar{a} = \bar{e}\bar{a}$.

(2)⇒(3). We prove it for a finitely generated right ideal $T \subseteq R$. The right ideal $\bar{T} = (T + J)/J$ of $\bar{R} = R/J$ is finitely generated, so $\bar{T} = \bar{a}\bar{R}$, where $\bar{a}^2 = \bar{a}$ by von Neumann's lemma (Lemma B.43). We may assume that $a \in T$. Hence by (2) and Lemma B.4 there exists $e^2 = e \in T$ such that $e - a \in J$. By Bass' lemma it remains to show that $T \cap (1 - e)R \subseteq J$. But if $t \in T \cap (1 - e)R$ then, since $\bar{T} = \bar{a}\bar{R} = \bar{e}\bar{R}$, we have $\bar{t} = \bar{a}\bar{t} = \bar{e}\bar{t} = \bar{0}$.

(3)⇒(1). This is clear by Lemma B.40. □

Clearly, semiperfect and regular rings are semiregular, and Utumi's theorem (Theorem 1.26) shows that every right continuous ring (and hence every right quasi-injective ring) is semiregular. In fact Theorem 1.25 shows that if M_R is continuous (in particular if M_R is quasi-injective) then $E = end(M_R)$ is semiregular and $J(E) = \{\alpha \in E \mid ker(\alpha) \subseteq^{ess} M\}$. We are going to prove a "dual" to this result (Theorem B.46) that reveals a condition under which the endomorphism ring of any projective module is semiregular. In particular, it enables us to show that $M_n(R)$ is semiregular whenever this is true of R and so, by Corollary B.42, to see that semiregularity is a Morita invariant.

We require a lemma and, with no extra effort we can do it for any quasi-projective module. Here we say that a module M is *quasi-projective* if, for every epimorphism $\alpha : M \to N$, and every R-linear map $\beta : M \to N$, there

$$\begin{array}{ccc} & M & \\ {}^{\gamma}\swarrow & \downarrow \beta & \\ M \stackrel{\alpha}{\to} & N \to 0 & \end{array}$$

exists $\gamma : M \to M$ such that $\beta = \alpha\gamma$. Clearly, projective and semisimple modules are quasi-projective.

For convenience, we often write composition of maps as juxtaposition in this appendix.

Lemma B.45. *If M_R is a quasi-projective module and $E = end(M)$, then $J(E) = \{\alpha \in E \mid \alpha M \subseteq^{sm} M\}$.*

Proof. Write $A = \{\alpha \in E \mid \alpha M \subseteq^{sm} M\}$, an ideal of E. If $\alpha \in A$, the fact that $\alpha M + (1-\alpha)M = M$ shows that $1-\alpha$ is epic. Then $1-\alpha$ has a right inverse by quasi-projectivity, and it follows that $A \subseteq J(E)$. Conversely, if $\alpha \in J(E)$, let $\alpha M + X = M$, where X is a submodule of M; we must show that $X = M$. If $\theta : M \to M/X$ is the coset map, then $\theta\alpha : M \to M/X$ is epic, so there exists $\beta \in E$ such that $\theta\alpha\beta = \theta$, again because M is quasi-projective. Thus $\theta(1-\alpha\beta) = 0$, so $\theta = 0$ because $\alpha \in J(E)$. Hence $X = M$, as required. □

Theorem B.46. *Let M_R be a quasi-projective module, and write $E = end(M)$. The following conditions are equivalent:*

(1) E is a semiregular ring.
(2) For all $\alpha \in E$, $M = P \oplus K$ with $P \subseteq \alpha M$ and $\alpha M \cap K \subseteq^{sm} K$.

In particular, if M is projective, then E is semiregular if and only if $M/\alpha M$ has a projective cover for all $\alpha \in E$.

Proof. (1)⇒(2). If $\alpha \in E$ choose $\pi^2 = \pi \in \alpha E$ such that $(1-\pi)\alpha \in J(E)$. Then $\pi M \subseteq \alpha M$, and $\alpha M \cap (1-\pi)M = (1-\pi)\alpha M \subseteq^{sm} M$ by Lemma B.45. So (2) follows with $P = \pi M$ and $K = (1-\pi)M$.

(2)⇒(1). If $\alpha \in E$ choose $M = P \oplus K$ as in (2), and let $\pi^2 = \pi \in E$, where $\pi M = P$ and $(1-\pi)M = K$. Then $\pi\alpha : M \to P$ is epic and so, since M is quasi-projective, there exists $\beta \in E$ such that $\pi\alpha\beta = \pi$. Write $\tau = \alpha\beta\pi$. Then $\tau^2 = \tau \in \alpha E$, so it remains to show that $(1-\tau)\alpha \in J(E)$. But this follows from Lemma B.45 because $(1-\tau)\alpha M = \alpha M \cap (1-\tau)M \subseteq^{sm} M$.

Finally, the last sentence follows by Bass' lemma (Lemma B.15). □

Hence, to show that semiregularity is a Morita invariant, we must show that R^n has the property in (2) of Theorem B.46 whenever R is semiregular. To do this, we need a module theoretic version of Lemma B.40 which, it turns out, will also enable us to extend Theorem B.44.

If M is a module, an element $q \in M$ is called a *regular element* if $q\lambda(q) = q$ for some $\lambda \in M^*$ [where $M^* = hom(M, R)$ denotes the dual of M]. Thus $a \in R$ is regular if and only if a is regular in R_R (or ${}_RR$).

Lemma B.47. *Let $q \in M_R$ be regular, say $q = q\lambda(q)$, where $\lambda \in M^*$. Then $e = \lambda(q)$ is an idempotent in R, $qR \cong eR$ is projective, and $M = qR \oplus S$, where $S = \{s \in M \mid q\lambda(s) = 0\}$.*

Proof. We have $e = \lambda(q) = \lambda[q\lambda(q)] = e^2$, so $q = qe$. Then $\lambda : qR \to eR$ is epic, and it is monic because $\lambda(qr) = 0$ means $qr = q\lambda(q)r = 0$. Hence $qR \cong eR$ is projective. Since $m - q\lambda(m) \in S$ for every $m \in M$, we have $M = qR + S$. This is direct because $qr \in S$ means $0 = q\lambda(qr) = qer = qr$. $\square$

If P_R is projective, let $P \oplus Q = F$ be free on a basis $\{b_i \mid i \in I\}$, and define $\pi_i \in F^*$ by $\pi_i(\Sigma_j b_j r_j) = r_i$. For each $i \in I$, write $b_i = p_i + q_i$, where $p_i \in P$ and $q_i \in Q$, and let $\xi_i = (\pi_i)_{|P}$ denote the restriction of π_i to P. Then $\xi_i \in P^*$ for each i, and the system $\{(p_i, \xi_i) \mid i \in I\}$ is called a *dual basis* for P because $p = \Sigma_i p_i \xi_i(p)$ holds for every $p \in P$. (In fact a module has a dual basis if and only if it is projective.)

With this the module version of Lemma B.40 is as follows. Note the relationship to Bass' lemma (Lemma B.15).

Lemma B.48. *The following are equivalent for $m \in M_R$:*

(1) $M = P \oplus K$, where P is projective, $P \subseteq mR$, and $mR \cap K \subseteq^{sm} K$.
(2) There exists $\lambda \in M^*$ such that $\lambda(m) = e = e^2$ and $m - me \in rad(M)$.
(3) There exists a regular element $q \in M$ such that $m - q \in rad(M)$.
(4) There exists $\gamma^2 = \gamma \in end(M)$ such that $\gamma(M) \subseteq mR$, $\gamma(M)$ is projective, and $m - \gamma(m) \in rad(M)$.

Proof. (1)$\Rightarrow$(2). Let $M = P \oplus K$ as in (1). Hence $P \subseteq^{\oplus} mR$, so P is finitely generated and thus has a finite dual basis $\{(p_i, \xi_i) \mid i = 1, 2, \ldots, n\}$. Write $p_i = mr_i$, $r_i \in R$, and define $\lambda \in P^*$ by $\lambda(p) = \Sigma_{i=1}^n r_i \xi_i(p)$. Extend λ to M by defining $\lambda(K) = 0$. If $m = p + k$, $p \in P$, $k \in K$, then $\lambda(m) = \lambda(p)$, so $m\lambda(m) = m\lambda(p) = \Sigma_{i=1}^n p_i \xi_i(p) = p$. If we write $\lambda(m) = e$, it follows that $e^2 = e$ and $m - me = m - p = k \in mR \cap K$. Hence $m - me \in rad(M)$.

(2)⇒(3). If $q = me$ as in (2), then q is regular because $q\lambda(q) = qe^3 = q$.

(3)⇒(4). Let $m - q \in rad(M)$, where $q \in M$ is regular, say $q\alpha(q) = q$ with $\alpha \in M^*$. If we write $e = \alpha(q)$ then $e^2 = e$ and $m - me = (m - q)(1 - e) \in rad(M)$; we claim that me is regular. In fact $e - \alpha(m) = \alpha(q - m) \in rad(R_R) = J$, so let $b[1 - e + \alpha(m)] = 1$, $b \in R$. Then $\beta = b\alpha$ is in M^* and $\beta(m) = 1 - b(1 - e)$. It follows that $\beta(me) = e$ and hence that $me\,\beta(me) = me$. Thus me is regular and so, replacing q by me, we may assume that $q \in mR$ in (3).

Now Lemma B.47 gives $M = qR \oplus S$, where $qR \subseteq mR$, $q = q\lambda(q)$ for some $\lambda \in M^*$, and $S = \{s \in M \mid q\lambda(s) = 0\}$. Let $\gamma : M \to M$ be the projection with $\gamma(M) = qR$ and $ker(\gamma) = S$. Then $\gamma(M)$ is projective by Lemma B.47, and $(1 - \gamma)(q) = 0$. Hence $m - \gamma(m) = (1 - \gamma)(m) = (1 - \gamma)(m - q) \in rad(M)$, proving (4).

(4)⇒(1). Given (4), we have $M = \gamma(M) \oplus (1 - \gamma)(M)$, where $\gamma(M) \subseteq mR$, so it remains to show that $mR \cap (1 - \gamma)(M)$ is small in $(1 - \gamma)(M)$. But $mR \cap (1 - \gamma)(M) \subseteq (1 - \gamma)(mR)$ because $\gamma^2 = \gamma$, and $(1 - \gamma)(mR)$ is small in M by hypothesis. Since $(1 - \gamma)(M) \subseteq^{\oplus} M$, we are done. □

An element m in a module M is called *semiregular* in M if the conditions in Lemma B.48 are satisfied, and M is called a *semiregular module* if every element is semiregular. Thus if R is a semiregular ring then R_R and ${}_RR$ are both semiregular modules. A module is called *regular* if every element is regular, so every regular module is semiregular. In fact we have

Corollary B.49. *A module M is regular if and only if M is semiregular and $radM = 0$.*

Proof. If M is regular it is semiregular by Lemma B.48. To see that $radM = 0$, let $0 \neq q \in M$. Then $M = qR \oplus S$ for some $S \subseteq M$ by Lemma B.47, so choose a submodule $N \subseteq M$ maximal such that $S \subseteq N$ and $q \notin N$. Then N is maximal and so $q \notin radM$. The converse follows by Lemma B.48. □

It is clear from Bass' lemma and Lemma B.48 that a projective module M is semiregular if and only if M/mR has a projective cover for all $m \in M$. We are going to improve on this; to do so we will need the following immediate consequence of (3) of Lemma B.48.

Corollary B.50. *If $m \in M_R$, and if $m - m_1 \in rad(M)$, where m_1 is semiregular, then m is semiregular.*

Lemma B.48 has two more important consequences. The first is a characterization of semiregular modules.

Theorem B.51. *The following are equivalent for a module M:*

(1) M is semiregular.
(2) If $N \subseteq M$ is finitely generated there exists $\gamma : M \to N$ such that $\gamma^2 = \gamma$, $\gamma(M)$ is projective, and $(1-\gamma)(N) \subseteq rad(M)$.
(3) If $N \subseteq M$ is finitely generated then $M = P \oplus K$, where P is projective, $P \subseteq N$, and $N \cap K$ is small in K.

Proof. (1)⇒(2). Write $N = m_1R + m_2R + \cdots + m_kR$ and proceed by induction on k, the case $k = 1$ being Lemma B.48. In general, use Lemma B.48 again to obtain $\beta : M \to m_1R$, where $\beta^2 = \beta$, $\beta(M)$ is projective and $(1-\beta)(m_1) \in rad(M)$. Write $K = (1-\beta)(\Sigma_{i=2}^{k} m_i R)$ and, by induction, choose $\delta^2 = \delta : M \to K$ such that $\delta(M)$ is projective and $(1-\delta)(K) \subseteq rad(M)$. Then $\beta\delta = 0$, so $\gamma = \beta + \delta - \delta\beta$ is an idempotent in $end(M)$, and $\gamma(M) = \beta(M) \oplus \delta(M)$ is projective. Finally, $(1-\beta)(N) \subseteq K + (1-\beta)(m_1R) \subseteq K + rad(M)$, so

$$\begin{aligned}(1-\gamma)(N) = (1-\delta)(1-\beta)(N) &\subseteq (1-\delta)(K) + (1-\delta)[rad(M)] \\ &\subseteq rad(M).\end{aligned}$$

(2)⇒(3). If $N \subseteq M$ is finitely generated, choose γ as in (2) and write $P = \gamma(M)$ and $K = (1-\gamma)(M)$. Then $N = P \oplus (N \cap K)$, so $N \cap K$ is finitely generated. Since $N \cap K \subseteq (1-\gamma)(N) \subseteq rad(M)$, this shows that $N \cap K$ is small in M and hence in K (as $K \subseteq^{\oplus} M$).

(3)⇒(1). This is clear by Lemma B.48. □

Corollary B.52. *A projective module M is semiregular if and only if M/N has a projective cover for every finitely generated (respectively principal) submodule N of P.*

Proof. If N is finitely generated then M/N has a projective cover by (3) of Theorem B.51 and Bass' lemma (Lemma B.15). Conversely, if this is true whenever $N = mR$, $m \in M$, then m is a semiregular element of M by Lemma B.48. □

The second fundamental consequence of Lemma B.48 is that direct sums of semiregular modules are semiregular. We need the following lemma.

(homological) proof; the fact that this is equivalent to the DCC on finitely generated left ideals is due to Björk [20]. Jonah [114] showed that these rings have the ACC on principal right ideals.

Von Neumann's lemma [156] (Lemma B.43) was proved in 1936. Theorems B.44 and B.56 were given in 1971 by Oberst and Schneider [176]; the present treatment follows [157]. Semiregular modules were introduced in [157] and generalize the regular modules of Zelmanowitz [236].

Regarding Corollary B.55, it is worth mentioning that $end(F)$ may not be semiregular if F is a free module. Indeed, if a countably generated free right R-module is semiregular it can be shown that R is right perfect (see [157, Theorem 3.9]). Semiregular rings with J right T-nilpotent have been studied by Stock [210].

C
The Camps–Dicks Theorem

In this brief appendix we give a self-contained treatment of an important theorem about semilocal rings, settling (affirmatively) an open question whether the endomorphism ring of every artinian module is semilocal. The following lemma will be needed. If $M = {}_SM_R$ is a bimodule and $a \in R$, let $\mathtt{l}_M(a) = \{m \in M \mid ma = 0\}$, an S-submodule of M.

Lemma C.1. *Let $M = {}_SM_R$ be a bimodule and let $a, b \in R$.*

(1) $\mathtt{l}_M(a - aba) = \mathtt{l}_M(a) \oplus \mathtt{l}_M(1 - ab)$ and $\mathtt{l}_M(1 - ab) \cong \mathtt{l}_M(1 - ba)$.
(2) If ${}_SM$ is artinian and $\mathtt{l}_M(a) = 0$ then $Ma = M$.

Proof. (1) $\mathtt{l}_M(a) \subseteq \mathtt{l}_M(a - aba)$ and $\mathtt{l}_M(1 - ab) \subseteq \mathtt{l}_M(a - aba)$ are clear, as is the fact that $\mathtt{l}_M(a) + \mathtt{l}_M(1 - ab)$ is a direct sum. If $m \in \mathtt{l}_M(a - aba)$ then $m(a - aba) = 0$, so $m = (m - mab) + mab$ is in $\mathtt{l}_M(a) + \mathtt{l}_M(1 - ab)$. Next, right multiplication by a induces an S-linear map $\cdot a : \mathtt{l}_M(1 - ab) \to \mathtt{l}_M(1 - ba)$, and it is routine to verify that $\cdot b$ is the inverse map.

(2) We have $M \supseteq Ma \supseteq Ma^2 \supseteq \cdots$, so $Ma^n = Ma^{n+1}$ for some n because ${}_SM$ is artinian. Hence $M = Ma$ because $\mathtt{l}_M(a) = 0$. □

We say that a set of submodules has the *ACC on summands* if no infinite sum of nonzero modules in the set is direct.

Theorem C.2 (Camps–Dicks Theorem). *The following are equivalent for a ring R:*

(1) R is semilocal.
(2) There exists a ring homomorphism $\varphi : R \to S$, where S is semisimple artinian, and a is a unit of R whenever $\varphi(a)$ is a unit of S.

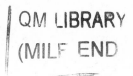